Dyes in History and Archaeology 21

Editor's note

Dyes in History and Archaeology is a refereed journal; the papers included are largely drawn from those presented at the annual meetings of *Dyes in History and Archaeology* and are accepted for publication, after peer review, on the assumption that a paper submitted does not contain essentially the same material as in a previous publication, and is not under consideration for publication elsewhere.

The journal was formerly published by Textile Research Associates, York (issue 7, 1989 to issue 15, 1997).

Dyes in History and Archaeology 21

**Including papers presented at the 21st Meeting,
held at Avignon and Lauris, France, 10–12 October 2002**

Edited by Jo Kirby
The National Gallery, London

with the assistance of Chris Cooksey
Dominique Cardon, *CNRS, Laboratoire d'Histoire et d'Archéologie, Lyon*
Maarten van Bommel, *Instituut Collectie Nederland, Amsterdam*
Anita Quye, *National Museums of Scotland, Edinburgh*
and Robert Atkinson

Archetype Publications

First published 2008 by Archetype Publications Ltd.

Archetype Publications Ltd.
6 Fitzroy Square
London W1T 5HJ
www.archetype.co.uk

Tel: 44(207) 380 0800
Fax: 44(207) 380 0500

ISBN: 978-1-904982-07-4

British Library Cataloguing in Publication Data
A catalogue record for this book is available from the British Library.

Typeset by Kate Williams, Swansea
Printed and bound in Great Britain by Latimer Trend & Co. Ltd., Plymouth

Contents

Editorial

The 21st Meeting of *Dyes in History and Archaeology* was held in Avignon and at Couleur Garance, Lauris, 10–12 October 2002, and was organised by Dominique Cardon, CNRS CIHAM-UMR 5648, Lyon, Claude Andary, Laboratoire de Botanique et Phytochimie, Faculté de Pharmacie, Université de Montpellier I, and Michel Garcia, Association Couleur Garance, Lauris. Avignon, seat of the papacy in the 14th century, is rich in historical monuments and works of art, including the Pont d'Avignon (Pont St-Bénézet) of the well-known song: a magnificent and atmospheric centre for a conference. The delegates were wonderfully entertained and treated to an extremely full and varied conference programme, amply demonstrated by the contributions published in the present volume; these represent about half of the total number of papers and posters presented. They also demonstrate the truly international nature of the field of natural dyes. Perhaps the highlight of the Meeting was the visit to Association Couleur Garance, Le Château, Lauris, visiting the old Moulin des Taillades, once used to grind madder roots, on the way; Lauris is in the Vaucluse department, formerly the heart of the French madder-growing region. Here the delegates walked in the garden of dye plants, in glorious weather (Plate 1), and watched demonstrations of dyeing. This was a splendid day, crowning a truly memorable conference and the organisers are thanked most warmly for giving the participants a most entertaining time.

I must apologise for the delay in the publication of this volume. Its preparation has been disrupted by other commitments, including assistance on the production of the English edition of Dominique Cardon's authoritative work on natural dyes, *Natural Dyes: Sources, Tradition, Technology and Science* (Archetype Publications, 2007). Preparation of the next volume in the series is, however, already under way, and further delays are not anticipated. My thanks are due to my co-editors, Chris Cooksey, Dominique Cardon, Maarten van Bommel and Anita Quye, whose advice and support during the production of this volume have, as always, been invaluable, and particularly to Robert Atkinson, who did much of the work inserting corrections, redrawing molecular diagrams and generally helping in the preparation of the final texts. I am particularly grateful to Jim Black and Lyn Corson of Archetype Books, for their patience and cheerful encouragement.

Note: all plant names have been checked using The International Plant Names Index (2004). Published on the Internet http://www.ipni.org [accessed 3 January 2008] and other sources accessed by way of the Royal Botanic Gardens, Kew, website. Similar checks have been made on the names of animal species, fungi and lichens where these occur.

Max Saltzman: A Recollection

Jo Kirby

At the beginning of the Meeting, some words were said by Ran Boytner, University of California, Los Angeles, Dominique Cardon, CNRS, and Jo Kirby, The National Gallery, London, in memory of the eminent colour scientist Max Saltzman (1917–2001), who had died in February the previous year. One of these recollections is given below.

In 1991, the 10th Meeting of *Dyes in History and Archaeology* was held at the National Gallery in London. My colleague Raymond White and I sent out the invitations and then sat back to wait.

One morning, there was a telephone call. When I answered it, a voice said, 'Hullo, this is Max Saltzman. I've received your invitation and I'd like to come ...' The National Gallery Scientific Department works on many aspects of pigments and paint, including colour change. One of the best-known text books – indeed, a bible – in the field of colour science is *Principles of Colour Technology* by Fred Billmeyer and Max Saltzman [3rd edn by Roy S. Berns, Wiley-Interscience, 2000], thus when that telephone rang, we were suitably awe-struck.

Max came to London and sat for the whole afternoon having tea in our departmental library – 'No', he said, 'I don't want to see your department, thank you: I'm sure it all works fine' – and we talked for a very long time. We were extremely impressed by how well he knew our work: many of Raymond's publications have been in the field of natural resins, but Max knew them all.

After that DHA meeting, he managed to attend several others in spite of declining health, and he and I had many long telephone conversations after that first one.

Max Saltzman was a scientist whose opinions we valued highly. Even though he had retired, he kept his finger on the pulse, to be aware of people, publications and research in the field of colour. His comments on experimental procedure, for example, and on the importance of the correct standards for the experiment to be performed or the sample to be investigated – textile, dyestuff or whatever it might be – were very much to the point. I cannot begin to list the great number of his very readable publications in the fields of dyestuffs and paint technology. And they were eminently readable: that was important to him. Max could be gruff, grumpy and demanding, like an old grandfather; he spoke his mind, and he responded to charm. For myself, I will always be grateful for his kindness and generosity with his knowledge. I treasure his memory and I am lucky to have known him. Our debt to him is incalculable.

Contributors

Stylianos Arghyros, School of Plant Sciences, University of Reading, Reading RG6 6AS, United Kingdom.

Dr ing. Maarten R. van Bommel, Instituut Collectie Nederland, Department of Conservation Research, PO Box 76709, 1070 KA Amsterdam, The Netherlands.

Dr René Breniaux, Département Chimie analytique, L'Oréal Recherche, 1 avenue Eugène-Schueller, 93601 Aulnay-sous-Bois Cédex, France.

Dr Dominique Cardon, CIHAM/UMR 5648, 18 quai Claude-Bernard, 69365 Lyon Cédex 07, France.

Christopher J. Cooksey, 59 Swiss Avenue, Watford, Herts WD18 7LL, United Kingdom.

Dr Brian H. Davies, Department of Information Studies, Aberystwyth University, Aberystwyth, SY23 3AS, United Kingdom.

Dr François Delamare, École des Mines de Paris, CEMEF (Centre de mise en forme des matériaux), BP 207. F–06904 - Sophia Antipolis Cédex, France.

Professor Alan T. Dronsfield, School of Education, Health and Science, University of Derby, Kedleston Road, Derby DE22 1GB, United Kingdom.

Dr Cara L. Dunford, New Zealand Institute for Industrial Research, PO Box 31310, Lower Hutt, New Zealand.

Dr Glenn A. Fenton, New Zealand Institute for Industrial Research, PO Box 31310, Lower Hutt, New Zealand.

Jennifer Giaccai, Walters Art Museum, 600 North Charles Street, Baltimore, MD 21201, USA.

Dr Valery Golikov, Centre of Historical and Traditional Technologies, Russian Research Institute for Cultural and Natural Heritage, 2 Kosmonaut Street, 129366 Moscow, Russia.

Kathryn Hallett, Conservation and Collection Care, Historic Royal Palaces, Hampton Court Palace, Surrey, KT8 9AU, United Kingdom.

Dr Regina Hofmann-de Keijzer, University of Applied Arts Vienna, Institute for Conservation Science and Restoration-Technology, Department Archaeometry, Salzgries 14/1, A-1013 Vienna, Austria.

Professor Philip John, School of Plant Sciences, University of Reading, Reading RG6 6AS, United Kingdom.

Jo Kirby, Scientific Department, The National Gallery, Trafalgar Square, London, WC2N 5DN, United Kingdom.

Professor Zvi C. Koren, The Edelstein Center for the Analysis of Ancient Textiles and Related Artifacts, Shenkar College of Engineering and Design, 12 Anna Frank Street, 52526 Ramat-Gan, Israel.

Marie Marquet, UMR 5648, Université Louis Lumière - Lyon II, 18, quai Claude Bernard, F-69365 Lyon, Cédex 07, France.

Dr Maria João Melo, Departamento de Conservação e Restauro e REQUIMTE/CQFB, Universidade Nova de Lisboa, Quinta da Torre, 2829-516 Monte de Caparica, Portugal.

Dr Ian Miller, Carina Chemicals, PO Box 30366, Lower Hutt, New Zealand.

Dr Bernard Monasse, École des Mines de Paris, CEMEF (Centre de mise en forme des matériaux), BP 207. F–06904 - Sophia Antipolis Cédex, France.

Sonja Nicholson, School of Plant Sciences, University of Reading, Reading RG6 6AS, United Kingdom.

Dr Mari Omura, Gango-ji Institute for Research of Cultural Property, Nara 630-0257, Japan.

Dr Anthony Pinto, 71, rue des Sardes-Grignon, 73200 Albertville, France.

Cheryl Porter, Thesaurus Islamicus Foundation, 21 Misr Helwan al-Ziraa'i Street, al-Ma'adi, Cairo, Egypt.

Witold Przewoźny, The Ethnographic Museum, Division of the National Museum in Poznań, ul. Grobla 25, 61-858 Poznań, Poland.

Dr Anita Quye, Conservation and Analytical Research, National Museums Scotland, Chambers Street, Edinburgh, EH1 1JF, Scotland.

Dr Ghislaine Richard, Département Chimie analytique, L'Oréal Recherche, 1 avenue Eugène-Schueller, 93601 Aulnay-sous-Bois Cédex, France.

Dr Jana Sanyova, Institut Royal du Patrimoine Artistique/ Koninklijk Instituut voor het Kunstpatrimonium/ Royal Institute for Cultural Heritage, Parc du Cinquantenaire 1/ Jubelpark 1, B-1000 Brussels, Belgium.

Dr Ken Sasaki, Department of Biomolecular Engineering, Kyoto Institute of Technology, Hashigamicho, Matsugasaki, Sakyo-ku, Kyoto 606-8585 Japan.

Dr Yoshiko Sasaki, Museum and Archives, Kyoto Institute of Technology, Matsugasaki, Sakyo-ku, Kyoto 606-8585, Japan.

Dr Masanori Sato, Conservation and Restoration Science Laboratory, Nara National Research Institute for Cultural Properties, 2-9-1, Nijo-cho, Nara 630-8577, Japan.

Dr Katarzyna Schmidt-Przewoźna, Instytut Włókien Naturalnych (Institute of Natural Fibres), ul. Wojska Polskiego 71b, 60-630 Poznań, Poland.

Dr Gerald J. Smith, New Zealand Institute for Industrial Research, PO Box 31310, Lower Hutt, New Zealand.

Rangi Te Kanawa, Conservation Textiles, PO 256, Te Kuiti, New Zealand.

Ina Vanden Berghe, Koninklijk Instituut voor het Kunstpatrimonium/ Institut Royal du Patrimoine Artistique/ Royal Institute for Cultural Heritage, Jubelpark 1/ Parc du Cinquantenaire 1, B-1000 Brussels, Belgium.

Dr Arie Wallert, Rijksmuseum, Hobbemastraat 21, P.O. Box 74888, 1070 DN Amsterdam, The Netherlands.

Dr Jan Wouters, Conservation Scientist, Consultant, Neerbroek 54, B-2070 Zwijndrecht, Belgium.

Zoë Zharikova, Centre of Historical and Traditional Technologies, Russian Research Institute for Cultural and Natural Heritage, 2 Kosmonaut Street, 129366 Moscow, Russia.

Dr I. Irving Ziderman, Tekhelet Foundation, 20 Brody Street, 92502 Jerusalem, Israel.

Dye Analysis of Selected Textiles from Three Roman Sites in the Eastern Desert of Egypt: A Hypothesis on the Dyeing Technology in Roman and Coptic Egypt

Jan Wouters,* Ina Vanden Berghe, Ghislaine Richard, René Breniaux and Dominique Cardon

Introduction

Since 1994, a programme of systematic excavations in the Eastern Desert of Egypt has focused on the small Roman fortresses (*praesidia*) scattered along the two caravan roads leading from Koptos (Quft) to either Myos Hormos (Quseir) or Berenike, two important Red Sea coastal harbours at the beginning of the Christian era (see Fig. 1).[1] These excavations have elucidated the standard architecture and the double function of these fortresses (Fig. 2): to keep the roads safe from the desert Bedouin; and to provide travellers, and especially caravans, with an adequate supply of water, thanks to a clever system of connected wells, cisterns and watering troughs.[2]

The sites have produced interesting objects, including evidence of the long-distance traffic in industrial products that connected the Mediterranean world with Asiatic countries via these caravan roads. The rubbish heaps preserved on some of these sites have yielded great numbers of *ostraca*, inscribed pieces of broken pottery that were used as cheap 'note-pads'. These have provided much information about everyday life in the *praesidia* of the Eastern Desert.

The rubbish heaps have also produced huge quantities of textile fragments, which reflect the development of techniques achieved in the ancient world during the three first centuries AD and the variety of textile products circulating along these roads.

These finds have proved particularly useful to historians due to the meticulous excavation techniques employed by the archaeologists in charge of these sites, one of whom, Jean-Pierre Brun from the Centre National de la Recherche Scientifique, France (CNRS), has studied very precisely the stratigraphy of the deposits, in most cases enabling fairly secure datings by periods of 20 years.

After a brief examination in 1994 by the Danish textile specialist Ulla Mannering from the University of Copenhagen, the study of the textiles discovered in 1994 and 1995 at Maximianon was entrusted to Dominique Cardon in 1997. To this huge group, the study of which is still unfinished, was then added the less abundant corpus of textiles from Krokodilô, discovered in 1997 and completely studied in 1998 during a mission on the new excavations at Didymoi. In the same year, it was also possible to begin the study of textiles being discovered at Didymoi. For these first two missions, in 1997 and 1998, the Egyptian government had given permission to take a limited number of samples for dye analysis. Since then, the study of the textiles from Didymoi has continued under the direction of Dominique Cardon with the collaboration of Hero Granger-Taylor, author of

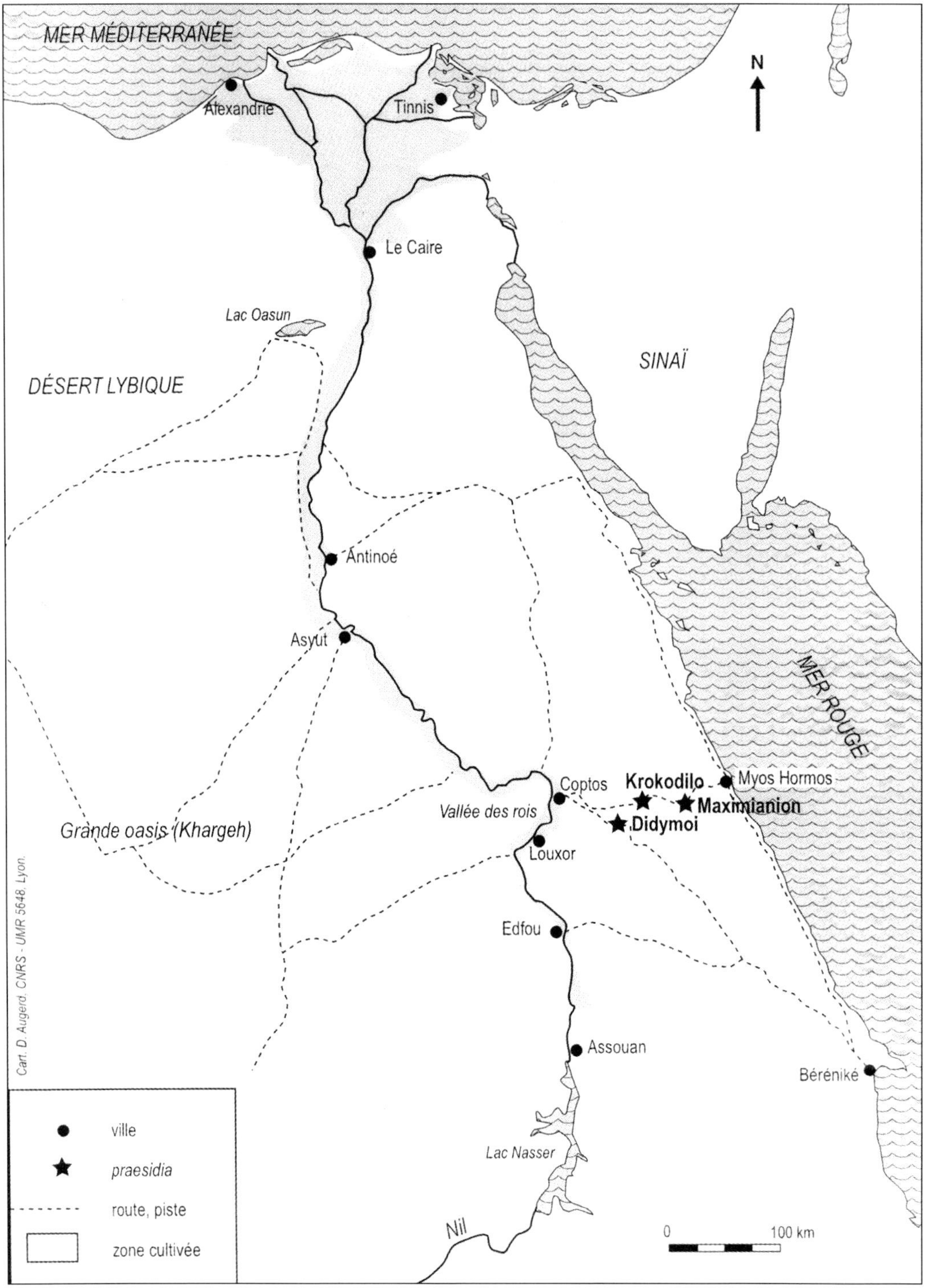

Figure 1 Location of Maximianon, Krokodilô and Didymoi, the only *praesidia* on the caravan roads from the Nile Valley to the Red Sea harbours where rubbish heaps with textiles have been preserved.

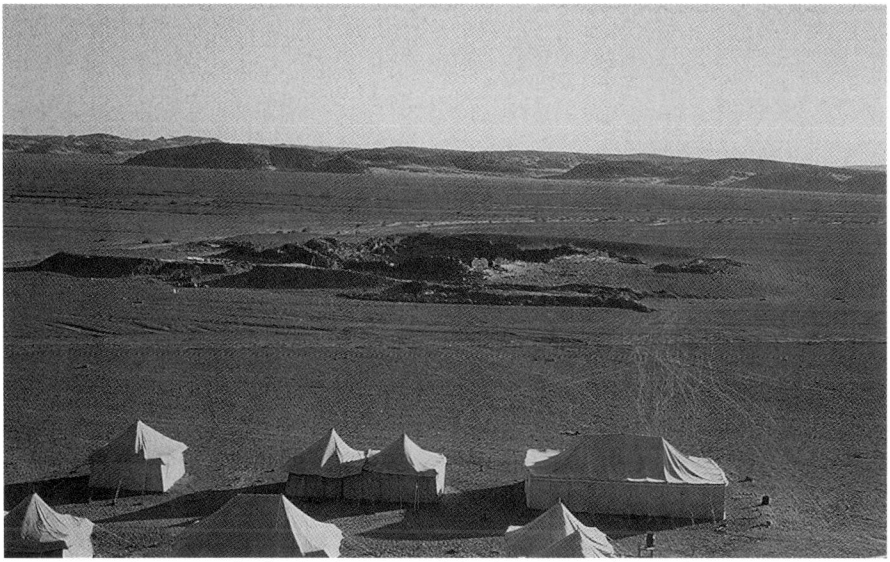

Figure 2 Didymoi: at the back, the ruins of the fortress; in front of the ruins, the enormous rubbish heap. In the foreground, the camp of the archaeological mission. (Photo: D. Cardon)

several publications on groups of archaeological textiles from eastern Mediterranean countries of similar date to, or slightly earlier than, those of the *praesidia*.[3]

The selection of samples for dye analysis

Small groups of textiles from Maximianon, Krokodilô and Didymoi, selected for their importance to the history of textile techniques, have already been published and a chapter on the textiles from Maximianon and Krokodilô, as well as a selected catalogue, was included in the recent publication of the Roman sites of the road to Myos Hormos.[4] However, only one dye analysis of a textile from the *praesidia*, performed by Witold Nowik at the Laboratoire de Recherche des Monuments Historiques in Champs-sur-Marne, has been published so far: that of a very fine purple band on a piece of tunic in damask weave from Didymoi.[5]

There was no doubt that a programme of carefully planned dye analyses of such a large group of securely dated textiles would contribute to the knowledge of dyeing technology in the eastern Mediterranean in antiquity. Moreover, broader sampling would also offer an opportunity to compare the dyeing techniques in Roman Egypt with those of Coptic Egypt.[6]

Seventy-four dye analyses from 54 textiles were performed at the Laboratory for Materials and Techniques of the Royal Institute for Cultural Heritage (KIK/IRPA), Brussels, Belgium, thanks to a generous sponsorship from L'Oréal, an international firm specialising in cosmetics. A further group of 20 red samples was analysed at L'Oréal's own Laboratoire de Chimie Analytique (Laboratory of Analytical Chemistry). Altogether, 94 dyes from 74 different textiles, found at three different, but similar, sites occupied during the first three centuries of the present era, were analysed.

The selection of six different groups of textiles (Table 1) was made with the aim of revealing new information and perhaps answer some of the remaining historical questions on both ancient weaving and dyeing technology.

The first three groups of samples were selected from types of textiles that are of particular interest from the point of view of the history of weaving technology. Group 1, the weft-faced compound tabbies (usually called by their technical name in French, *taquetés*; see Fig. 3) and group 2, the damask twills from the three sites, figure among the earliest examples (in the case of the *taquetés*) or, in the case of the damasks, actually represent the earliest examples discovered so far of these two types of complex weaves. Both techniques first appear at the beginning of our era. They are found first in wool textiles from a few archaeological sites, mainly in the eastern Mediterranean world, but are then used more and more commonly for weaving elaborate patterns on precious textiles made of wool or silk. Because of this historical importance, samples of all the red yarns and of some of the yellow or green yarns from all the *taquetés* and damasks discovered on the three sites up to 1998 were analysed, with the exception of the fragments that were too small to allow samples to be taken without damaging the general pattern of the textile.

Group 3 consists of a very unusual collection of tubular weaves with warp stripes in 2/1 chevron twill. This type of textile appears in the three sites of Maximianon, Krokodilô and Didymoi, but has not been described from anywhere else, including the contemporary sites in Egypt and in the neighbouring regions of the eastern Mediterranean. In this case too, the red warp threads of all the textiles from this group were analysed.

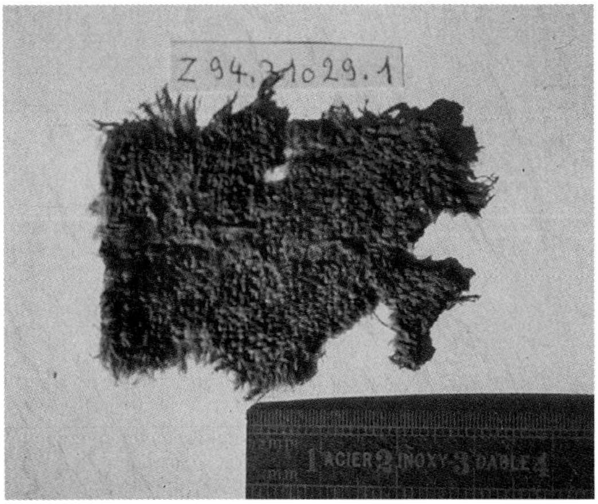

Figure 3 A typical example of the geometrical, reversible designs, alternately in red and yellow and yellow and blue, on the early wool *taquetés* from the *praesidia* of the Eastern Desert of Egypt. Fragment Z94.31029.1 from Maximianon. (Photo: D. Cardon)

The fourth group is composed of samples from different types of black or brown/black textiles, or parts of textiles. The contemporary written sources show that two main ways to dye black were known at the time: the tannin-plus-iron mordant process, and the process consisting of top-dyeing a dark blue ground with a red dye, a mordant being applied before the second dye-bath. Dye analyses were planned to determine which of these technical solutions was preferred by dyers for the different types of textiles.

Lastly, the high proportion and great variety of red and purple/violet textiles included in this selection were meant to contribute not only to technological questions, such as the imitation of the most precious dyes, extracted from insects and seashells, but also to broader issues of economic and social history: would highly priced dyestuffs such as shellfish purple or kermes be present at all in textiles from these sites, inhabited mostly (if not only) by middle-class or lower-class people? Effectively, the *ostraca* depict the *praesidia* as small communities of ordinary soldiers, Roman and non-Roman, commanded by low- to medium-rank officers, some accompanied by their families; all served and surrounded by a satellite group of people, the *familia*, providing commodities ranging from fresh fruit, vegetables and wine, to prostitutes!

Methods

All descriptions and classifications of textiles and fragments were performed by Dominique Cardon. All dye analyses were carried out by high-performance liquid chromatography with photodiode array detection (HPLC–PDA) either at the Laboratory for Materials and Techniques of the Royal Institute for Cultural Heritage (KIK/IRPA), Brussels, Belgium, or at the Laboratoire de Chimie Analytique, L'Oréal, Aulnay-sous-Bois, France. The analytical conditions were as follows.

KIK

To a few millimetres of dyed yarn, 250 μL of water/methanol/37% hydrochloric acid (1/1/2, v/v/v) was added. The mixture was heated for 10 min at 105 °C in open Pyrex tubes in a heating block. After cooling, any particulate matter was removed by filtering through a porous polyethylene frit (Vac-Elut, Analytichem, USA). The clear filtrate was dried in an evacuated desiccator over sodium hydroxide. The dry residues were taken up in 50–100 μL methanol/water (1/1, v/v) and 20 μL of this solution was injected into the chromatograph. The HPLC equipment consisted of a high-pressure pump programmable for flow rate, time periods and composition of the eluting solvents (Model M615, Waters, USA); a column with renewable cartridges of 4.0 × 125 mm of Lichrosorb-RP18, 5 μm (Merck, Overijse, Belgium); a photodiode array detector (model 996, Waters, USA); a system for data storage, manipulation and retrieval (Millennium, Waters, USA). Three solvents were used: (a) water, (b) methanol, and (c) 5% (w/v) phosphoric acid in water. The elution program was 60A/30B/10C: 3 min; linear gradient to 10A/80BA/10C: 26 min; flow rate: 1.2 mL min^{-1}, creating a system back-pressure of 18 to 24 MPa. The temperature in the chromatography laboratory was stabilised between 20 °C and 22 °C. Any dye component was identified according to two criteria: the retention time and the UV-visible spectrum (recorded between 200 and 800 nm).

L'Oréal

A sample of coloured wool (0.1–10 mg, depending on the amount of textile available) was extracted with a mixture of methanol/water/hydrochloric acid (0.5/0.5/0.5 mL) by heating at 85 °C for 10 min with stirring in a 5 mL capped vial using a ReactiTherm (Pierce Chemical Co., Rockford, Illinois, USA). After cooling for 5 min in a cold water bath, the content of the vial was transferred to a 60 mL separating funnel with 2 × 10 mL water. The extraction of the dyes from the aqueous media was achieved with 4 × 5 mL portions of ethyl acetate. The upper organic layer was dried under vacuum at 30 °C and dissolved in 50% methanol before HPLC analysis.

Chromatography was performed using a Waters 625 LC Pump, Waters 717 plus injector, Waters PDA 996 detector and Waters Millennium software (Waters, Milford, USA). A 20 μL volume of the extract was injected and the components separated on a 4.6 × 150 mm Alltima C18 column, particle size 5μm (Alltech, ref. 88050). The separation of madder dyes was achieved within 40 min with an isocratic elution using methanol/1% acetic acid (75/25, v/v). For the separation of insect dyes a gradient was employed: the initial conditions were methanol/1% acetic acid (40/60 v/v). A linear increase to 80% methanol was performed within 40 min from the initial time and then elution with methanol/1% acetic acid (80/20, v/v) was maintained for 10 min (total run time 60 min).

Results

General

All results are given in Table 1 as a function of fabric characteristics and colour. All figures refer to the relative presence of dyestuff components calculated from integration data obtained at the wavelength indicated.

All the results of Table 1 are summarised in Table 2. Of the 94 analytical results, 85 contain components indicating the presence of a type of madder. Thirty-six out of the 39 results that point to the presence of an indigoid dye (not including true shellfish purple), show combinations with a madder type. These served mainly to obtain shades of purple and blacks, but also to achieve special shades of green (any influence of ageing on the original hue is unknown): in group 6c, Z95.14011.2, the green ground of a textile decorated by a beautiful 'shaded band' in shades of purplish pink-to-mauve and dark purple (see Plate 2); in group 6d, K.3732.2, the almond-green warp of a fine half basket weave with a violet weft. They could even be used to

Table 1 Results of dye analyses as a function of fabric characteristics and colour. One result per sample involving acid treatment; ag: anthragallol; al: alizarin; ea: ellagic acid; in: indigotin; ir: indirubin; ka: kermesic acid; law: lawsone; lu: luteolin; lu': a component displaying the same UV-VIS spectral characteristics as luteolin, but eluting at a different retention time; mbi: 6-monobromoindigotin; mu: munjistin; pca: protocatechuic acid; pu: purpurin; xp: xanthopurpurin.

Site	Description	Yarn	Dye composition (λ in nm)	Dyestuff(s)	Laboratory
Group 1: Taqueté (weft-faced compound tabby)					
Maximianon	Z95.13004.3: geometrical design in red and yellow	red weft	17 al, 83 pu (255)	madder type	KIK/IRPA
		yellow weft	53 lu, 27 al, 20 pu (255)	weld + madder type	
	Z94.31029.1: geometrical design in red, yellow and blue	red weft	51 al, 49 pu (255)	madder type	KIK/IRPA
		yellow weft	43 pca, 7 lu, 27 al, 10 in, 13 pu (255)	weld + madder type + indigoid	
Didymoi	D98.13901.1: design of waves and branches in red, blue and yellow	red weft	51 al, 49 pu (255)	madder type	L'Oréal
	D98.1410.2: design of flowers and birds in red and yellow	red weft	30 al, 70 pu (255)	madder type	L'Oréal
	D98.2305.2: geometrical design in blue and yellow and the beginning of a band with a pink ground	pink weft	14 al, 86 pu (255)	madder type	L'Oréal
	D98.3410.1: geometrical design in red, blue and yellow	red weft	14 al, 86 pu (255)	madder type	L'Oréal
Group 2: Damask twills					
Krokodilô	K2406.1: dark-pink *clavus* on a cream-coloured ground	pink weft	+ ea, + lu, 3 al, 4 in, 92 pu (255)	madder type + indigoid + trace of tannin and weld	KIK/IRPA
	K2550.3: plain pinkish orange	orange	4 pca, 6 ea, + lu, 61 al, 28 pu (255)	madder type + tannin + trace of weld	KIK/IRPA
	K2415.1: plain orange	orange	15 lu, 1 ap, 42 al, 3 xp, 39 pu (255)	madder type + weld	KIK/IRPA
	K3737.26: plain green	green	83 lu, 4 ap, 8 in, 2 ir, 3 lu' (255)	weld + indigoid	KIK/IRPA
Maximianon	Z95.23012B.7: plain light greyish green	green	16 lu, 8 al, 76 pu (255)	madder type + weld	KIK/IRPA
	Z95.23012B.5: plain orange-red	red	+ law, 1 mu, 15 al, 83 pu (255)	madder type + trace of lawsone	KIK/IRPA
	Z95.33042.13: plain pinkish red	red	12 al, 88 pu (255)	madder type	KIK/IRPA
Didymoi	D98.2409.1B: plain red	red	21 al, 79 pu (254)	madder type	L'Oréal
	D98.2309.1: plain red-orange	red	30 al, 70 pu (254)	madder type	L'Oréal
	D98.2305.1: plain light red	red	14 al, 86 pu (254)	madder type	L'Oréal
	D14.HS3: plain orange-red	red warp	14 al, 86 pu (254)	madder type	L'Oréal
Group 3: Tubular 2/1 chevron twills					
Krokodilô	K3425.1: coloured chevron	red warp	+ ag, + mu, 50 al, 1 xp, 48 pu (255)	madder type	KIK/IRPA
Maximianon	Z95.33029.10: coloured chevron	red warp	3 mu, 29 al, 2 xp, 66 pu (255)	madder type	KIK/IRPA
Didymoi	D98.3417.1: coloured chevron	red warp	83 al, 17 pu (254)	madder type	L'Oréal
Group 4: Textiles with black bands or a black ground					
Krokodilô	K2416.7: cream-coloured tabby ground with black *clavus*	black weft	+ ag, 31 al, 7 in, 55 pu, 7 ir (255)	madder type + indigoid	KIK/IRPA

	K.H.S.15: cream-coloured tabby ground with black *clavus* in weft-faced plain weave	black weft	+ ag, 36 al, 6 in, 55 pu, 3 ir (255)	madder type + indigoid	KIK/IRPA
	K4713.1: red tabby ground with black *clavus*	red ground	41 al, 59 pu (255)	madder type	KIK/IRPA
		black weft	21 al, 4 in, 73 pu, 2 ir (255)	madder type + indigoid	KIK/IRPA
	K2405.1: red tabby ground with black *clavus*	red ground	1 mu, 35 al, 64 pu (255)	madder type	KIK/IRPA
		black weft	8 al, 35 in, 53 pu, 4 ir (255)	madder type + indigoid	KIK/IRPA
	K5701.1: red tabby ground with black *clavus*	red ground	47 al, 53 pu (255)	madder type	KIK/IRPA
		black weft	16 ea, 19 lu, 9 al, 44 in, 5 pu, 7 ir (255)	indigoid + weld + madder type + tannin	KIK/IRPA
	K2416.6: plain black	black	100 ea (255)	tannin	KIK/IRPA
	K2418.1: red tabby ground with very dark blue-green *clavus*	red ground	+ ag, 2 mu, 34 al, 64 pu (255)	madder type	KIK/IRPA
		blue-green weft	10 al, 5 in, 84 pu, 1 ir (255)	madder type + indigoid	KIK/IRPA
Maximianon	Z95.28012.B11: red half basket ground with black *clavus*	red ground	1 ag, 33 al, 66 pu (255)	madder type	KIK/IRPA
Group 5a: Plain red or orange tabby or half basket weave ground					
Didymoi	D1418.1: thin red tabby	red	39 al, 61 pu (254)	madder type	L'Oréal
	D3415.3: thin crimson-red tabby	red	67 al, 33 pu (254)	madder type	L'Oréal
Group 5b: Red clavus, *purple* clavus, *tapestry pattern, stripes or checks on tabby ground*					
Krokodilô	K2407.3: cream-coloured tabby with striped *clavus*	purple weft	2 pca, 15 al, 18 in, 65 pu (255)	madder type + indigoid	KIK/IRPA
		dark purple weft	1 pca, + ag, 1 mu, 34 al, 3 in, 61 pu (255)	madder type + indigoid	KIK/IRPA
Maximianon	Z94.32030.1: cream-coloured tabby with purple *clavus*	purple weft	7 al, 15 in, 77 pu, 1 ir (255)	madder type + indigoid	KIK/IRPA
	Z95.33029.11: cream-coloured tabby with purple *clavus*	purple weft	3 al, 7 in, 90 pu (255)	madder type + indigoid	KIK/IRPA
	Z95.13008.B10: cream-coloured tabby with purple *clavus*	purple weft	5 al, 7 in, 88 pu (255)	madder type + indigoid	KIK/IRPA
	Z95.23020.B1: cream-coloured tabby with reddish-purple *clavus*	purple weft	7 al, 4 in, 89 pu (255)	madder type + indigoid	KIK/IRPA
Didymoi	D1412.1: tabby with multicoloured checks	red	42 al, 58 pu (254)	madder type	L'Oréal
	D1431.5: cream-coloured tabby ground with dark red *clavus*	red weft	41 al, 59 pu (254)	madder type	L'Oréal
	D.HS.10: brown 2/2 twill with red thread on sewn border	red thread	1 al, 99 pu (254)	madder type	L'Oréal
	D4409.1: red design of a fragment of tapestry	red weft	14 al, 86 pu (254)	madder type	L'Oréal
	K.H.S.1: cream-coloured tabby with purple *clavus*	purple weft	+ mu, 7 al, 2 in, 90 pu, + ir (255)	madder type + indigoid	KIK/IRPA

(contd.)

Site	Description	Yarn	Dye composition (λ in nm)	Dyestuff(s)	Laboratory
Group 5c: Thin, dense twills, with a plain red ground					
Didymoi	D1412.20: diamond twill	red	39 al, 61 pu (254)	madder type	L'Oréal
	D2305.10: 2/1 twill	red	57 al, 43 pu (254)	madder type	L'Oréal
	D23000.HS.1: 2/2 diamond twill	red warp	59 al, 41 pu (254)	madder type	L'Oréal
Group 5d: Coarse 2/2, z/z twills with a plain orange to salmon pink ground					
	D1431.1: salmon pink	pink weft	3 al, 97 pu (254)	madder type	L'Oréal
Group 6a: Textiles with a ground in a purplish hue					
Krokodilô	K5701.5: basket weave	purple-brown	+ ag, 1 mu, 32 al, 6 in, 61 pu (255)	madder type + indigoid	KIK/IRPA
Maximianon	Z94.22030.13.: 2/2 twill, zS2/z spinning system	purple-red	+ ag, 43 al, 58 pu (255)	madder type	KIK/IRPA
	Z94.31030.88.33: tabby	red	21 al, 79 pu (255)	madder type	KIK/IRPA
	Z94.32016.76.3: tabby	red	1 mu, 55 al, 44 pu (255)	madder type	KIK/IRPA
	Z94.32005.194.1: half basket weave, z-spun warp, z-spun weft	orange	6 ea, 51 al, 43 pu (255)	madder type + tannin	KIK/IRPA
	Z95.33029.3: dense 2/2 twill, dark purplish-brown ground	purplish brown	5 pca, 95 ea (255)	tannin	KIK/IRPA
Group 6b: Textiles with an undyed or striped ground and with purple tapestry bands or geometrical patterns					
Krokodilô	K3709.1: uncoloured ground with purplish-pink *clavus*	purple weft	6 al, 6 in, 87 pu, 1 ir (255)	madder type + indigoid	KIK/IRPA
Maximianon	Z95.14011.1: tubular knitting with white and purple stripes	purple	5 pca, 7 al, 5 in, 82 pu, 1 ir (255)	madder type + indigoid	KIK/IRPA
	Z95.33044.6: cream-coloured tabby ground with purple *clavus*	purple weft	4 al, 15 in, 81 pu (255)	madder type + indigoid	KIK/IRPA
	Z95.23024.1: cream-coloured tabby ground with purple gamma	purple weft	8 al, 8 in, 82 pu, 2 ir (255)	madder type + indigoid	KIK/IRPA
	Z94.32029.2: cream-coloured tabby ground with purple gamma	purple weft	7 al, 10 in, 83 pu (255)	madder type + indigoid	KIK/IRPA
	Z95.24007.1: cream-coloured tabby ground with purple gamma	purple weft	11 al, 5 in, 83 pu, 1 ir (255)	madder type + indigoid	KIK/IRPA
	Z94.22034.100.19: cream-coloured tabby ground with red *clavus*	red weft	1 ea, + ag, 37 al, 62 pu (255)	madder type + tannin	KIK/IRPA
	Z94.22010.5: cream-coloured half basket weave with purple *clavus*	purple weft	72 in, 28 mbi (288)	royal purple	KIK/IRPA
Group 6c: Textiles with purple bands on a brown, green or red ground					
Krokodilô	K5701.6: purple *clavus* on red tabby ground	purple weft	13 al, 7 in, 80 pu (255)	madder type + indigoid	KIK/IRPA
		red ground	35 ea, 45 al, 20 pu (255)	madder type + tannin	KIK/IRPA
	K2418.3: purple *clavus* on brown tabby ground	purple weft	1 ea, + mu, 9 al, 4 in, 86 pu (255)	madder type + indigoid + tannin	KIK/IRPA
		brown ground	100 ea (255)	tannin	KIK/IRPA

	K2315.1: purple *clavus* on blue-green tabby ground	purple weft	16 al, 8 in, 76 pu (255)	madder type + indigoid	KIK/IRPA
		green ground	93 in, 7 ir (288)	indigoid	KIK/IRPA
	K3737.1: purple *clavus* on red tabby ground	purple weft	79 ka, 7 al, 9 in, 4 pu, 1 mbi (255)	kermes + madder type + royal purple	KIK/IRPA
		red ground	1 ea, 1 ag, 50 al, 48 pu (255)	madder type + tannin	KIK/IRPA
Maximianon	Z95.14011.2: *clavus* in multicoloured tapestry weave on green tabby ground	purple weft	+ ag, 2 mu, 49 al, 6 in, 43 pu (255)	madder type + indigoid	KIK/IRPA
		pink weft	9 al, 8 in, 83 pu (255)	madder type + indigoid	KIK/IRPA
		greyish-purple weft	7 al, 15 in, 78 pu (255)	madder type + indigoid	KIK/IRPA
				madder type + indigoid	
		green ground	10 al, 87 in, 3 pu (255)	indigoid + madder type	KIK/IRPA
	Z95.33024.5: pink *clavus* on basket-weave green ground	pink weft	3 lu, 8 al, 15 in, 74 pu (255)	madder type + indigoid + weld	KIK/IRPA
		green ground	27 lu, 5 al, 66 in, 2 pu (255)	indigoid + weld + madder type	KIK/IRPA
	Z95.33029.9: pink angle in tapestry weave on dark blue-green tabby ground, fragment 1	pink weft	77 ka, 18 in, 5 mbi (255)	kermes + royal purple	KIK/IRPA
			52 ka, 39 in, 9 mbi (288)		
		green ground	95 in, 5 ir (288)	indigoid	KIK/IRPA
	Id., fragment 2	pink weft	73 ka, 19 in, 8 mbi (255)	kermes + royal purple	KIK/IRPA
			43 ka, 46 in, 11 mbi (288)		
	Z94.271003.5: dense 2/2 twill, plain red ground	red	+ ag, 1 mu, 45 al, 1 xp, 53 pu (255)	madder type	KIK/IRPA
	Z95.26.007.9: dense 2/2 twill, plain red ground	red	1 mu, 11 al, 88 pu (255)	madder type	KIK/IRPA
	Z94.22010.7: 2/2 twill, plain red ground	red	1 mu, 11 al, 88 pu (255)	madder type	KIK/IRPA
	Z95.33044.8: purple *clavus* on brown tabby ground	purple weft	17 ea, 3 al, 10 in, 70 pu (255)	madder type + indigoid	KIK/IRPA
Didymoi	D3410.2: purple *clavus* on red tabby ground	red warp	34 al, 66 pu (254)	madder type	L'Oréal
Group 6d: Miscellaneous					
Krokodilô	K3732.2: half basket weave	purple weft	2 al, 4 in, 94 pu (255)	madder type + indigoid	KIK/IRPA
		green warp	85 in, 8 pu, 7 ir (255)	indigoid + madder type	KIK/IRPA
	K4709.1: coarse 2/2 twill, z-spun warp, z-spun weft, plain orange ground	orange ground	44 pca, 45 al, 11 pu (255)	madder type	KIK/IRPA
Maximianon	Z94.22012.191.7: coarse 2/2 twill, z-spun warp, z-spun weft, salmon-pink ground	pink ground	62 al, 38 pu (255)	madder type	KIK/IRPA

Table 2 Summary of dye analysis results.

Source	Number of detections
madder type	39
madder type + indigoid	28
madder type + tannin	4
tannin	3
madder + weld	3
madder + indigoid + tannin	3
madder + indigoid + weld	3
indigoid	2
kermes + murex purple	2
madder type + indigoid + tannin + weld	2
murex purple	1
madder type + lawsone	1
weld + indigoid	1
kermes + madder type + murex purple	1
madder + tannin + weld	1
TOTAL	94

obtain a particular shade of yellow (group 1, Z94.31029.1). In four cases, components that relate to shellfish purple were found: once as a pure dye, twice combined with kermes and once with kermes and a type of madder. All flavonoid compositions found indicated the use of weld. The detection of a small amount of lawsone in one case remains unexplained.

Type of madder

Analyses executed within this project and those resulting from earlier work on Coptic textiles,[7] on medieval European textiles[8] and on 16th–17th-century tapestries from Bruges,[9] together with the data resulting from a fundamental study of the composition of madder dyes in plant root extracts and on dyed wool,[10] formed a solid basis for an evaluation of madder compositions in a wide geographical and historical context. Moreover, many of the results pointing to madder were obtained in two different laboratories. All these data are represented in Table 3. The most important features of the *Rubia peregrina* L. references were the low content of alizarin (between 0 and 21% when taking into account the deviations) and the relatively high content of rubiadin measured in several samples at KIK/IRPA, but this contrasts with the absence of this anthraquinone in the single reference sample analysed in the L'Oréal laboratory. In *Relbunium hypocarpium* (L.) Hemsl. preparations, alizarin was always absent. In *Rubia tinctorum* L. reference samples, alizarin and purpurin were found in approximately equal amounts (and with statistically overlapping integration values). When compared to historical samples, the following observations could be made. High relative amounts of rubiadin were not found in any historical sample. Statistically equal values for alizarin and purpurin occurred in samples from Coptic Egypt that were not purple or black and in those from the Bruges tapestries. Remarkably low amounts of purpurin were calculated from the medieval European fabrics. On the other hand, all samples from Roman Egypt and the purple and black colours from the Coptic era showed statistically lower values for alizarin than would be expected from *Rubia tinctorum*.

These observations formed the basis of another way of representing the results from red, purple and black colours. Taking into account the statistical spread of purpurin in *Rubia peregrina* reference samples ranging from 0 to 21% and the more or less equal mean amounts of alizarin and purpurin in *Rubia tinctorum*, these data were grouped according to relative amounts of alizarin falling into the following ranges: 0–20%; 21–50%; 51–80%; and more than 80%. The results of these calculations are represented in Figure 4. It is more obvious now that the mean results from European medieval textiles resulted from a distribution of alizarin

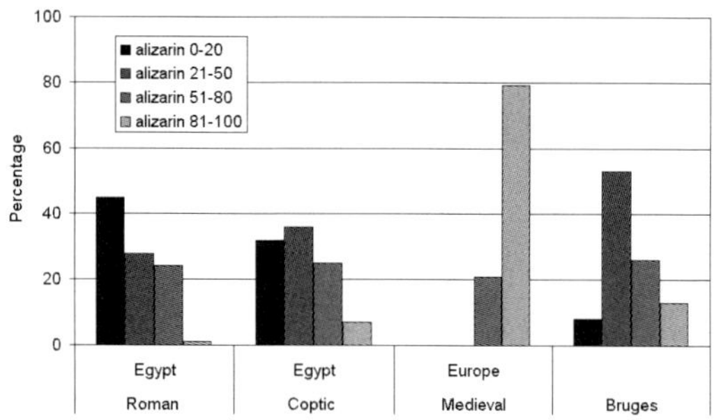

Figure 4 Madder compositions, grouped according to the relative amounts of alizarin.

Table 3 Summary of madder compositions found in laboratory reference and historic samples.

	<alizarin> (sd)	<purpurin> (sd)	<rubiadin> (sd)
References			
Rubia peregrina (KIK/IRPA), n= 35	9 (12)	59 (26)	32 (23)
Rubia peregrina (L'Oréal), n = 1	1	99	0
Rubia tinctorum (KIK/IRPA), n = 5	54 (15)	45 (13)	1 (1)
Relbunium hypocarpium (KIK/IRPA), n = 16	0 (0)	99 (18)	1 (1)
Roman Egypt (1st–2nd century AD)			
Didymoi reds (L'Oréal), n = 20	34 (22)	66 (22)	0 (0)
Others (KIK/IRPA), n = 66	29 (23)	71 (23)	0 (0)
Coptic Egypt (3rd–8th century AD)			
Purple/black (KIK/IRPA), n = 7	36 (35)	64 (34)	0 (0)
Others (KIK/IRPA), n = 21	52 (31)	48 (30)	0 (1)
Medieval Europe (10th–15th century AD)			
(KIK/IRPA), n = 24	91 (10)	9 (10)	0 (0)
Bruges tapestries (16th–17th century AD)			
(KIK/IRPA), n = 38	49 (24)	50 (24)	2 (3)

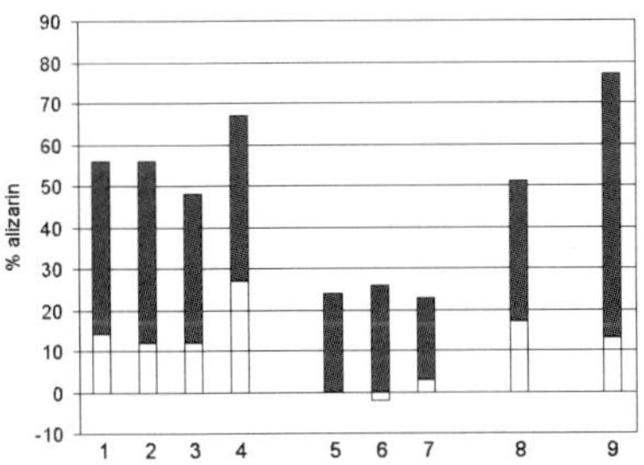

Figure 5 Madder compositions in Roman Egyptian samples, grouped according to the relative amounts of alizarin, the actual colour and the site of provenance: (1) all reds; (2) reds from Didymoi; (3) reds from Maximianon; (4) reds from Krokodilô; (5) all purples; (6) purples from Maximianon; (7) purples from Krokodilô; (8) all blacks; (9) all other colours. Grey blocks represent the spread of % alizarin found.

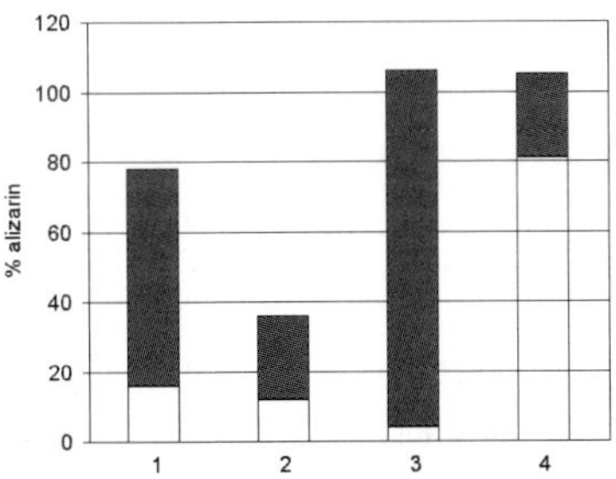

Figure 6 Madder compositions in Coptic Egyptian samples, grouped according to the relative amounts of alizarin and to the actual colour: (1) all reds; (2) all purples; (3) all blacks; (4) all other colours. Grey blocks represent the spread of % alizarin found.

and purpurin completely different from that in those of the Egyptian area. Indeed, the dyes with a low alizarin content were more predominant in the latter.

This observation then formed the basis for the representation of the alizarin groups, but now as a function of colour and also according to site of provenance. The results of this are represented in Figure 5 for the Roman Egyptian samples and in Figure 6 for those from Coptic Egypt. From these two figures it is obvious that the madder composition in purples is different from that in the reds, blacks and any other colour, in that all alizarin ratios of the purples belong to the sub-20% group. Since these purples were produced by mordanting and top-dyeing madder over vat-dyed indigo, it was necessary to find out whether or not the procedure of top-dyeing as such influenced the anthraquinoid component ratios in the madder dye. Therefore, all results from Roman and Coptic Egypt, where components from both an indigoid and a madder dye source were found, were reconsidered as follows. Indigotin and indirubin on the one hand and alizarin and purpurin on the other were integrated at 255 nm and the results of both recalculated to show the relative presence of an indigoid and a madder-type dye, respectively. All calculations were statistically evaluated. The results are presented in Table 4.

Table 4 The relative presence of components from indigoid and madder dye sources in Roman and Coptic Egyptian textiles. All calculated values result from integrations at 255 nm.

Colour	Indigotin + indirubin (sd)	Alizarin + purpurin (sd)
Roman Egyptian reds (n = 5)	8 (4)	92 (4)
Roman Egyptian purples (n = 22)	8 (5)	92 (5)
Roman Egyptian blacks (n = 5)	26 (25)	74 (25)
Roman Egyptian others (n = 6)	63 (40)	37 (40)
Roman Egyptian purples from Maximianon (n = 13)	9 (4)	91 (4)
Roman Egyptian purples from Krokodilô (n = 9)	7 (5)	93 (5)
Coptic Egyptian purples (n = 4)	9 (5)	91 (5)
Coptic Egyptian blacks (n = 3)	34 (23)	66 (23)

Considering the group of Roman Egyptian colours, it seems that the relative ratio of the two sources is the same in the reds and the purples, but different to those in the blacks and the other colours. The results from Coptic Egypt also show a substantially different source ratio in the purples as compared to the blacks. Moreover, the *biological* source ratios of purples are similar in the Roman Egyptian sites Maximianon and Krokodilô. All this means that top-dyeing to produce purple does not seem to compromise the ratio between the indigoid and madder dye sources, because this ratio is exactly the same as that found in reds. This then leads to the suggestion that the lower alizarin – and, hence, higher purpurin – content of the madder source to produce purple may have been desired because of the more effective contribution made to purple by purpurin as compared to alizarin. The madder composition in the blacks was comparable to that of the reds, i.e. it had a higher alizarin content. The relative proportions of alizarin and purpurin should be considered less critical for the production of black, where, indeed, the indigoid source was found to be more prominent.

It is impossible to say at present whether the low-alizarin-content madder in the Egyptian purples may be ascribed to a different biological source or to a different dyeing procedure. However, it can be proposed that the madder dyeing technology for producing purple by top-dyeing madder over indigo was different to that for producing any other colour involving madder.

Purple

Apart from top-dyeing madder over indigo, purple hues may also be obtained from lichens, alkanet roots and certain marine snails (royal purple and shellfish purple). Lichens and alkanet dyes were not found in Roman or Coptic Egyptian textiles. Royal purple was detected in four samples from the Roman period. The information given in Table 5 should be interpreted taking into account certain factors in the analysis of indigoid compounds. Indigoid dyestuff components, including the brominated ones that occur in the different muricid species, may be analysed by HPLC.[11] However, experience built up over the years at the laboratory of KIK/IRPA suggests that the detection of these components on archaeological samples is not straightforward. Indeed, the properties of indigoid products, especially in terms of solubility, contradict their apparent 'availability' to be analysed in aqueous HPLC systems. Obviously, it should be taken into account that preparations and HPLC separations of indigoid dye components, especially the brominated ones, represent a high degree of instability that may lead to low reproducibility of analytical results and to the loss of components with a prominent diagnostic value. Therefore, when the presence of shellfish purple in an archaeological sample is suspected, several approaches are always combined: one involving the normal acid treatment procedure, in order not to disregard any other possible dye component; one involving extraction in hot dimethylformamide (DMF); one repeating one or both of the former; and one involving different integration wavelengths so as to detect components of purple at their optimum wavelength of 288 nm. A conclusion is then drawn based on all the results.

The result from an acid-treated sample of K3737.1 (see Plate 3) showed components from kermes (ka: kermesic acid), madder (al: alizarin, pu: purpurin) and royal purple (in: indigotin, mbi: monobromoindigotin) when integration was carried out at 255 nm. Due to their low concentration, alizarin and purpurin disappeared in the result of integration at 288 nm, but the relative presence of the purple components became more prominent. Surprisingly, mbi was absent in the DMF extract and indirubin (ir) appeared. It was concluded that this purple was composed of kermes (*Kermes vermilio* Planchon), madder (*Rubia tinctorum* or equivalent) and royal purple (no species determined).

The results from acid treatment and DMF extraction of sample Z94.22010.5 were highly consistent. Only indigoid components of purple were found, albeit at different ratios. This sample contained a pure royal purple, but the exact murex species was not identified.

Textile sample Z95.32029.9 consisted of two fragments with a purplish-pink coloured tapestry-woven angle (Plate 4). From fragment 1, two independent samples were acid treated and analysed ((AH)a and (AH)b). In each case, components from kermes and royal purple were found in reproducible relative ratios. Surprisingly again, only indigotin and indirubin appeared in the DMF extract. One acid-treated sample from fragment 2 also showed components of kermes and royal purple, moreover in relative ratios perfectly comparable to those of fragment 1. In this case, the same indigoid components of purple were found in the DMF extracted sample. It was concluded here that the two fragments contained purple of the same composition (kermes and royal purple) and, except for madder components, were very closely related to the purple from sample K3737.1.

Conclusion

Ninety-four samples, taken from 74 textiles discovered at the archaeological sites of Didymoi, Krokodilô and Maximianon and belonging to the Roman Egyptian historic period, were analysed for the presence of dyes. This number of samples and fabrics was considered relevant enough for suggesting remarkable features of dyeing technology in a historical context.

An overwhelming number of samples contained components indicating the presence of a red dye from a plant belonging to the Rubiaceae family. The most prominent species of this family is *Rubia tinctorum* L., commonly called madder. A statistical evaluation of relative ratios of alizarin and purpurin was conducted as a function of the colour produced and relying

Table 5 Royal purple in Roman Egyptian textiles. (AH) acid hydrolysis; (DMF) dimethylformamide extraction.

Sample	Site	Yarn	Result	λ (nm)
K3737.1	Krokodilô	purple weft of a purplish-greyish *clavus*	(AH): 79 ka, 7 al, 9 in, 4 pu, 1 mbi	255
			52 ka, 43 in, 5 mbi	288
			(DMF): 63 in, 37 ir	288
Z94.22010.5	Maximianon	purple weft of purple *clavus*	(AH): 72 in, 28 mbi	288
			(DMF): 10 in, 85 mbi, 5 dbi	288
Z95.32029.9	Maximianon	purplish-pink weft of tapestry-woven angle, fragment 1	(AH)a: 76 ka, 20 in, 4 mbi	255
			(AH)b: 77 ka, 18 in, 5 mbi	255
			52 ka, 39 in, 9 mbi	288
			(DMF): 77 in, 23 ir	288
		purplish-pink weft of tapestry-woven angle, fragment 2	(AH): 73 ka, 19 in, 8 mbi	255
			43 ka, 46 in, 11 mbi	288
			(DMF): 84 in, 16 mbi	288

upon former research on *Rubia* species and on results of dye analyses on textiles from the Coptic Egyptian period. It appeared that madder compositions in purple colours, produced from top-dyeing madder on yarns dyed with an indigoid source, were substantially different from those used to produce reds, blacks and other colours. Moreover, this phenomenon seemed to be the same in the Roman and the Coptic Egyptian periods. Therefore it may be concluded that the use of madder to produce purple, in combination with an indigoid dye, was based on grounds technologically different to those used for reds and blacks. Given the fact that the main difference was the relatively low content of alizarin in the madder component of purple, it was tempting to ascribe the difference in technological approach to the use of another plant species, for instance *Rubia peregrina* L.: former research has indicated that the alizarin/purpurin content of dyeings with this plant may indeed vary from 0/100 to 21/79. However, supplementary evidence would be needed to confirm this hypothesis since *Rubia peregrina* is not present in Egypt according to recent research: the only *Rubia* species reported on Egyptian territory at present – *R. tenuifolia* d'Urv. – is quite rare, with a woody root, very poor in colorants and its composition is similar to that of *R. tinctorum*.[12] The remaining explanations therefore are either the use of imported *R. peregrina* or the development by Egyptian dyers of a process of selective extraction of colorants from dyer's madder that will have to be reconstituted through laboratory dyeing experiments.

The great majority of tapestry bands or geometrical signs decorating tunics and mantles that have been preserved from the three sites of Maximianon, Krokodilô and Didymoi are dyed in a broad range of purple shades, from pinkish violet to very dark bluish purple. Even though only 28 samples from these sites were selected with the aim of identifying the source(s) of their 'purple' ground or decoration, it is remarkable that murex dyes were detected in three textiles: in a pure form in the purple band of a cream-coloured tunic; combined with kermes in two of the angle-shaped purple decorations of a dark greenish-blue mantle; and combined with both kermes and madder in the mauve band of a blood-red tunic. All these refer to expensive production, because of the presence not only of royal purple, but also kermes. Such combinations have not been reported in the results of dye analyses of ancient textiles so far. However, written sources do mention the combination of true purple with kermes: it was known in antiquity as 'hysgin purple' and Pliny the Elder, in his *Natural History*, also mentions that paler, mauve shades (perhaps similar to that of the fragment illustrated in Plate 4), called 'conchylian colours', were also produced by diluting the purple dye-baths with water and urine.[13] A mixture of kermes with a vegetable indigoid dye has been identified by H. Böhmer and R. Karadag as the colouring matter present in a dark purple arrow-design on a textile from Palmyra in Syria.[14]

Equally remarkable is the absence of any insect dye in the red samples analysed; in particular, in contrast to the Coptic period, no Indian lac (*Kerria lacca* Kerr, 1782) was found in textiles from the Roman Egyptian era.

The use of tannins, and the degradation processes that must be associated with them, remains controversial in the study of ancient textile technology. Moreover, the techniques for dye analysis are not yet at a point where the detection of any component contributing to colour can be guaranteed. These observations may underlie the basis for the detection of tannin alone in a brownish-black colour of a plain, rather coarse textile (group 4, K2416.6) and in a purplish brown present on an extremely fine twill (group 6a, Z95.33029.3). Tannins alone have not been detected in blacks. Obviously, the standard way to obtain blacks was the three-step process consisting of dyeing with an indigoid dye, followed by mordanting and top-dyeing with madder. Indeed, all the black tapestry bands decorating white or red tunics – two very popular colour combinations in Maximianon, Krokodilô and Didymoi – are the result of just such a combination of indigo and madder. Even more sophisticated processes for dyeing black, not described in the technical literature until much later (in the 18th and 19th centuries AD) are shown by two dye analyses to have already been practised by Egyptian dyers of the first centuries of our era; namely, a black dye using indigo, madder and tannin (Z95.25012.B11) and another adding weld to the former ingredients (K5701.1).

Lastly, the results of these dye analyses of textiles from the Roman *praesidia* must be examined from the point of view of their contribution to current knowledge on some special groups of textiles. In the case of the tubular chevron twills (group 3), nearly always decorated with red warp stripes, all the dye analyses have shown the presence of a similar type of madder dye with a significant alizarin content. This will have to be taken into account for further comparisons with similar textiles if any happen to be discovered in the future.

The analyses of shades of pink, red and orange, the most common colours used for the designs of the *taquetés* or for the plain ground of many damask twills, show far less consistency. Considering the high quality of these two types of textiles, it is remarkable that no insect dye has been identified, only madder types, both alizarin-rich and purpurin-rich. The high price and prestige of both these groups of textiles was, however, revealed by the identification of rather complicated processes to give special shades to these textiles, as in the yellow weft of a *taqueté* from Maximianon (Fig. 3, Z94.31029.1), resulting from a combination of weld, madder and indigo, or in the dark pink *clavus* of an ivory-coloured, fine damask tunic from Krokodilô (K2406.1), obtained by a subtle combination of purpurin-rich madder, indigo and traces both of tannin and of weld (Plate 5).

The analyses of the colourful textiles found in the rubbish heaps left by the Roman garrisons in the Egyptian desert are thus giving us interesting insights into the diversity and range of qualities of the textiles that could find their way to the farthest outposts of the empire. Moreover, when comparing dyeing technologies in both Roman and Coptic Egypt, there seems to have been a long-lasting transmission of specific skills for almost an entire millennium.

Editor's note

*Author for correspondence: Jan Wouters, Conservation Scientist, Consultant, Neerbroek 54, B-2070 Zwijndrecht, Belgium (jan.j.wouters@telenet.be). Paper received 28 November 2003. When the research was carried out the author was employed at the Royal Institute for Cultural Heritage, Jubelpark 1, B-1000 Brussels, Belgium.

Notes and references

1. Archaeological excavations were conducted under the direction of Hélène Cuvigny, CNRS/ Institut de recherche sur l'Histoire des Textes, Paris; funding was provided by the Ministère des Affaires Etrangères, France, and logistics by the Institut Français d'Archéologie Orientale, Cairo.
2. Cuvigny, H. (ed.) (2003) *La route de Myos Hormos: l'armée romaine dans le désert Oriental d'Égypte*, 2 vols, Cairo.
3. Granger-Taylor, H. and Sheffer, A. (1994) 'The textiles: a preliminary selection', in J. Aviram, G. Foerster and E. Netzer (eds), *Masada IV: The Yigael Yadin Excavations 1963–5 Final Reports*, Jerusalem, pp. 151–255; Granger Taylor, H. (2000) 'The textiles from Khirbet Qazone (Jordan)', in D. Cardon and M. Feugère (eds), *Archéologie des textiles des origines au Ve siècle: Actes du colloque de Lattes, Oct. 1999*, Montagnac (Monographie Instrumentum 14), pp. 149–62; Granger-Taylor, H. (2006) 'Textiles from Khirbet Qazone and the Cave of Letters, two burial sites near the Dead Sea: similarities and differences in find spots and textile types', in S. Schrenk (ed.), *Textiles In Situ: Their Find Spots in Egypt and Neighbouring Countries in the First Millennium CE* (*Riggisberger Berichte* 13), Riggisberg.
4. Cardon, D. (2003) 'Chiffons dans le désert: textiles de Maximianon et Krokodilô', in Cuvigny 2003 (see note 2 above), vol. 2, pp. 619–59, pl. IV–VI; Cardon, D. (1998) 'Textiles de Maximianon/al-Zerqa: les plus anciens exemples de décors par réserve sur laine', *Bulletin du CIETA* 74, pp. 15–25; Cardon, D. (1999) 'Les damassés de laine de Krokodilô, 100–120 ap. J.-C.', *Bulletin du CIETA* 75, pp. 6–21; Cardon,

D. (2002) 'Chiffons dans le désert: textiles des *praesidia* de Maximianon, Krokodilô et Didymoi', in M. Durand and F. Saragoza (eds), *Egypte, la trame de l'histoire: textiles pharaoniques, coptes et islamiques*, Paris, pp. 40–43.

5. Cardon, D. (2001) 'On the road to Berenike: a piece of tunic in damask weave from Didymoi', in P. Walton Rogers, L. Bender Jorgensen and A. Rast-Eicher (eds), *The Roman Textile Industry and its Influence: A Birthday Tribute to John Peter Wild*, Oxford (Oxbow), pp. 12–20 and plate 1.
6. Wouters, J. (1993) 'Kleurstofanalyse van Koptisch textiel (Dye analysis of Coptic textiles)' in A. De Moor (ed.), *Koptisch Textiel/ Coptic Textiles*, Zottegem (Publicaties van het Provinciaal Archeologisch museum van Zuid-Oost-Vlaanderen, site Velzeke), pp. 53–64.
7. Wouters 1993 (see note 6 above).
8. Wouters, J. (1993) 'L'analyse des colorants naturels par chromatographie liquide haute performance: méthodologie et application à une série de textiles médiévaux conservés en Languedoc-Roussillon ou exposés dans le cadre de l'exposition "Fils Renoués – Trésors textiles du Moyen Age en Languedoc-Roussillon"' in D. Cardon (ed.), *Fils Renoués: Trésors textiles du Moyen Age en Languedoc-Roussillon*, Carcassonne, pp. 158–65.
9. Wouters, J. (1987) 'Analyse des colorants des tapisseries brugeoises des XVIe et XVIIe siècles/ Kleurstofanalyse van Brugse wandtapijten uit de 16de en de 17de eeuw', in G. Delmarcel and E. Duverger, *Bruges et la tapisserie/ Brugge en de tapijtkunst*, Bruges, pp. 515–26.
10. Wouters, J. (2001) 'The dye of *Rubia peregrina*', *Dyes in History and Archaeology* 16/17, pp. 145–57.
11. Wouters, J. and Verhecken, A. (1991) 'High performance liquid chromatography of blue and purple indigoid natural dyes', *Journal of the Society of Dyers and Colourists* 107, pp. 266–9.
12. Böhmer, H. (2002) *Koekboya: Natural Dyes and Textiles*, Ganderksee (REMHÖB-Verlag), p. 119; Boulos, L. (1999–2005) *Flora of Egypt*, 4 vols, Cairo.
13. Cardon, D. (2003) *Le monde des teintures naturelles*, Paris, pp. 440–41.
14. Böhmer, H. and Karadag, R. (2003) 'New dye-research on Palmyra textiles', *Dyes in History and Archaeology* 19, pp. 89–90.

Dyestuff Analysis of Two Textile Fragments from Late Antiquity

Regina Hofmann-de Keijzer* and Maarten R. van Bommel

Abstract To gain more knowledge about the dyes in two textile fragments from Late Antiquity from the Museum of Applied Arts in Vienna, Austria, dyestuff analysis was carried out by high-performance liquid chromatography (HPLC).

The first fragment is from an Egyptian garment dating from the 7th or 8th century AD. Two multicoloured tapestry-woven textiles, the *clavi* and the neckbands, are attached to a linen ground. In 13 woollen samples, 11 dyestuffs, originating from four dyes, were identified: indigotin and indirubin from indigo or woad; alizarin and purpurin from madder; luteolin and apigenin from weld; and five laccaic acids from lac dye. In addition, unknown compounds were found with spectra similar to known flavonoids or anthraquinones, but with retention times that could not be compared with reference material. Differences in dyestuff composition between the *clavi* and the neckbands were found. An unknown yellow component was detected in indigotin dyeings. The identification of lac dye supports the dating of the fragment according to art-historical research.

The second fragment represents a purple medallion, presumably dated to the 2nd or 3rd century AD. It is one of the few textiles from Late Antiquity in which shellfish purple has been detected. The decorative metal thread is made of pure gold.

Introduction

The Austrian Museum of Applied Arts (Museum für angewandte Kunst (MAK)) in Vienna (formerly the Museum of Art and Industry) has a collection of so-called Coptic textiles consisting of about 1400 objects. This collection includes textiles from the Coptic period as well as textiles of Late Antiquity, which, strictly speaking, do not belong to the Coptic period. The collection was begun in the year 1882, when the museum bought about 700 Coptic textiles from the Viennese art dealer Theodor Graf.[1]

In order to obtain more technological and cultural-historical information, an investigation into the dyestuffs present in two remarkable textile fragments of Late Antiquity was carried out using high-performance liquid chromatography with photodiode array detection (HPLC–PDA). The first textile is a fragment of an Egyptian garment decorated with well-preserved multicoloured *clavi* and neckbands. The aim of the analysis was to identify the dyes, to examine the correlation between the dyes used and the colours present, to look for differences between the *clavi* and the neckbands, and to get to know more about the history of the object. The second textile is a purple medallion with golden and white ornaments; in this case the purple and the golden thread were analysed.

A fragment of an Egyptian garment (Tunic, T 8528/ 1928)

Description of the object

In 1928 the fragment of a garment came to the museum from the private collection of Johanna Nadherny-Borotin. The fragment is part of an Egyptian tunic and measures 36 × 128.5 cm. The linen ground fabric consists of an S-spun warp and an S-spun weft.[2] Two different tapestry-woven textiles, about 8 cm wide, are attached to the linen ground: the vertical units of *clavi* and the transverse neck decoration, the neckbands. Both tapestry ornaments are made of a linen warp, S-spun and Z-plied (S_2Z). The weft is made of dyed wool, S-spun, and of wool and linen of natural colour, S-spun. Four woollen yarns of different hues were used for the neckbands and eight woollen yarns of different hues for the *clavi*. The colours of the neckbands are deeper than those of the *clavi*, which seem to have faded.

The ornaments of the neckbands consist of small medallions and birds. Seven different scenes are present on the ten fields of *clavus* 1 and the nine fields of *clavus* 2 (see Plate 6):

- Scene 1: Adam and Eve, fields 2-1 and 2-9;
- Scene 2: a shepherd, fields 1-1 and 1-10;
- Scene 3: a female figure with a child, fields 1-2, 1-9, 2-2 and 2-8;
- Scene 4: a male figure and a greenish-blue animal with a long neck, fields 1-3, 1-8, 2-3 and 2-7;
- Scene 5: a male figure ('camel driver') with an animal bearing a saddle, fields 1-4, 1-7 and 2-6;
- Scene 6: a female figure on a four- or six-wheeled wagon, fields 1-5, 1-6 and 2-5;
- Scene 7: a female figure ('rider') with two blue animals (mules or horses?), field 2-4.

The interpretation of the almost completely preserved *clavus* decoration allows us to place the Egyptian tunic in the Late Antique–Early Islamic period, most probably in the 7th or 8th century AD.[3] In the iconographical study of this tunic carried out recently, some of the scenes are described as Old Testament themes and the textile fragment was identified as a Christian garment, though its decorative elements seem to be influenced by Jewish and Islamic ideas.[4]

Investigation

The 12 woollen samples taken from the *clavi* and the neckbands were first examined under a stereomicroscope at a magnification of 20× to observe the colour and to see whether the yarns contain fibres of different colours. One blue sample of the neckbands consisted of two different threads; therefore this was separated into a black-blue (K12-bk) and a dark-blue (K12-db) sample.

Dyestuff analyses were carried out on all 13 woollen threads in order to answer the following questions:

- Which dyestuffs (that is, dye constituents) can be found and which dyes can be identified?
- What is the relationship between the identified dyes and the colours?
- Are there differences between the dyes identified in the *clavi* and in the neckbands?
- Can a dye be found which provides information about the history of the object?

The organic dyestuffs of the 13 samples were analysed by HPLC–PDA according to the method described by Wouters and Rosario-Chirinos (1992) with some minor modifications, using a 2 mm Luna C18 column.[5] The sample preparation was carried out in two different ways. All samples that might contain indigoids were first extracted with dimethylformamide at 140 °C for 10 min. These extracts can be directly injected onto the HPLC column. Hydrochloric acid was then added to the fibres and the remaining dyestuffs were extracted using the normal procedure. The other samples were hydrolysed directly with hydrochloric acid at 100 °C for 10 min in order to extract the dyestuffs. The extracts were then evaporated to dryness, the dyestuffs were dissolved in dimethylformamide and analysed by HPLC. The HPLC procedure used has been described in detail in an earlier publication.[6]

Results and discussion

While in two documents from late Roman Egypt, the *Papyrus Graecus Holmiensis* and the *Papyrus X Leidensis*, a wide range of possible dyes are mentioned,[7] so far only eight dyes have been identified by analysis in Coptic Egyptian textiles:[8] madder (*Rubia tinctorum* L.), wild madder (*Rubia peregrina* L.), lac dye (*Kerria lacca* Kerr, 1788), kermes (*Kermes vermilio* Planchon) and Armenian cochineal (*Porphyrophora hameli* Brandt) were used for dyeing red; weld (*Reseda luteola* L.) for dyeing yellow; and indigo (*Indigofera* sp.) or woad (*Isatis tinctoria* L.) for dyeing blue. The wide range of different hues found on Coptic textiles, or on textiles from Late Antiquity, could be obtained by using these dyes in different concentrations and also by using combinations of the dyes and different mordants. Other, still unidentified, dyes might have been used as well, however.

In the 13 woollen samples of the tunic, four out of the eight dyes mentioned above were identified: madder, lac dye, weld and indigo or woad (see Table 1). The relationship between the colours and the identified dyes is given in the same table. Six woollen yarns were dyed with a single dye, five yarns with two dyes and two yarns with three dyes. The red hue of the *clavi* is dyed only with madder. Blue, from light blue to a dark, nearly black blue, is given by indigotin

Table 1 Relationship between the colour and the identified dyes and dyestuffs present in blue, green, orange, red and yellow yarns of the *clavi* and the neckbands. * The relative ratio of dyestuffs present is calculated at 275 nm for indigotin (int), indirubin (inr), alizarin (ali) and purpurin (pur), and at 350 nm for luteolin (lut) and apigenin (api).

Sample location	Sample	Sample colour	Dye	Ratio of dyestuffs*	Number of dyes
Clavi					
2-9, Adam and Eve: background	K02	red	madder	ali:pur = 80:20	1
2-7, male figure: contour	K04	dark blue	indigo or woad	int:inr = 87:13	1
2-7, male figure: column-shaped element with blue and light blue squares	K05	blue	indigo or woad	int:inr = 82:18	1
2-7, male figure: column-shaped element with blue and light blue squares	K06	light blue	indigo or woad	int:inr = 97:3	1
2-2, female figure: animal	K08	beige-yellow (turquoise in textile)	indigo or woad	trace of int	1
2-9, Adam and Eve: border	K01	blue (greenish blue in textile	indigo or woad and weld	int:inr = 99:1	2
2-9, Adam and Eve: border	K07	yellow	weld and indigo or woad	lut:api = 100:0; trace of int	2
2-7, male figure: vegetable from behind the animal	K03	yellow-orange	madder and weld	ali:pur = 85:15; lut:api = 82:18	2
Neckbands					
A, background	K12-bk	black-blue	indigo or woad	int:inr = 90:10	1
A, background	K12-db	dark blue	indigo or woad and weld	int:inr = 91:9; lut:api = 93:7	2
B, bird	K10	red-orange	madder and weld	ali:pur = 79:21; lut:api = 94:6	2
A, bird	K11	green	indigo or woad, weld and madder	int:inr = 95:5; lut:api = 94:6; ali:pur = 76:24	3
A, medallion	K09	bluish red	madder, lac dye and indigo or woad	ali:pur = 78:22; int:inr = 84:16	3

in different concentrations. The turquoise of the *clavi* is produced by a very low concentration of indigotin. In the blue sample from the *clavi*, which looks greenish on the textile, and in the dark blue sample from the neckbands weld was identified as well as an indigotin dye. In the yellow sample of the *clavi* luteolin was detected, together with a trace of indigotin. In this case indigotin might be contamination caused by a small fibre dyed with an indigotin dye situated next to the area from where the sample was taken.

In two orange samples, mixtures of madder and weld were found. Mixtures of madder and weld seem to be rare in Coptic textiles. During the research on Coptic textiles from Flemish private collections Wouters (1993) identified mixtures of madder and weld in one yellow and five orange samples, mentioning that these mixtures have not been identified with certainty in Coptic textiles so far.[9] This mixture was also found in three samples from Coptic textiles in the Abegg-Stiftung: two yellow and one orange.[10]

Mixtures of three dyes seldom occur in Coptic textiles. A mixture of an indigotin dye, weld and madder was found by Wouters (1993) in a green sample in the Coptic textile 34/6.[11] This result corresponds with the analysis of the green yarn from the neckbands, which also contains a mixture of these three dyes. Furthermore, the ratio of weld and madder to the amount of indigo (89/11) is approximately the same as that found by Wouters. Lac dye in combination with madder and an indigotin dye were identified in a Coptic textile for the first time. In samples of Coptic textiles from Flemish private collections, lac dye, kermes and Armenian cochineal were detected, in each case alone and not in combination with other dyes.[12] Insect dyes mixed with other dyes were, however, identified in textiles from the Abegg-Stiftung: lac dye, madder and sappanwood in a bluish-violet sample; kermes, madder and a flavonoid dye in a purple sample.[13]

An overview of the 18 dyestuffs identified in the 13 samples from the *clavi* and the neckbands is given in Table 2. Eleven of these could be associated with a particular colouring matter. Indigotin, originating from indigo or woad, was identified in blue, green, yellow and red samples. With the exception of samples containing indigotin in low concentration, indirubin was found as well. The ratio of the dyestuffs varies between 82 and 99% indigotin and 18 and 1% indirubin (see Table 1). Alizarin and purpurin, in a ratio varying from 76 to 85% alizarin and 24 to 15% purpurin (Table 1), originating from madder,

Table 2 Dyestuffs analysed and dyes identified in the blue, green, orange, red and yellow yarns of the *clavi* and neckbands.

Dyestuff	Dye	Number of yarns							
		Clavi				Neckbands			
		Blue	Orange	Red	Yellow	Blue	Green	Orange	Red
indigotin	indigo or woad	4			2	2	1		1
indirubin	indigo or woad	4				2	1		
alizarin	madder		1	1			1	1	1
purpurin	madder		1	1			1	1	1
alizarin equivalent	madder?		1					1	
five laccaic acids	lac dye								1
luteolin	weld	1	1		1	1	1	1	
apigenin	weld					1	1	1	
apigenin equivalent	weld?	1	1		1				
unknown anthraquinone	?				1				
unknown anthraquinone	?								1
two unidentified flavonoids	?						1		
unknown yellow compound, λ_{abs} = 338 nm and 385 nm	?	3				1			

were present in red and orange samples. Some of these samples also contain an alizarin equivalent. This compound has almost the same UV-visible spectrum as alizarin, but elutes with a different retention time. Lac dye could be identified in a bluish-red sample of the neckbands by the presence of five laccaic acids. Luteolin, the main component of weld, was found in yellow, orange, green and blue samples. Apigenin, the minor component of weld, was only found in samples from the neckbands containing a high concentration of luteolin. The ratio between these dyestuffs was 82–100% luteolin and 18–0% apigenin (see Table 1). In the samples of the *clavi* an apigenin equivalent was detected, which has the same spectrum as apigenin, but a different retention time.

In some samples, unidentified flavonoids and anthraquinones were present. These compounds have spectra similar to flavonoids, absorbing at 350–370 nm, or anthraquinones, absorbing at 480–490 nm. The retention times of these unknown flavonoids and anthraquinones could not be related to any reference material.

In four samples from the tunic, containing only indigotin and indirubin, a yellow compound with absorption bands at 338 nm and 385 nm was detected (see Fig. 1). The origin of this unidentified compound is unknown. It may be a degradation product or it may originate from the indigo plant or from another plant used in the dyeing process. The authors have found the same unknown compound in indigotin samples from other objects.[14]

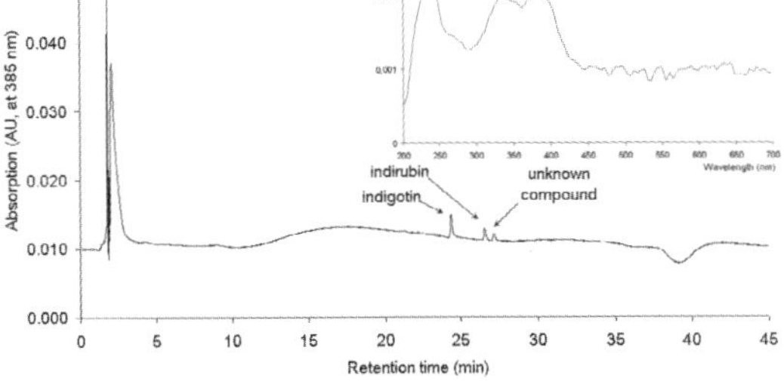

Figure 1 HPLC chromatogram of sample K12-bk, showing indigotin, indirubin and the unknown yellow compound found in four indigotin samples from the tunic. Analysis performed at the Netherlands Institute for Cultural Heritage (ICN), Amsterdam.

The analysis revealed some differences between the dyes used for the *clavi* and the neckbands. While mixtures of two dyes were found in both tapestry ornaments, mixtures of three dyes were only found in the neckbands. In all the weld-dyed samples from the neckbands, both luteolin and apigenin were found; in the samples from the *clavi* only luteolin, or luteolin and an apigenin equivalent, were present. The lack of apigenin in samples from the *clavi* is correlated with a low concentration of luteolin. These results may be due to the fading of the colours in the *clavi* or may indicate that low amounts of weld were used for dyeing. Lac dye was only found in the neckbands.

Whereas madder, weld and an indigotin dye (which could have been woad or an *Indigofera* species) were common in Egypt from the 14th century BC until the Coptic period,[15] this is not the case with lac dye, found in the bluish red of the neckbands together with madder and indigotin. Forbes (1964) believed that in Egypt lac dye appeared when the connections with Armenia, Byzantium and the north were interrupted by the Arab invasions of the 7th century AD and hence Armenian cochineal and kermes were unobtainable; lac dye may have been known in earlier times but it was hardly used.[16] The detection of lac dye by Wouters (1993) in a Coptic textile which, according to radiocarbon dating, should be dated between the second half of the 8th century AD and the end of the 10th century AD confirms Forbes's statement.[17] Wouters (1993) also found pure Indian lac dye in another Coptic textile and Eastwood (1984) found lac dye in three out of 57 samples, but on textiles dated from the 13th to the 16th century AD.[18]

De Jonghe and Verhecken-Lammens (1993) mention that Coptic tapestry ornaments on plied linen warp, like the *clavi* and neckbands of this tunic, are probably mostly later than those in which the linen warp is not plied. They are usually dated after the late 6th century AD.[19]

Conclusion

The colours of the neckbands are deeper than the colours of the *clavi*, which seem to have faded. Dyestuff analysis indicates differences between the two tapestry-woven ornaments. The yellow dyeings of the neckbands contain apigenin correlated to a high concentration of luteolin. Luteolin as a main compound and apigenin as a minor compound are characteristic for dyeings with weld. In the yellow dyeings of the *clavi*, luteolin and an apigenin equivalent were found therefore it is not certain that weld was used. For the yarns of the neckbands, more complicated dyeing techniques, using three dyes for some colours, were employed. The differences in the colours as well as those revealed by dye analysis suggest different histories for these two tapestry ornaments. When the tunic was made, the *clavi* and the neckbands were attached to the linen ground fabric at the same time: the possibility that the neckbands were fixed on the tunic at a later period can be excluded. The differences may therefore be due to different dyeing techniques, which may be correlated to a different light sensitivity. It might also be true that some fading of the *clavi* took place during a previous use of the *clavi* elsewhere.

Lac dye, found in the neckbands, helps to date the cloth to within a period after the Arab invasion in the second half of the 7th century AD.[20] The plied linen warp used in the *clavi* and the neckbands suggests that they were not made before the late 6th century AD.[21] It can be concluded that the results of the scientific research support the dating to the 7th or the 8th century AD, according to art-historical research.

A fragment of a purple medallion (T 6025/ 1905 = T 10069/ 1953)

Description of the object

In 1905, two textile fragments fixed on the front and the back of the same linen support fabric came to MAK, originating from Theodor Graf's inheritance. Both are medallions. In the museum, the decision was made to separate them. In the year 1953, one fragment was stitched to another linen support fabric and given a new inventory number (T 10069). This tapestry-woven medallion measures 15 × 18 cm; as Plate 7 shows, small, four-leafed motifs made of a golden metal thread are arranged on a purple ground, also decorated with white geometrical lines. The warp is made of natural-coloured wool, S-spun and the weft is made of purple wool, Z-spun and S-plied (Z_2S). Linen of natural colour, S-spun, was used for the geometrical white lines. The metal thread, Z-spun, has a silk core, Z-spun.[22] The medallion is believed to be dated to the 2nd or 3rd century AD; however, this has not been confirmed by scientific research.

In the 1930s Rudolf Pfister chose this textile for dyestuff analysis. From 1934, the French chemist and pioneer in the field of the analysis of dyestuffs in antique textiles analysed many purple textiles from Egypt, but he could find real purple in only two objects. This medallion is one of them. In *Nouveaux textiles de Palmyre* (Paris, 1937), he describes the analysis of the purple wool sample:[23]

> ... l'acide chlorhydrique laisse la teinte intacte, l'acide reste incolore et n'abandonne rien aux solvants habituels. La fibre (après traitement à l'acide) rincée à l'eau, traitée à l'hydrosulfite ammoniacal, se décolore lentement en jaune; un fil de coton se teint en bleu-violet après oxydation à l'air et à la lumière. C'est donc bien la pourpre véritable; il s'agit d'un tissu particulièrement riche puisque les fils d'or, également, sont très rares en Égypte.

> [... hydrochloric acid leaves the colour intact, the acid remains colourless and does not yield anything to the usual solvents. The fibre (after acid treatment) rinsed in water, treated with ammonium hydrosulfite, discolours slowly to yellow; a cotton thread is dyed blue-violet after oxidation in the air and light. Thus it is the real purple; it is a textile particularly rich because the gold thread, as well, is very rare in Egypt.]

Investigation

The purple yarn was analysed in order to verify Pfisters's results with modern chromatographic techniques. The yarn was extracted with 50 μL dimethylformamide at 140 °C for 10 min. The dyestuff analysis was performed by HPLC with a water/acetonitrile/phosphoric acid gradient, using a photodiode array (PDA) detector. Additionally scanning electron microscopy with energy-dispersive X-ray-analysis (SEM–EDX) was used for the examination of the metal thread.

Results and discussion

Pfisters's result of real purple was confirmed by the analysis. With HPLC–PDA five different components were found in the purple dyeing of the woollen sample: mono-bromoindigotin, dibromoindigotin, monobromoindirubin, dibromoindigotin and a trace of indigotin (Fig. 2). Furthermore, the SEM–EDX analysis showed that the metal used for the metal thread consists of pure gold (Fig. 3).

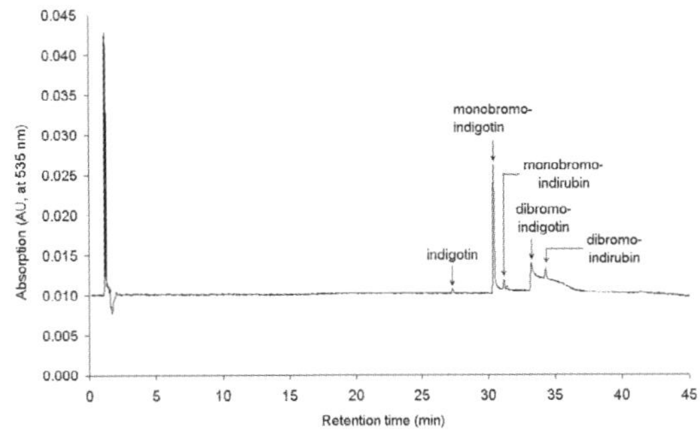

Figure 2 HPLC chromatogram of the purple thread. Analysis performed at the Netherlands Institute for Cultural Heritage (ICN), Amsterdam.

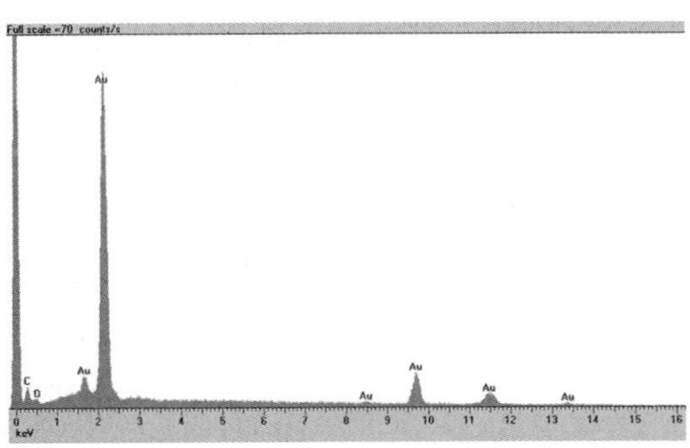

Figure 3 SEM–EDX spectrum of the metal thread. Analysis performed at the University of Applied Arts Vienna, Department of Archaeometry.

An attempt was made to find out if the detected indigoid compounds could reveal something about the source of the purple dyeing. In his experiments, Verhecken (1990) produced purple stains and dyeings by single species of snails.[24] Wouters (1992) dissolved the dyestuffs of these stains and dyeings in pyridine and analysed them by HPLC. As a result of this investigation Wouters (1992) found that a distinction might be possible between two types of purple: type 1, from *Phyllonotus trunculus* (now *Hexaplex trunculus* (Linnaeus, 1758)) contains indigotin and monobromoindigotin as main compounds and no compound X, which may be a brominated indirubin; type 2, from *Phyllonotus brandaris* (now *Bolinus brandaris* (Linnaeus, 1758)) and *Thais haemastoma* (now *Stramonita haemastoma* (Linnaeus, 1766)) contains dibromoindigotin as the main compound and the compound X.[25]

The ratios of the dyestuffs found in the purple thread of the medallion differ from the ratios found in purple types described by Wouters, therefore it is not possible to correlate the purple dyeing with one of these two purple types. It is possible that type 2 snails were used and that vatting parameters, causing debromination, influenced the composition of the dye, or that snails from both types were used together. Further ageing effects also have to be taken into account, as well as the possible influence of the use of dimethylformamide as a solvent. Before being able to identify the source of a purple dye more research is necessary; we need more knowledge about the composition of purple dyes, which are produced by glandular secretion of different snail species. Ageing studies would be necessary, as well as studies concerning the influence of different vatting parameters, such as temperature, alkalinity, the reductive system, irradiation and duration.

Purple hues occur very frequently on Coptic textiles. In most of these yarns, so-called 'Coptic purple' can be detected. This was used for imitating shellfish purple and it consists of two different dyes: a red anthraquinone dye, generally that of madder, and a blue indigotin dye. Shellfish purple was obtained from the hypobranchial glands of Mediterranean sea snails: the spiny dye-murex (*Bolinus brandaris*, syn. *Phyllonotus brandaris*, *Murex brandaris*), the banded dye-murex (*Hexaplex trunculus*, syn. *Phyllonotus trunculus*, *Murex trunculus*) and the rock-shell (*Stramonita haemastoma*, syn. *Thais haemastoma*, *Purpura haemastoma*).[26] Real purple made from marine snails was the most expensive dye of antiquity. Stauffer (1996) mentions the value and the rarity of purple textiles in Late Antiquity: textiles dyed with shellfish purple were status symbols, estimated in rank as equivalent to, or even higher than, gold brocaded textiles. Textiles dyed with real purple have been found very seldom. Most real purple-dyed textiles originate from regions near the habitat of the purple snails, in Syria, Phoenicia, Palestine and later Byzantium; sporadically they have been found in Egypt. Those found in Egypt are always decorated with golden threads and are dated from the 3rd to the 4th century AD.[27] In previous research, real purple has been detected in five objects found in Egypt: in the purple medallion from MAK and in 'un petit carré de style hellénistique' from the Rudolf Pfister collection;[28] in a tapestry-woven medallion, worked in purple wool, dated to about the 5th century AD;[29] in a woollen ornament of an Egyptian mantle in the British Museum, London, dated to the 3rd century AD;[30] and in the Coptic textile no. 139/2, a tapestry band in coloured wool and dyed linen on a plied linen warp.[31]

Conclusion

The purple medallion in the MAK collection is one of the few textiles of Late Antiquity in which shellfish purple has been identified, as well as a metal thread made of pure gold. This textile fragment is, therefore, among the most precious textile objects of Late Antiquity. The dating to the 2nd or 3rd century AD is not based on scientific research, but in any case it can be placed in the period between the 3rd and 4th century AD because all textiles found in Egypt dyed with shellfish purple and decorated with golden threads are dated to this period.[32]

Acknowledgements

We would like to express our gratitude to all those who helped us realise this project, especially Dr Angela Völker, Blanda Winter, Martina Dax and Christine Stührenberg (MAK – Austrian Museum of Applied Arts Vienna) for the textile historical and technological discussions. We thank Lecturer Eng. Octaviana Marincas and Dipl. Ing. Rudolf Erlach (University of Applied Arts Vienna, Institute for Conservation Science and Restoration-Technology/Department of Archaeometry) for the fibre analysis.

Editor's note

*Author for correspondence: University of Applied Arts Vienna, Institute for Art and Technology, Department of Archaeometry, Salzgries 14/1, A-1013 Vienna, Austria (Regina.Hofmann@uni-ak.ac.at). Paper received 6 May 2003; revised version received 5 September 2005.

Notes and references

1. Egger, G. (1967) *Koptische Textilien*, Vienna (Österreichischen Museums für angewandte Kunst), p. 3.
2. The technological description of the fragment was carried out by Christine Stührenberg and Martina Dax, MAK, Vienna.
3. Egger 1967 (see note 1 above), p. 21; Pillinger, R. (2002) 'Gewandfragment mit menschlichen Figuren und anderen Darstellungen im Museum für angewandte Kunst in Wien. Versuch einer ikonographischen Deutung', *Mitteilungen zur christlichen Archäologie* (Vienna, Österreich-

ischen Akademie der Wissenschaften) 8, pp. 29–30.
4. Pillinger 2002 (see note 3 above), pp. 18–31.
5. Wouters, J. and Rosario-Chirinos, N. (1992) 'Dye analysis of pre-Columbian Peruvian textiles with high-performance liquid chromatography and diode-array detection', *Journal of the American Institute of Conservation* 31, pp. 237–55.
6. Hofmann-de Keijzer, R. and Bommel, M.R. van (2005) 'TLC and HPLC analysis on red and violet cotton yarns of Indonesian textiles', *Dyes in History and Archaeology* 20, pp. 70–79.
7. Wouters, J. (1995) 'Dye analysis in a broad perspective: a study of 3rd- to 10th-century Coptic textiles from Belgian private collections', *Dyes in History and Archaeology* 13, pp. 38–45; see p. 38.
8. Wouters 1995 (see note 7 above) pp. 39–40; Schweppe, H. (1993) *Handbuch der Naturfarbstoffe: Vorkommen; Verwendung; Nachweis*, Landsberg/Lech, p. 60.
9. Wouters, J. (1993) 'Dye analysis of Coptic textiles', in A. De Moor (ed.), *Koptisch textiel uit Vlaamse privé-verzamelingen/ Coptic Textiles from Flemish Private Collections*, Zottegem, pp. 53–64; see p. 62.
10. Schrenk, S. (2004) *Textilien des Mittelmeerraumes aus spätantiker bis frühislamischer Zeit*, Riggisberg, p. 479.
11. Wouters 1993 (see note 9 above), p. 61.
12. Wouters 1995 (see note 7 above), p. 41.
13. Schrenk 2004 (see note 10 above), pp. 480–81.
14. Bommel, M.R. van and Hofmann-de Keijzer, R., unpublished results, Instituut Collectie Nederland, Amsterdam, The Netherlands.
15. Wouters 1995 (see note 7 above) p. 39.
16. Forbes, R.J. (1964) *Studies in Ancient Technology*, vol. IV, Leiden, p. 107.
17. Wouters 1993 (see note 9 above), p. 60.
18. Wouters 1993 (see note 9 above), p. 60; Eastwood, G.M. (1984) 'Egyptian dyes and colours', *Dyes on Historical and Archaeological Textiles* 3, pp. 9–19.
19. De Jonghe, D. and Verhecken-Lammens, C. (1993) 'Technological discussion', in A. De Moor (ed.), *Koptisch textiel uit Vlaamse privé-verzamelingen/ Coptic Textiles from Flemish Private Collections*, Zottegem, p. 38.
20. Forbes 1964 (see note 16 above), p. 107; Wouters 1993 (see note 9 above), p. 60.
21. De Jonghe and Verhecken-Lammens 1993 (see note 19 above), p. 38.
22. The fibre analysis of the yarns and of the metal thread was carried out by Ass. Prof. Eng. Octaviana Marincas, MSc, during her research internship at the University of Applied Arts Vienna. The technological description of the fragment was done by Martina Dax, MAK, Vienna.
23. Pfister, R. (1937) *Nouveaux textiles de Palmyre*, Paris, p. 12.
24. Verhecken, A. (1990), 'Experiences with mullusc purple', *La Conchiglia* 22, pp. 250–52.
25. Wouters, J. (1992) 'A new method for the analysis of blue and purple dyes in textiles', *Dyes in History and Archaeology* 10, pp. 17–21; see p. 19.
26. Schweppe 1993 (see note 8 above), pp. 315–18.
27. Stauffer, A. (1996) *Spätantike, frühchristliche und islamische Textilien aus Ägypten*, Bern (Bernisches Historisches Museum), p. 10; information kindly provided by Annemarie Stauffer, 2005.
28. Pfister 1937 (see note 23 above), p. 12.
29. Walton, P. (1985) 'Shellfish purple in a Coptic textile', *Dyes on Historical and Archaeological Textiles* 4, pp. 33–4.
30. Wouters 1992 (see note 25 above), p. 20.
31. Wouters 1993 (see note 9 above), p. 61.
32. Stauffer 1996 (see note 27 above), p. 10; information kindly provided by Annemarie Stauffer, 2005.

A New HPLC-PDA Method for the Analysis of Tyrian Purple Components

Zvi C. Koren

Introduction

The chemical fingerprinting of molluskan Tyrian Purple pigments has advanced by quantum leaps in recent years. Until about a decade ago, the detection of a true purple pigment on an ancient dye vat potsherd, or the identification of a real purple dye on an archaeological textile, required the detection of the singular chemical 'biochemical-marker' to indicate the *bona fide* pigment.[1] This genuine purple 'trademark', the dye common to all authentic purples, is the famous doubly brominated indigo, a dark purple substance that can only be produced in nature from a molluskan source and not from a plant source. Ancient vegetal chicanery has 'faked' this purple hue by double-dyeing wool with the blue plant-produced indigo dye and with a red dyestuff of plant or insect origin, or by dyeing an iron salt-mordanted woolen with such a red dyestuff.[2] However, only a marine creature possesses the biochemical ability to produce dibromoindigo, commonly abbreviated as DBI, as well as other brominated colorants. The dye's initial fame is due to Friedländer who in 1909 synthesized it and also isolated it from various purple-producing mollusks.[3]

Recent analyses have shown that modern and archaeological purple pigments from various molluskan species contain more components than the primary DBI.[4] These colorants can be classified as isatinoids, indigoids, and indirubinoids, whose molecular structures are given in Figure 1. Thus, in order to obtain a more complete profile of that dyestuff, it is necessary to identify all the detectable colorants contained in the purple pigment. Such a multicomponent analysis will be able to produce a more detailed chemical characterization that should help in the determination of the geographical and zoological provenance of purple dyestuffs.

In the course of the current method development, it was found that while the HPLC method previously used for the detection of indigoidal dyes was satisfactory for detecting a number of key Tyrian Purple components,[5] it was nevertheless not suitable for the identification of other important fingerprinting compounds. Specifically, this method lacked the ability to identify various isatinoids, and it did not possess the resolution required to produce a clear separation between the singly brominated indirubins and the dibrominated indigo molecule.

This paper presents a new reverse-phase high-performance liquid chromatographic (HPLC) method for the well-resolved separation and identification of ten brominated and unbrominated indigoid dyes in 25 minutes. The linear gradient elution scheme is composed of a ternary aqueous acidic methanolic mobile phase, and the component detection is performed by way of a photodiode array (PDA) detector. This

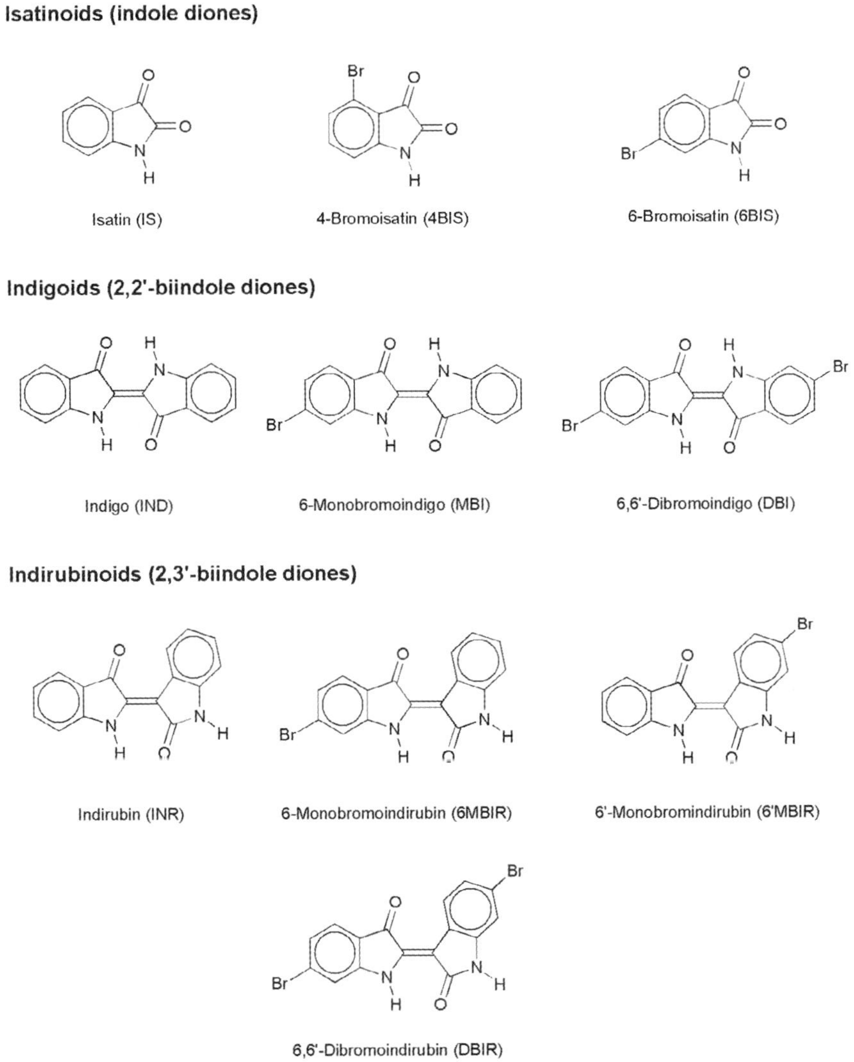

Figure 1 Molecular structures and common and abbreviated names of the isatinoids, indigoids, and indirubinoids.

is the first time that all of these components have been analyzed by a single elution method.

Phoenician purple pigments: a brief look

The biogenesis and chemistry of the molluskan purple dyestuff, though not completely understood, have been studied,[6] and also reviewed.[7] Recently, natural dyeings were produced with dried *Hexaplex trunculus* (*Murex trunculus*) pigments reduced by cockles,[8] as well as with fresh *H. trunculus* glands using an all-murex dye-bath.[9] The pigment is produced from the colorless precursors contained in the hypobranchial glandular fluid of the living animal. These precursors consist of substituted and unsubstituted indoxyl sulfates, as well as their brominated and unbrominated counterparts. Purpurase, the enzyme required for the hydrolysis of these indoxyl sulfates, is also present in the gland, although not in the same compartment as the precursors. When the gland is punctured, as in the excision process, or when the animal expires, the enzyme comes into contact with the precursors and hydrolyzes them. These hydrolyzates then undergo a series of spontaneous oxidative reactions to form indigoids and other products, such as tyrindoxyls and tyriverdins. Solar radiation can photolyze the tyriverdins to produce brominated indigoids. The increased temperature caused by the infrared portion of the solar radiation will accelerate all the reactions, and the rate of the photochemical reactions is proportional to the radiation intensity.

Along the Levantine Mediterranean coasts of present-day Israel and Lebanon, three molluskan species have been associated with purple dyeing.[10] These Muricidae sea snails, whose common familial approbation is – or was – *Murex*, consist of *Hexaplex trunculus* (Linnaeus, 1758, formerly *Murex trunculus*, *Phyllonotus trunculus*), *Bolinus brandaris* (Linnaeus, 1758, formerly *Murex brandaris*) and *Stramonita haemastoma* (Linnaeus, 1766, formerly *Purpura haemastoma*, *Thais haemastoma*). The industrial biochemical production of the royal purple pigment from these sea snails originated from as early as perhaps 4000 years ago.[11]

HPLC analyses of brominated colorants

Chromatography is a far superior instrumental tool for performing a multicomponent analysis than is a spectrometric technique alone. The former is a separatory technique, while the latter only provides an overlap of information, a total picture that is the sum of its parts.

Past HPLC analyses of brominated indigoid colorants have been applied to only some of the dye components that may be present in a shellfish (Tyrian) purple pigment. Table 1 (below) lists those molluskan-related colorants that have until recently been analyzed by HPLC, with the first such analysis performed nearly two decades ago. The commonly used elution method was originally published in a Chinese article describing the identification of vegetal indigo and indirubin;[22] it was subsequently applied to brominated indigoids of a molluskan source.[23] The method consists of varying methanol and water compositions linearly whilst maintaining 5% w/v H_3PO_4 solution at a constant 10% composition. For methanol, the cumulative times (and concentrations) are: 0–1 min (30%), 3 min (60%), 15 min (90%).

This method has been excellent for the detection of several key components of shellfish purple pigments. However, recent analyses of these dyestuffs have clearly indicated that other colorants, such as isatin and possibly other substituted isatinoids, are also present in purple *Hexaplex trunculus* pigments.[24] Hence, a new

Table 1 Molluskan-related indigoidal dyes analyzed via HPLC.

Researchers (reference)	Year	Method used (ref.)	Dyes investigated									
			IS	4BIS	6BIS	IND	INR	MBI	6MBIR	DBI	6′MBIR	DBIR
Wouters and Verhecken (12)	1991	(13)				✓	✓	✓		✓		
Wouters (14)	1992	(12)				✓	✓	✓		✓		✓*
Koren (15)	1994	(12)				✓	✓	✓		✓		✓
Koren (16)	1995	(12)				✓		✓		✓		✓
Koren (17)	1997	(12)				✓		✓		✓		✓
Clark and Cooksey (18)	1997	(not reported)				✓	✓	✓	✓	✓	✓	✓
Cooksey (19)	2001	(not reported)				✓	✓	✓	✓	✓		✓
Cooksey & Withnall (20)	2001	(12)				✓	✓	✓	✓	✓		✓
Withnall, Patel, Cooksey, Naegel (21)	2005	(21)				✓	✓	✓		✓		✓
Koren (current work)		(see text)	✓	✓	✓	✓	✓	✓	✓	✓	✓	✓

*A 'brominated indirubin' was postulated, without precise molecular identification.

HPLC method was needed for performing more detailed multicomponent analyses of purple snail pigments.

Experimental

Reagents and dyes

The ternary mobile phase consisted of methanol, water, and 5% w/v phosphoric acid solution. The latter was prepared to yield a pH of 1.50 at 25 °C. The standard dyes were synthesized and provided by Chris Cooksey (formerly of University College, London), except for indigo, which was obtained as 'Indigo rein' (BASF Chemicals).

HPLC system

The chromatographic system used consisted entirely of Waters™ equipment. The pump was a model 600E Controller and the detector was a 996 PDA, both controlled by Millennium-32 software. The stationary phase consisted of a 3.0 × 150 mm C_{18} Symmetry column with 5 μm particle and 100 Å pore diameters, respectively. Sample solutions were injected manually through a 20-μL sample loop.

Extraction and sample preparation

Each dye was dissolved in dimethyl sulfoxide (DMSO) and processed under subdued lighting conditions as brominated indigoids can undergo photo-debromination. The sample was heated for 5 min at 100 °C, allowed to cool to room temperature for 15 min, and then filtered in an Alltech 0.45-μm micro-spin polypropylene centrifuge tube with nylon filter.

Elution methods

Two linear gradient elution methods with differing flow rates were applied, as indicated in Table 2; these are identical for the first 21 minutes. The first method is more conventional in that it uses a constant flow rate (0.8 mL min^{-1}) throughout, while the second method developed consists of an accelerated flow rate from 21 minutes in order to expedite the elution of the last remaining dye.

Results and discussion

Extracting solvent

Various solvents, such as dimethyl formamide (DMF, used warm[25] and at 150 °C[26]) and pyridine (100 °C),[27] have been used for extracting indigoids. In the course of the author's investigations into the 'purple problem' it was found that an excellent solvent for the dissolution of these blue, red, and purple colorants is dimethyl sulfoxide (DMSO) at 100 °C or higher. It has certain advantages over these other solvents, as it can be used at higher temperatures (its boiling point is 189 °C, as compared with 153 °C and 115 °C for DMF and pyridine, respectively)[28] and it does not possess the noxious pungent odor of pyridine. It has also been found to be

Table 2 Linear HPLC gradient elution methods for the analyses of indigoidal dyes: constant (Method 1) and increasing (Method 2) flow rate methods.

Method	Time (min)	Flow rate (mL/min)	Methanol (%)	Water (%)	5% H_3PO_4 (pH = 1.50) (%)
Methods 1 & 2	0 → 3	0.8	30 → 75	60 → 15	10
	3 → 20	0.8	75	15	10
	20 → 21	0.8	75 → 100	15 → 0	10 → 0
Method 1	21 → 30	0.8	100	0	0
Method 2	21 → 24	0.8 → 1.4	100	0	0
	24 → 25	1.4 → 2.1	100	0	0
	25 → 27	2.1	100	0	0

the optimal extracting solvent for the red dye carthamin in red safflower dyeings.[29]

Constant flow-rate method

The flow rate of 0.8 mL min^{-1} was selected for this chromatographic system so that the initial column pressure was close to, but not higher than, about 3000 PSI, and did not increase during the run by more than about 10%. This method (see Table 2) produced the chromatograms shown in Figure 2 for a DMSO-solution containing all of the ten standard dyes. The detection wavelengths are chosen to provide for the best visualizations of the peaks. The figure shows that this chromatographic method produces well-resolved peaks throughout.

The elution order conforms to a logical progression due to molecular structure (polarity) and size. The three relatively small isatinoid molecules are the first to elute between about 3 and 7 min, with the small polar isatin (IS in Table 1) eluting first, but retained in the column for about 3 min, far away from the solvent front or hold-up time. The less polar and heavier isomeric bromoisatins are more retained and, though they elute close to each other, they are nevertheless clearly separated. The 4-bromoisatin (4BIS) is not a component of a molluskan purple pigment, but has been used in this study to show that the developed method is responsive enough to separate it from the isomeric component substituted in the 6-position. Isatin and its 6-bromoisatin (6BIS) isomer may be found in the final pigment, as they are intermediates in the photo-oxidative pathway leading to the purple product.[30] The advantage of this method over those previously published is that previous indigoid studies did not address the chromatographability of the isatinoids.

The second series of dyes elutes between about 10 and 13 min, and consists of dimerized isatinoids, namely indigo (IND), indirubin (INR, the asymmetric isomer of indigo) and the singly brominated indigo (MBI), respectively.

The third group, eluting between about 18.5 and 20.5 min, consists of more brominated, and thus higher molecular weight, dimers specifically the singly brominated asymmetric indirubinoids and the doubly brominated indigo. The dibromoindigo (DBI) component is sandwiched between the 6- and the 6′-monobromoindirubins (6MBIR and 6′MBIR), respectively. The method developed in this study produced a much better separation of these three 'stubborn' dyes than previously reported.[31] The 6MBIR, DBI, and 6′MBIR dyes were separated by a total of about 2 min, compared with only 1 min as published

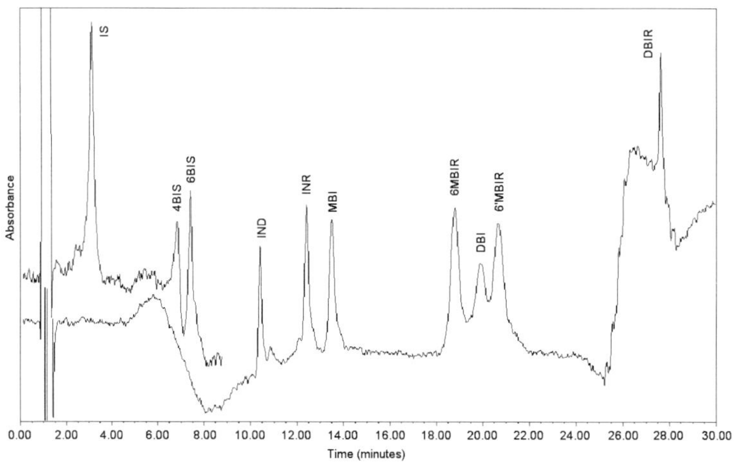

Figure 2 Chromatograms of a DMSO solution of the ten indigoidal dyes visualized at 400 nm for the isatinoids (top) and at 594 nm for the indigoids and indirubinoids (bottom).

earlier. It was found that the use of an acidic elution system maximized the separation of the latter two dyes, which in many chromatographic studies tend to elute together, to an average of about 45 seconds. Thus, though these two components elute close to one another, nevertheless, their respective peaks are both still clearly resolvable, as seen from the chromatogram (Fig. 2). Further, as the molecular weights of DBI and 6′MBIR are substantially different, additional mass spectrometric detection of these colorants would be helpful for future work.

The last eluting dye, the bulky asymmetric, doubly brominated indirubin (DBIR), is the most retained. Its chromatographic positioning far away from the other components was necessitated by the need to separate clearly previously eluting components from each other. It was found that DBIR's elution could be hastened by maximizing the methanol composition to 100% in the mobile phase, with a parallel reduction in the aqueous acidic phase from 10% to 0% phosphoric acid, as soon as the next-to-last dye (6′MBIR) washed out of the column (see Table 2). This condition prompted DBIR to elute at less than 28 min, as shown in the chromatogram (Fig. 2).

For each component, *i*, the average retention time (symbolized as RT or as t_R) and relative retention time ($RRT = RT_i/RT_{indigo}$) are given in Table 3 together with the respective standard deviations (SD) and % relative standard deviations (RSD). The RSD is an important statistical property as it is a measure of the precision or reproducibility of the experimental results. It is the standard deviation relative to the average value (AVG), and as a percentage is defined as $\%RSD = 100 \times (SD/AVG)$.

It can be seen from Table 3 that the precisions in the average retention times of the dyes range from 0.11 to 0.83% for the six runs performed in the same day. As expected, the reproducibilities of the relative retention times of the dyes are even better. The RRT has been widely used as a characteristic retention quantifier in HPLC studies; in the current work, the RRTs have resulted in the highest degree of reproducibility of all the most common retention parameters. The precisions in the average RRTs range from 0.04 to 0.76%.

UV-visible spectra

The ultraviolet/visible spectra of these dyes, as produced by the photodiode array detector of the chromatograph, are presented in Figures 3–5. For each spectrum, the wavelengths with maximal absorptions, λ_{max}, are given together with the relative absorbances (or absorptivities,

Table 3 Average retention times (RT) and relative retention times (RRT) for the indigoidal dye components and corresponding standard deviations (SD) and % relative standard deviations (RSD) for the two flow-rate methods (Table 2).

	Method 1: Constant flow rate (Same day, $n = 6$)						Method 2: Increasing flow rate (Same day, $n = 6$)					
	Retention times			Relative retention times			Retention times			Relative retention times		
Dye component	AVG	SD	%RSD	AVG	SD	%RSD	AVG	SD	%RSD	AVG	SD	%RSD
IS	3.11	0.0110	0.35	0.305	0.0008	0.25						
4BIS	6.76	0.0107	0.16	0.663	0.0003	0.04						
6BIS	7.32	0.0093	0.13	0.718	0.0006	0.09						
IND	10.20	0.0161	0.16	1	0	0	10.15	0.0748	0.74	1	0	0
INR	12.19	0.0375	0.31	1.195	0.0019	0.16						
MBI	13.26	0.0491	0.37	1.300	0.0030	0.23						
6MBIR	18.58	0.1227	0.66	1.821	0.0093	0.51						
DBI	19.70	0.1590	0.81	1.931	0.0124	0.64						
6′MBIR	20.47	0.1691	0.83	2.006	0.0153	0.76						
DBIR	27.46	0.0292	0.11	2.693	0.0031	0.11	25.04	0.0909	0.36	2.468	0.0092	0.37

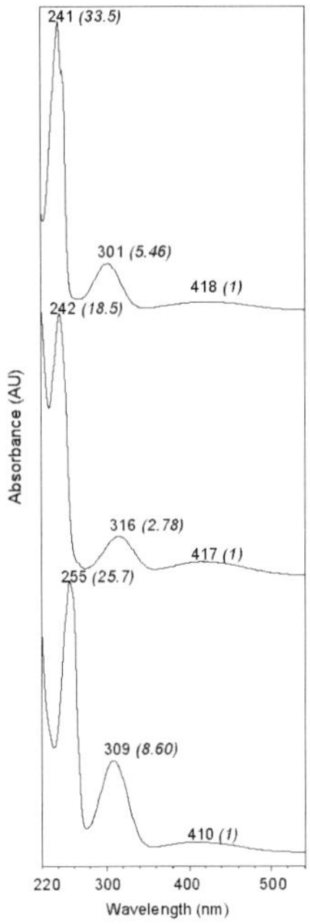

Figure 3 UV-visible spectra of the isatinoids as determined by the PDA detector with local λ_{max} values (and corresponding relative absorptivities): (a) IS; (b) 4BIS; (c) 6BIS.

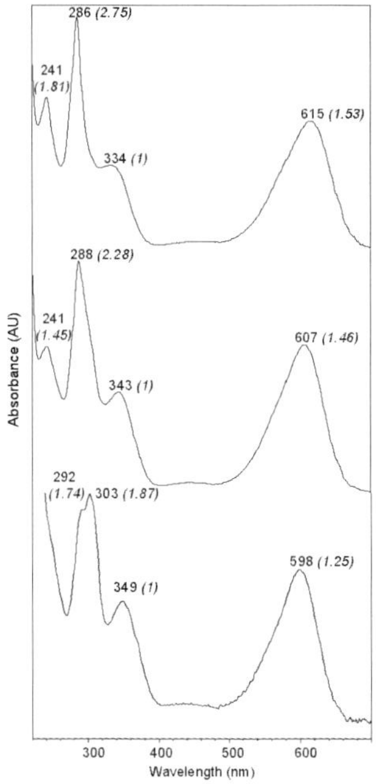

Figure 4 UV-visible spectra of the indigoids as determined by the PDA detector with local λ_{max} values (and corresponding relative absorptivities): (a) IND; (b) MBI; (c) DBI.

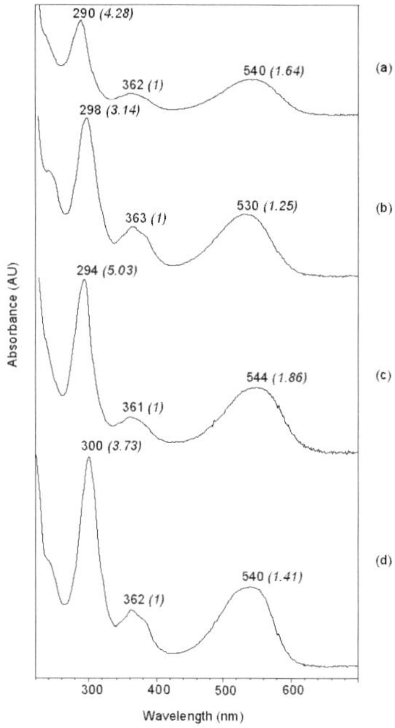

Figure 5 UV-visible spectra of the indirubinoids as determined by the PDA detector with local λ_{max} values (and corresponding relative absorptivities): (a) INR; (b) 6MBIR; (c) 6′MBIR; (D) DBIR.

usually symbolized as ε or *a*) at those wavelengths. These values can be used in future work for the Beer-Lambert quantifications of the absolute concentrations of these dyes.

Accelerated flow-rate method

The retention time of DBIR was more than 27 min with the constant flow rate method discussed above. In order to shorten its retention time further, the mobile phase flow rate was increased after 21 min, which is the time for the complete elution of the previously eluting component, 6′MBIR. As Table 2 indicates, the flow rate was increased from a value of 0.8 to 1.4 mL min^{-1} over a 3-min interval, and then accelerated to 2.1 mL min^{-1} over a 1-min period. These flow rates and intervals were chosen so that the column pressure would be less than 4000 PSI, which is near the upper pressure limit of most HPLC instruments. Figure 6 shows the variation of pressure throughout the run. For the first 21 min the two elution methods are identical. In the accelerated method, the pressure rises to about 3800 PSI, which is still a tolerable high-pressure value.

As indicated in Table 3, DBIR elutes at about 25 min with this new method, more than 2 min faster than with the conventional constant flow rate. The average retention time of DBIR as well as its RRT, relative to indigo, are highly reproducible with %RSD values of 0.36% and 0.37%, respectively, for six same-day run-to-run experiments.

This chromatographic method can be summarized in Figure 7, which shows group separations according to chemical functionality as a time-dependent function of the percentage compositions of the eluents and of the flow rate. The isatinoid, indigoid, and indirubinoid dyes

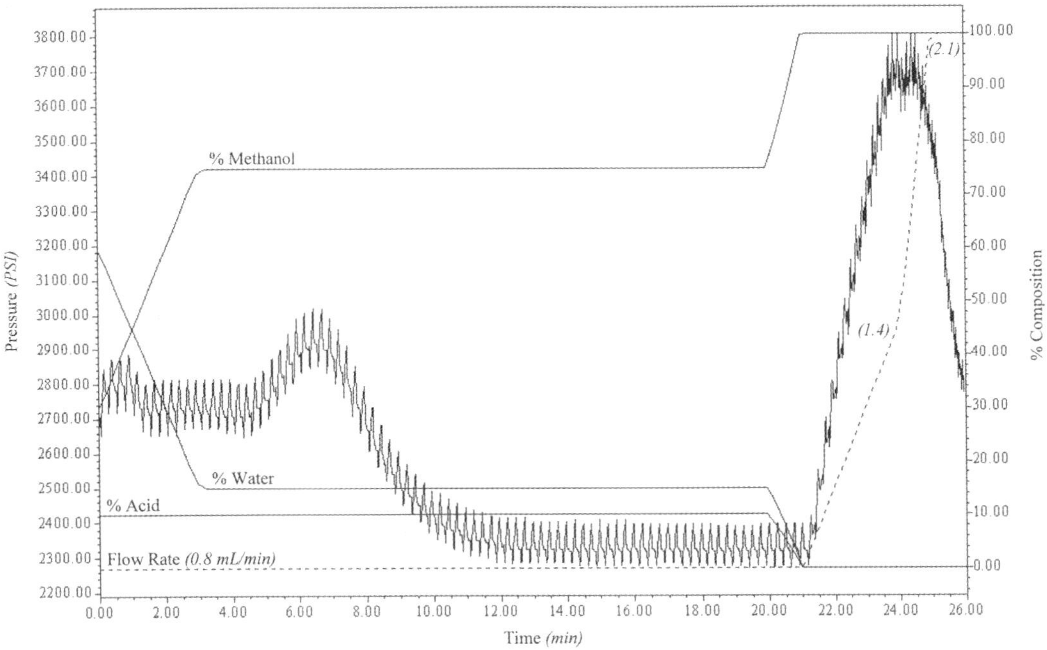

Figure 6 Variation of pressure with time for the increasing flow rate method.

investigated belong to the following chemical groups respectively: indole diones, 2,2′-biindole diones, and 2,3′-biindole diones. The small indole diones are the first to elute followed by the second group, composed of relatively simple biindole diones (indigo and indirubin) and the singly brominated indigo. The complex biindole diones, consisting of the doubly substituted indigo sandwiched between the singly substituted indirubinoids, is the third group to elute. The last eluting high molecular weight biindole dione, DBIR, is more asymmetric than its isomer DBI, and is the sole member of the last group to elute out of the chromatographic column. No other colorant has yet been detected to elute after DBIR.

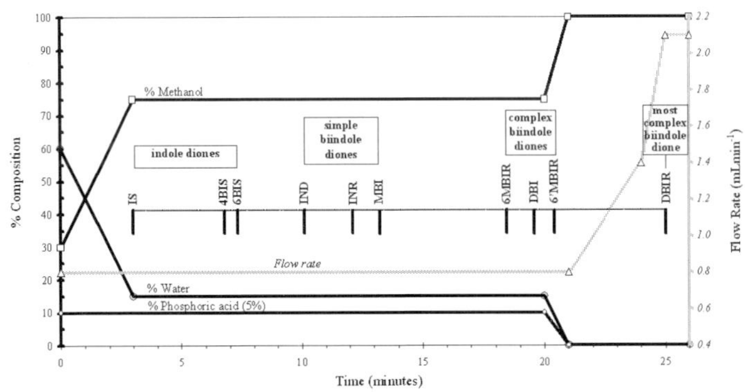

Figure 7 Group elutions of the isatinoids, indigoids, and indirubinoids based on the increasing flow rate method.

Conclusion

This work has shown that more indigoidal and related components than previously analyzed can be chromatographed in about 25 min by a single elution method, which yields well-resolved peaks, homologous group separations, and quantifiable retention parameters. The method should therefore be able to predict and detect the chromatographic positions of other colorants that may be present in various molluskan purple pigments. For example, based on the 'road map' of Figure 7, the logical position for other monomeric colorants, such as brominated and unbrominated sulfur-substituted isatinoids, would be after their unsulfurated complements, but before the dimeric indigo molecule.

This analytical scheme would thus be able to provide qualitatively and quantitatively a more detailed 'biochemical fingerprint' of modern pigments produced from various purple-yielding marine species collected at different geographical locations around the world. Such a full multicomponent chemical characterization would thus aid in determining the zoological and geographical provenance of the sea snails used in producing the ancient royal and sacred pigments found on archaeological dye vat potsherds and on ancient textile dyeings. This method was recently applied to the analysis of the purple paint pigment on a 2,500-year-old marble jar belonging to King Darius I of ancient Persia.[32]

Acknowledgements

The author is deeply grateful to the Dr. Sidney M. Edelstein Foundation for support of this research. In addition, the samples provided by Chris Cooksey and the assistance with the chromatographic runs given by Efrat Konstantini, Eva Eich and Angelika Rubinov have been much appreciated.

Editor's note

This article is based, in part, on a paper titled 'The purple question reinvestigated: just what is really in that purple pigment?' and delivered by the author at the 20th Annual Meeting of Dyes in History and Archaeology, Netherlands Institute for Cultural Heritage, Amsterdam, 1–2 November 2001.

Address for correspondence: The Edelstein Center for the Analysis of Ancient Textiles and Related Artifacts, Shenkar College of Engineering and Design, 12 Anna Frank Street, 52526 Ramat-Gan, Israel (zvi@shenkar.ac.il).
Paper received 27 August 2003; revised version received 5 January 2004.

Notes and references

1. Koren, Z.C. (1993) 'Methods of dye analysis used at the Shenkar College Edelstein Center in Israel', *Dyes in History and Archaeology* 11, pp. 25–33.
2. Koren, Z.C. (1994) 'Analysis of the Masada textile dyes', in J. Aviram, G. Foerster, and E. Netzer (eds), *Masada IV. The Yigael Yadin Excavations 1963–1965. Final Reports*, Jerusalem, pp. 257–64.
3. Friedländer, P. (1909) 'Über den Farbstoff des antiken Purpurs aus Murex brandaris', *Berichte der Deutschen Chemische Gesellschaft* 42, pp. 765–70; Friedländer, P., Bruckner, S. and Deutsch, G. (1912) 'Über Brom- und Methoxyderivate des Indigos', *Annalen* 388, pp. 23-49; Friedländer, P. (1922) 'Über die Farbstoffe aus Purpura aperta and Purpura lapillus', *Berichte der Deutschen Chemische Gesellschaft* 55, pp. 1656–8.
4. Cooksey, C.J. (2001) 'Tyrian purple: 6,6′-dibromoindigo and related compounds', *Molecules* 6, pp. 736–69; Koren, Z.C. (2001) 'The purple question reinvestigated: just what is really in that purple pigment?', *20th Annual Meeting of Dyes in History and Archaeology Abstracts*, Netherlands Institute for Cultural Heritage, Amsterdam, p. 10.
5. Dai, F.B., Qiao, C.Z. and Li, L. (1986) 'Determination of indigo and indirubin in qingdai by HPLC', *Acta Pharmaceutica Sinica* 21, pp. 868–71; Wouters, J. and Verhecken, A. (1991)

'High-performance liquid chromatography of blue and purple indigoid natural dyes', *Journal of the Society of Dyers and Colourists* 107, pp. 266–9; Wouters, J. (1992) 'A new method for the analysis of blue and purple dyes in textiles', *Dyes in History and Archaeology* 10, pp. 17–21.
6. Fouquet, H. and Bielig, H.-J. (1971) 'Biological precursors and genesis of Tyrian purple', *Angewandte Chemie International Edition English* 10, pp. 816–17; Fouquet, H. (1970) 'Bau und Reaktionen natürlicher Chromogene indigoider Farbstoffe bei Purpurschnecken', PhD Thesis, University of Saarlandes; Baker, J.T. (1974) 'Tyrischer Purpur: ein antiker Farbstoff, ein modernes Problem' ('Tyrian purple: an ancient dye, a modern problem'), *Endeavour* 33, pp.11–17.
7. Cooksey 2001 (see note 4 above); Cardon, D. (2003) *Le monde des teintures naturelles*, Paris, pp. 421–67; Schweppe, H. (1993) *Handbuch der Naturfarbstoffe: Vorkommen; Verwendung; Nachweis*, Landsberg/Lech (Ecomed), pp. 304–18 (or Nikol Verlagsgesellschaft mbH & Co. KG, Hamburg).
8. Edmonds, J. (2000) *The Mystery of Imperial Purple Dye* [Cover title: *Tyrian or Imperial purple dye*] (Historic Dye series no. 7), Little Chalfont (published by the author).
9. Koren, Z.C. (2005) 'The first optimal all-*Murex* all-natural purple dyeing in the eastern Mediterranean in a millennium and a half', *Dyes in History and Archaeology* 20, pp. 136–49 and pls 15.1–15.5.
10. Spanier, E. and Karmon, N. (1987) 'Muricid snails and the ancient dye industries', in E. Spanier (ed.), *The Royal Purple and the Biblical Blue: Argaman and Tekhelet. The Study of Chief Rabbi Dr. Isaac Herzog on the Dye Industries in Ancient Israel and Recent Scientific Contributions*, Jerusalem, pp. 179–92.
11. Reese, D.S. (2000) 'Iron Age shell purple-dye production in the Aegean', Appendix 6.1, in J.W. Shaw and M.C. Shaw (eds), *Kommos IV. The Greek Sanctuary, Part 1*, Princeton, pp. 643–8; Stieglitz, R.R. (1994) 'The Minoan origin of Tyrian purple', *Biblical Archaeologist* 57, pp. 46–54.
12. Wouters and Verhecken 1991 (see note 5 above).
13. Dai *et al.* 1986 (see note 5 above).
14. Wouters 1992 (see note 5 above).
15. Koren, Z.C. (1994) 'HPLC analysis of the natural scale insect, madder and indigoid dyes', *Journal of the Society of Dyers and Colourists* 110, pp. 273–7.
16. Koren, Z.C. (1995) 'High-performance liquid chromatographic analysis of an ancient Tyrian Purple dyeing vat from Israel', *Israel Journal of Chemistry* 35, pp. 117–24.
17. Koren, Z.C. (1997) 'The unprecedented discovery of the Royal Purple dye on the two thousand year-old royal Masada textile', *AIC Textile Specialty Group Postprints*, vol. 7 (American Institute for Conservation of Historic and Artistic Works, 25th Annual Meeting, San Diego, CA), Washington DC, pp. 23–32.
18. Clark, R.J.H. and Cooksey, C.J. (1997) 'Bromoindirubins: the synthesis and properties of minor components of Tyrian purple and the composition of the colorant from *Nucella lapillus*', *Journal of the Society of Dyers and Colourists* 113, pp. 316–21.
19. Cooksey 2001 (see note 4 above).
20. Cooksey, C. J. and Withnall, R. (2001) 'Chemical studies on *Nucella lapillus*', *Dyes in History and Archaeology* 16/17, pp. 91–6.
21. Withnall, R., Patel, D., Cooksey, C. and Naegel, L. 'Chemical studies of the purple dye of *Purpura pansa*', *Dyes in History and Archaeology* 19, pp. 109–17.
22. Dai *et al.* 1986 (see note 5 above).
23. Wouters and Verhecken 1991 (see note 5 above); Wouters 1992 (see note 5 above).
24. Koren 2001 (see note 4 above).
25. Cooksey and Withnall 2001 (see note 20 above).
26. Koren 1997 (see note 17 above); Koren 1995 (see note 16 above); Koren 1994 (see note 15 above); Koren 1993 (see note 1 above).
27. Wouters and Verhecken 1991 (see note 5 above); Wouters 1992 (see note 5 above).
28. *The Merck Index: An Encyclopedia of Chemicals, Drugs and Biologicals* (1989), 11th edn, ed. S. Budavari, Rahway, NJ, pp. 511, 513–14, 1267.
29. Koren, Z.C. (2001) 'A successful Talmudic-flavored high-performance liquid chromatographic analysis of carthamin from red safflower dyeings', *Dyes in History and Archaeology* 16/17, pp. 158–66.
30. Fouquet and Bielig 1971, Fouquet 1970, Baker 1974 (see note 6 above).
31. Clark and Cooksey 1997 (see note 18 above).
32. Koren, Z.C. (2008) 'Archaeo-chemical analysis of Royal Purple on a Darius I stone jar', *Microchimica Acta* (in press).

The Biblical Dye *Tekhelet* and its Use in Jewish Textiles

I. Irving Ziderman

Historical introduction

Two purple textile dyes are referred to in ancient records: one had a blue hue and the other a red.[1] The bluish purple was named *tekhelet* in biblical and Talmudic Hebrew and 'hyacinthine purple' by the Romans. The reddish variety was called *argaman* in the Hebrew and 'Tyrian purple' by the Romans. Archaeological findings in the Mediterranean area have indicated that the ancients utilised three species of marine shellfish for dyeing, namely: banded dye-murex (*Hexaplex trunculus* (Linnaeus, 1758) = *Phyllonotus trunculus*, *Murex trunculus*), spiny dye-murex (*Bolinus brandaris* (Linnaeus, 1758) = *Murex brandaris*) and dogwinkle or rock-shell (*Stramonita haemastoma* (Linnaeus, 1766) = *Thais haemastoma*, *Purpura haemastoma*) (Fig. 1). All three are extant today and the dyes are prepared from their hypobranchial gland.

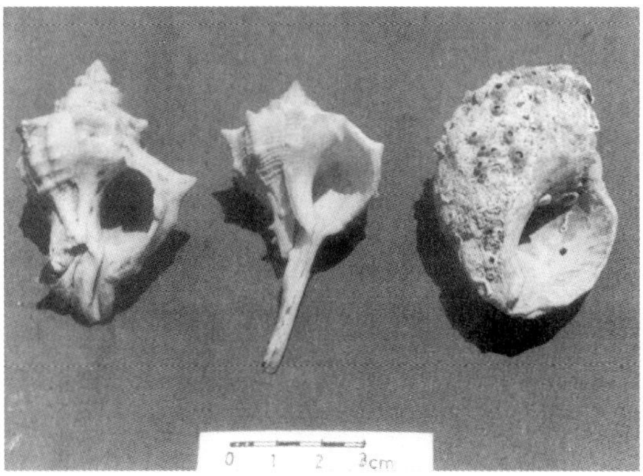

Figure 1 The three species of marine shellfish used in antiquity for manufacturing purple dyes. The shells are, left to right, banded dye-murex (*Hexaplex trunculus*), used for dyeing *tekhelet*/hyacinthine purple, and spiny dye-murex (*Bolinus brandaris*) and dogwinkle (*Stramonita haemastoma*), both used for dyeing *argaman*/Tyrian purple. The dogwinkle shell bears signs of marine wear and encrustation of biological fouling.

The broken shells found at the sites of ancient purple dye-houses are principally those of banded dye-murex.[2] This suggests that banded dye-murex was the main source of purple in antiquity. Now, there has been much confusion as to the colour and chemical composition of the dyestuff obtained from this shellfish.

Semantics of colour[3]

In order to facilitate a discussion of this issue, it is necessary first to clarify the semantics of the words that have been used to describe the various colours. Colour designations (e.g. blue, violet, red and purple) may be regarded as having one of three meanings:

- *aesthetic*: an abstract term for colour sensation or hue;
- *chemical*: name of dyestuff substance;

- *coloured material*: for example, dyed textiles (fibres, yarn, cord, fabrics or garments) as commercial commodities.

Ancient references – biblical, classical and Talmudic – should typically be understood in this third sense. Accordingly, identification of an ancient colour often means identifying a commercial textile, rather than a particular shade. The semantics of colour varies both within a language and between languages. Furthermore, there is a distinction between generic names and hue colour names. The cases of purple, violet and blue are discussed below.

Purple

In English, 'purple' is not only the name of a particular hue, but also a generic term describing a range of hues or a class of similar hues. It is generic in three senses:

- *Coloristic*: purple is the sensation of mixing spectral red and blue lights in various ratios; it is outside the spectrum.
- *Tinctorial*: in hyacinthine and Tyrian purples, a blue tint in the former gives a bluish purple, while red in the latter gives a reddish purple. Since a red tint characterises modern usage of 'purple' in English colour terminology, it would nowadays be sufficient to use 'violet' and 'purple' respectively for bluish purple and reddish purple.
- *Malacological*: dyeings with any species of shellfish, whatever the colour.

Violet

In English, 'violet' is a spectral division and very limited in its range of hue. However, in some other languages, such as German and French, 'violet' is a generic term that would translate to the English 'purple' as defined above, as does the Hebrew *segol*.

Blue

In Tyndale's 1529 rendition of the Bible into English, *tekhelet* is translated 'jacinth', a word signifying 'hyacinth'. The King James version of the Bible published in 1611 renders *tekhelet* as 'blue.' Tracey Rihll has discussed the changing meaning of 'blue' in English since the 17th century:[4]

> Blue was used in the 1800s ... to describe a pony, a pig, cows, milk, flint, a number of diseases in animals (e.g. 'blue-spald' in cattle), of numerous plants some of which we think of as blue but many others as green (e.g. ground ivy, marsh grass), of a lot of birds which we would call grey, brown or black (e.g. hedge sparrow, black tern, pigeon).
>
> It is also worth noticing, on the symbolic front, that it had and still has negative connotations in English: in Shakespeare's time it was the colour of servants' clothes and could be used as shorthand to refer to someone of low standing; a 'blue gown' in Scotland was a licensed beggar; a blue-belly was a Protestant dissenter; a blue day was one on which something bad had taken place (opposite to 'red letter' day); we still of talk of 'being blue' to mean melancholy. Given its socio-economic and symbolic focus, I cannot imagine that people had in mind either azure-type hues or expensive dyes.
>
> The change over to meaning blue-blue (i.e. not red-blue, green-blue, blue-black, blue-grey, blue-brown) seems to start in the 18th century (and takes longer to penetrate dialectical usage), after Newton anyway, and at the same time the range of the meaning of violet and indigo started shrinking – they have practically disappeared in modern English. So when Newton (or anyone before him speaking English) said 'blue', he did not mean what we now assume the word to mean.

Accordingly, in 1611, 'blue' was a generic term that designated a range of hues and would have included reddish blue, i.e. 'violet' in English. Indeed, 'violet' is now preferred for *tekhelet* in several dictionaries and modern translations of the Bible (*The Jerusalem Bible*, London 1966; *The New English Bible*, London 1970). The colours that are obtained from the various shellfish, as detailed in the following sections, confirm the validity of this translation as 'violet' in English.

The colours obtained from banded dye-murex

During the last 150 years, various investigators have made dyes from banded murex, but different colours were obtained. In 1832, Bizio obtained a bluish purple and found it to be a mixture of indigotin with a purple substance.[5] Accordingly, he concluded that banded dye-murex was the source of hyacinthine purple, and that Tyrian purple was the reddish purple that he had obtained from both spiny dye-murex and dogwinkle, and which he identified to be the same purple substance as he had found in the dye mixture from banded dye-murex. This purple compound is now identified as 6,6′-dibromoindigotin (DBI), the predominant ingredient of Tyrian purple dyeings. Bizio's identification of the two purple dyes was supported by his chemical research, which has since been corroborated in independent studies of these shellfish species by other investigators.[6]

But Lacaze-Duthiers reported in 1859 that, in his hands, banded dye-murex yielded not only violet, but occasionally either blue or purple.[7] He contended that it was only the purple colour which characterised the banded murex, just as it was for spiny murex and dogwinkle. The other colours from the banded murex would have been produced by modification of the primary purple, introducing varying amounts of blue. Accordingly, Lacaze-Duthiers rejected Bizio's assignment of banded dye-murex as the specific source for hyacinthine purple and of spiny dye-murex for Tyrian purple. As a result, he left no hypothesis regarding the source and nature of hyacinthine purple/*tekhelet*. The controversy between the two scholars on this issue remained unresolved.[8]

Gender segregation of precursors[9]

Elsner, too, observed that some individuals of banded dye-murex yield blue and others purple. He discovered that this difference was correlated to the snail's sex, males largely giving blue and females purple: in other words, there is a gender segregation of dyestuff precursors in the hypobranchial gland of the snails. The females contain the bromoindoxyl, while the males have predominantly the indoxyl (see Plate 8). Two molecules of indoxyl can couple together forming indigotin, the blue dye from males. Similarly, two molecules of bromoindoxyl can react to form DBI, the purple dye from females. Thus, natural segregation of precursors can account for the blue and purple colours that are obtained from banded murex individuals. Using a commingled population of males and females, the resulting mixture of purple DBI with blue indigotin will be a violet dye, as was described by Bizio for hyacinthine purple.

Furthermore, banded dye-murex males undergo sex reversal to functioning females (protandric sequential hermaphroditism) after one season. Since the penis is retained, this female may be erroneously identified as a male, even though it now produces only DBI.

The formation of 6-monobromoindigotin (MBI)

Another factor that contributes to the formation of a range of colours with banded dye-murex was revealed by the recent discovery that, together with DBI and indigotin, the dye contains a third ingredient, 6-monobromoindigotin (MBI).[10] The formation of MBI may be understood on considering the gender distribution of dyestuff precursors in the snails' hypobranchial gland.[11] An intimate admixing of glandular material from a catch of male and female snails would permit a mixture of the two precursors, indoxyl and bromoindoxyl, to react together chemically to form MBI (see Fig. 2). Thus, three chemical reactions would compete with each other, leading to variations in chemical composition

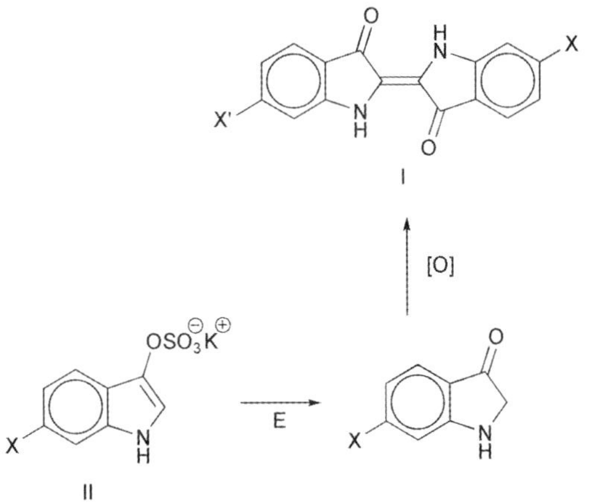

Figure 2 Major pathway for formation of indigotin dyes, I, from colourless indoxyl precursors, II, obtained from the murex hypobranchial gland. In indigotin, X = X′ = H. In MBI, X = Br and X′ = H. In DBI, X = X′ = Br. The precursor II is an indoxyl sulfate salt and its 6-bromo derivative, as found in banded dye-murex. There is no evidence that it is the potassium salt in the mollusc: the potassium comes from the alkali added to the separated indoxyl sulfuric acids that emerge from the chromatography column. E = enzymatic or acidic hydrolysis. [O] = aerobic oxidative coupling reaction. This transformation does not have a photolytic step dependent on sunlight, in contrast to the requirement with the precursors from all other species, since the latter contain a C-2 substituent that is absent in banded dye-murex. (Reproduced with permission from Ziderman 1981 (see note 14) © 1981 Society of Dyers and Colourists.)

of the dye formed. The colour of MBI was still unknown then, since it had not yet been isolated, but was presumed to be purple. Accordingly, it was proposed that depletion of the indigotin content, due to enhanced formation of MBI, would give redder colours that are purple rather than violet. Different colours would thereby be possible, depending upon the reaction conditions used.

There is considerable variation in the proportional ratio between the three compounds in the dye from banded dye-murex, as analysed by the different workers. These contrasting results are dependent on the gender composition of the snail catch and on the experimental conditions used by the investigator to prepare the dye.

The nature of *tekhelet*

Several textile remains have been found to be dyed with a mixture of DBI and indigotin that corresponds empirically to the product of banded dye-murex.[12] Furthermore, 'a *Ph. trunculus*-type' dye was found on a textile from a Sarmathian tomb of the first century BCE on the Black Sea coast, according to its analytical data – indigotin (70%), DBI (7%) and MBI 23%.[13] Accordingly, the bluish purple from the banded murex may be identified as biblical *tekhelet*, the hyacinthine purple of classical antiquity.[14] This corroborates Bizio's view in his disputation with Lacaze-Duthiers (see above). In Plate 8 the colour and source of hyacinthine purple are depicted, and compared with the other two precious dyes of antiquity, namely Tyrian purple and kermes (Hebrew *tola`at shani*, from the scale insect *Kermes vermilio* Planchon).

Ancient recipes for making imitation *tekhelet* substantiate that ancient *tekhelet* was a violet colour. Thus a 7th-century document recently discovered describes the imitation of hyacinthine purple by double dyeing with woad and madder.[15] This chemical composition has also been reported in a new analysis of the violet wool tassel from the Dead Sea Cave of the Letters,[16] previously considered to have been dyed with woad and kermes as an imitation of *tekhelet* for use on ritual tassels. Such a dye combination corresponds to the Talmudic description (Midrash *Sifrey*, on Numbers 14:41) of forging ritual *tekhelet*: 'Behold! I use red dye and indigo so that they resemble *tekhelet*: who could then expose me?'

The hyacinth connection

Tekhelet was translated 'hyacinth' in Greek by Hellenistic Jews and in Latin by the Romans. What does the term signify in this context?

In classical Greek, 'hyacinth' had meant a flower and not murex purple. In the Hellenistic period, however, several Jewish translations of the Bible into Greek appeared, in which *tekhelet* was consistently rendered 'hyacinth'.[17] This translation is surely authentic, having been made at a time when the use of *tekhelet* in the Second Temple was completely familiar to both the translators and their readers. The new usage of 'hyacinth' for a dyestuff would appear to have derived from its similar hue to *Hyacinthus orientalis* L., a violet-coloured flower native to the Phoenician hinterland.

Concurrent with this usage in Hellenistic Greek, 'hyacinth' was one of the two classes of purple dyeing described in Latin by the Elder Pliny (Book XXI.xxii.45–6; also called 'amethyst' and 'ianthine').[18] In an earlier passage (Book IX.lxi.130–lxiv.140), Pliny describes two hyacinthine dyeings: namely, hyacinthine purple, made from a mixture of *bucinum* and *pelagia* molluscs, and a paler one (also called *conchylia*), made from *pelagia* alone.[19] *Bucinum* appears to be today's dogwinkle, while *pelagia* is a category for the various dye-murexes.[20] Thus Pliny's description of 'hyacinth' is compatible with identification of banded dye-murex as the source for *tekhelet*.

In classical Greek, two terms describe the colour purple.[21] *Porphorous* is from *porphora*, murex shellfish. *Phoinikous* means red, scarlet or purple: it is of uncertain origin and was not used for murex. Did they perhaps correspond to the two purple products, hyacinthine and Tyrian?

Jewish textiles

Tekhelet was the most precious dye of the ancients, and was required for the most holy textiles of the religious rites in the biblical Tabernacle and in the Temple in Jerusalem. Detailed descriptions are given in the biblical book of Exodus (chapters 24–40) of the awnings, curtains and priestly garments. The Book of Numbers (chapter 4) describes coverings for the sacred vessels during the wanderings in the Wilderness.

The Talmudic tractate *Menahot* contains descriptions of how *tekhelet* was made from shellfish.[22] The details are remarkably similar to those of Pliny's description of shellfish dyeing, including treatment of wool with 'fresh' and 'clear' extract, and the test dyeing.[23] Furthermore, the Talmud there states that the colour of *tekhelet* resembles the sea and the sky, and details how to use it in tying ritual tassels. There is also the following test to distinguish *tekhelet* from indigo: liquid alum, juice of fenugreek and urine that had been kept for 40 days were mixed. The test sample was soaked in the mixture overnight. If the colour did not fade, it was genuine *tekhelet*. If it faded, it was baked inside a piece of hard leavened barley dough. If the colour then improved, it was genuine *tekhelet*, but if it deteriorated it was indigo.

Ritual tassels

In the Book of Numbers (ch. 15), there is a commandment that *tekhelet* cord be tied together with white cords to form a tassel (Hebrew *tsitsit*) attached to each corner of four-cornered garments. Incorporation of *tekhelet* in ritual corner-tassels was abandoned in the seventh century, when Moslem conquerors destroyed the purple dye-houses in the Holy

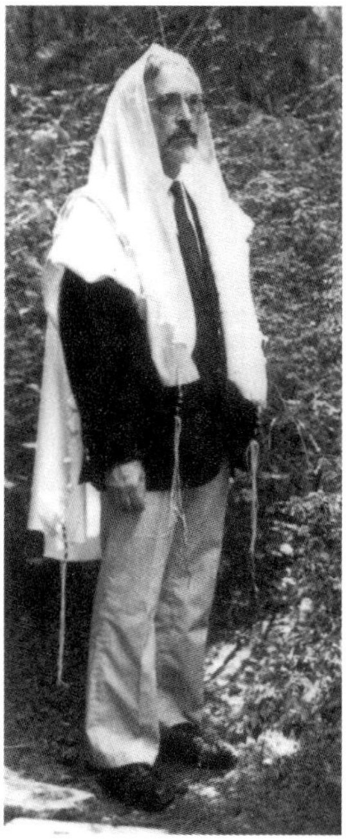

Figure 3 The author wearing a prayer shawl with *tekhelet* on the corner-tassels.

Land. However, it is now being revived there using hypobranchial glands from banded dye-murex.

In rabbinical and Karaite traditions of Jewish practice, different patterns are used for tying the tassels with *tekhelet*. A special four-cornered vestment (Hebrew *tallit*) was introduced to fulfil the biblical obligation. There are two such ritual garments: one is the Jewish prayer shawl (see Fig. 3); the other is a short cassock-like garment worn under or over the shirt.

Founded in the 8th-century, the Karaite Jewish sect still use blue cords in the tassels on their prayer shawls, since they do not require shellfish *tekhelet* and may use any blue dye. Each tassel is composed of eight white cords and eight blue cords that are doubled over and then tied together as a corner-tassel of 32 hanging cords (see Plate 9).

Among rabbinical Jews, the practice is to take only four threads, double them over and then tie them as a tassel of eight hanging cords. In the absence of *tekhelet*, white cords are used in their stead so as to preserve the required number of eight cords. In the Middle Ages, three different presumptions were made as to the way the ancients would have incorporated the long-lost *tekhelet* in ritual tassels. Accordingly, of the eight tassel cords, one, two or four would be *tekhelet*. One of the cords is much longer than the others and is wound around them repeatedly: when available, *tekhelet* is used for these windings, except for the first and last windings, nearest and farthest from the garment corner respectively, which must always be white (see Plate 10).

The prevalent pattern is to make 39 windings that are separated by double knots into sets of 7, 8, 11 and 13 windings. Furthermore, Talmudic law requires that, when genuine *tekhelet* is used, these windings must be arranged throughout in groups of three. Winding each group of three through an individual loop has been used to preserve the integrity of each set of three windings.

Colour instability of MBI

The revival of *tekhelet* dyeing in Israel has been undertaken in order to renew its use in ritual corner-tassels. The initiative began with a scientific study that determined the historical source of the required dye to be banded dye-murex, the colour to be violet and the chemical composition to be indigotin and DBI.[24] Experiments with shellfish glands confirmed that a violet colour may be made with banded dye-murex and a purple with spiny dye-murex and with dogwinkle.[25] But usually a purple colour was obtained, which was considered inappropriate as *tekhelet* since it was the same colour as the Tyrian purple from spiny dye-murex. The nature of this purple from banded dye-murex has been revealed through the discovery of the thermal instability of the colour.

Wool that had been dyed purple with banded dye-murex changed colour to blue on heating.[26] This intriguing finding may not be ascribed to a change occurring in any DBI present, since this compound is eminently stable. It was therefore hypothesised that this purple dyeing contains mainly MBI, and that MBI has this peculiar property of thermal colour instability.

When synthetic MBI became available later, it was found to have a violet colour.[27] On heating wool dyed with this MBI, it indeed changes colour to blue, thereby proving the validity of the hypothesis.[28] Pure MBI behaved similarly.[29]

But what is the mechanism of the colour transition in MBI? The hypothesis considered initially was that the colour change might be due to a possible thermal debromination of MBI forming indigotin. This view has since been shown to be untenable because chemical analyses of the blue product have established that no change in chemical structure or composition has occurred.[30]

Photolytic debromination

In order to produce a bluish colour that could be marketed as *tekhelet* for the ritual tassels, the purple from banded murex (largely MBI) has been photolytically debrominated to indigotin, which was vat-dyed *in situ* onto wool.[31]

Driessen discovered photolytic debromination of DBI in 1944.[32] An aqueous solution of the halogen-containing indigoid in the *leuco*-form is first prepared by reaction with the powerful reducing agent sodium dithionite. Then the solution is exposed to sunlight or an equivalent ultraviolet source, while held in a transparent glass vessel for effective irradiation. But dithionite is a modern synthetic reagent that was not known in antiquity, and neither were glass reaction vessels. Therefore it is questionable whether photolytic debromination was available in antiquity to make indigotin. Furthermore, it would have been unnecessary in that era to make indigotin from purple, considering the ready availability of inexpensive woad and/or indigo from plants, then used for producing indigotin dye. Besides, the highly precious value of the purple dye would have been squandered.

But it should now be possible to prove murex provenance of an indigotin dyeing by using analytical chemistry. A diagnostic test could be performed by high-pressure liquid chromatography (HPLC). This sensitive technique can detect the presence of residual traces of DBI and MBI that are too minute to affect the colour, but must be present in every dyestuff originating in shellfish. HPLC analysis of archaeological textiles dyed blue with indigotin has not hitherto detected a vestige of brominated indigotin.[33] Accordingly, they are vegetable dyes, either woad or indigo. Hence, analytical evidence has not yet been presented to suggest that shellfish were used for dyeing blue by debromination of purple to indigotin.

Conclusion

The new understanding of the status of MBI in the banded dye-murex dye, as described above, should lead to the development of a marketable *tekhelet* dye that is not merely indigotin. Firstly, the thermal transition of the purple to form blue would seem to be a more acceptable process, as an alternative to photolytic debromination to indigotin. Secondly, investigation of the chemical reactions of the natural dye precursors from banded dye-murex should reveal the empirical conditions required for direct synthesis of a stable dye mixture of indigotin and DBI that resembles the violet colour of the hyacinth flower.

Editor's note

Address for correspondence: Tekhelet Foundation, 20 Brody Street, 92502 Jerusalem, Israel (ziderman@smile.net.il). Paper received 9 February 2004; revised version received 20 May 2004.

Notes and references

1. Ziderman, I.I. (1986) 'Purple dyes made from shellfish in antiquity', *Review of Progress in Coloration and Related Topics* 16, pp. 46–52.
2. Reese, D.S. (1980) 'Industrial exploitation of murex shells: purple-dye and lime production at Sidi Khrebish, Benghazi', *Libyan Studies: The Society for Libyan Studies 11th Annual Report (1979–80)*, pp. 79–83; Reese, D.S. (1987). 'Palaikastro shells and Bronze Age purple-dye production in the Mediterranean basin', *Annual of the British School of Archaeology at Athens* 82, pp. 201–6; Reese, D.S. (2000) 'Iron Age shell purple-dye production in the Aegean', in J.W. Shaw and M.C. Shaw (eds.), *Kommos 14. The Greek Sanctuary*, Princeton, part 1, pp. 643–7.
3. Ziderman, I.I. (1990) 'Textile dyes of animal origin in ancient Israel and Phoenicia', paper presented at the International Workshop on History of Chemical Technology, Hebrew University of Jerusalem, 1990 (unpublished).
4. Tracey Rihll, pers. comm., 2001.
5. Bizio, B. (1832) *La porpora (degli antichi) rivocata entro i confini del rosso*, Venice, Dissertazione critica, Tipografia di Commercio; Grimaud de Caux, G. and Gruby, A. (1842) 'Description anatomique de l'organe qui fournit la liqueur purpurigène dans le Murex brandaris, et analyse microscopique de cette liqueur', *Comptes rendus hebdomadaires des séances de l`Académie des Sciences* 15, pp. 1007–9; Grimaud de Caux, G.

(1856) 'Sur l'animal de la pourpre des Anciens', *Revue et Magasin de Zoologie pure et appliquée* 8, pp. 34–42.

6. Friedländer, P. (1922) 'Über die Farbstoffe aus Purpura aperta und Purpura lapillus', *Berichte der Deutschen Chemischen Gesellschaft* 55, pp. 1655–8; Pfister, R. (1937) *Nouveaux textiles de Palmyre,* Paris, p. 11; Bouchilloux, S. and Roche, J. (1954) 'Sur les prochromogènes et les pigments purpuriques de Murex trunculus Linné' *Comptes rendus des Sciences, Sociétié de Biologie* 148, pp. 1732–4; Fouquet, H. and Bielig, H.-J. (1971) 'Biological precursors and genesis of Tyrian purple', *Angewandte Chemie International Edition* 10, pp. 816–17; Elsner, O. (1992) 'Solution of the enigmas of dyeing Tyrian purple and the biblical *tekhelet*', *Dyes in History and Archaeology* 10, pp. 11–16.
7. Lacaze-Duthiers, H. de (1859) 'Mémoire sur la pourpre', *Annales des Sciences Naturelles: Zoologie* 4, 12, pp. 5–84.
8. Bizio, B. (1860) 'Réclamation', *Cosmos* 16, p. 624; Lacaze-Duthiers, H. de (1860) 'Response', *Cosmos* 17, pp. 34–5; Bizio, B. (1861) 'Notizie concernenti alcune controversie circa la porpora degli antiqui', *Atti del' I. R. Istituto Veneto* 3, 6, pp. 795–802.
9. Elsner 1992 (see note 6 above). In our new paper (Ziderman, I.I. and Larush-Asraf, L., 'Biochemistry of *Phyllonotus trunculus* dyes', presented at the 26th Meeting of *Dyes in History and Archaeology*, Vienna, November 2007), we discuss the findings of later authors who, like Elsner, also report a sex segregation of banded dye-murex dyes but differ from him in finding that the males mostly give purple and the females blue! Moreover, the occurrence of imposex is considered to compromise the empirical criteria used for recognising the sex of the snails.
10. Wouters, J. and. Verhecken, A. (1991) 'High-performance liquid chromatography of blue and purple indigoid natural dyes', *Journal of the Society of Dyers and Colourists* 107, pp. 266–9; Michel, R.H., Lazar, J. and. McGovern, P.E. (1992) 'The chemical composition of the indigoid dyes derived from the hypobranchial glandular secretions of murex molluscs', *Journal of the Society of Dyers and Colourists* 108, pp. 145–50; Koren, Z.C. (1995) 'High-performance liquid chromatographic analysis of an ancient Tyrian purple dyeing vat from Israel', *Israel Journal of Chemistry* 35, pp. 117–24.
11. Ziderman, I.I. (2001) 'Revival of biblical *tekhelet* dyeing with banded dye-murex (*Phyllonotus trunculus*): chemical anomalies', *Dyes in History and Archaeology* 16/17, pp. xvii and 87–90.
12. Ziderman, I.I. (1990) 'Seashells and ancient purple dyeing', *Biblical Archaeologist* 52, pp. 98–101.
13. Wouters, J. (1992) 'A new method for the analysis of blue and purple dyes in textiles', *Dyes in History and Archaeology* 10, pp. 17–21.
14. Ziderman, I.I. (1981) 'Blue thread of the *tzitzit*: was the ancient dye a Prussian blue or Tyrian purple?' *Journal of the Society of Dyers and Colourists* 97, pp. 362–4; Ziderman, I.I. (1987) 'First identification of authentic *tekhelet*', *Bulletin of the American Schools of Oriental Research* 265, pp. 25–33.
15. Granger-Taylor, H. (2001) 'Dyes and coloured textiles of the Greek and Roman world', *Proceedings of the International Conference on Colours in Antiquity*, University of Edinburgh, UK.
16. Koren, Z.C. (2001) 'The fashionable colours of ancient Israel', *Proceedings of the International Conference on Colours in Antiquity*, University of Edinburgh, UK.
17. Herzog, I. (1987). 'Hebrew porphyrology' (D.Litt. dissertation, University of London, 1913), in E. Spanier (ed.), *The Royal Purple and the Biblical Blue,* Jerusalem, pp.17–131, esp. p. 78.
18. Pliny [Gaius Plinius Secundus] (1938–62) [*Naturalis Historia*] *Pliny Natural History*, tr. H. Rackham [and others] (Loeb Classical Library), 10 vols, Cambridge, Mass. and London, vol. III (1940), IX.liii.105–5; lx.124–lxiv.141; vol. VI (1951), XXI.xxii.45–6; Born, W. (1937) 'Purple', *Ciba Review* 4, pp. 112–13; Forbes, R.J. (1964) *Studies in Ancient Technology,* 2nd edn, 9 vols, Leiden, vol. 4, pp. 98–142; Ziderman, I.I. (1992) 'Molluscal source of biblical "blue" dyestuff', *Proceedings of the 9th International Malacology Congress,* Edinburgh, 1986, pp. 407–14.
19. Dedekind, A. (1898) 'La pourpre verte et sa valeur pour l'interprétation des écrits des Anciens', *Archives de Zoologie Expérimentale et Générale* 3, 6, pp. 467–80.
20. Ziderman 1992 (see note 18 above).
21. Fox, N. (2001) 'Concepts of colour in Greek literature', *Proceedings of the International Conference on Colours in Antiquity*, University of Edinburgh, UK.

22. Ziderman, I.I. (1987) 'Halakhic aspects of reviving the ritual *tekhelet* dye in the light of modern scientific discoveries', in E. Spanier (ed.), *The Royal Purple and the Biblical Blue*, Jerusalem, pp. 207–20.
23. Pliny 1940 (see note 18 above), IX.lx.126; lxii.133.
24. Ziderman 1981 (see note 14 above).
25. Ziderman, I.I. (1986) '3600 years of purple-shell dyeing: characterisation of hyacinthine purple (*tekhelet)*', in H.L. Needles and S. Haig Zeronian (eds) *Historic Textile and Paper Materials: Conservation and Characterization*, Washington DC (Advances in Chemistry Series 212), pp. 187–98.
26. Ziderman 2001 (see note 11 above).
27. Clark, R.J.H. and Cooksey, C.J. (1999) 'Monobromoindigos: a general synthesis, the characterization of all four isomers and an investigation into the purple colour of 6,6′-dibromoindigo', *New Journal of Chemistry* 23, pp. 323–8.
28. Ziderman, I.I. (2004) 'Purple dyeing in the Mediterranean world: characterisation of biblical *tekhelet*', in L. Cleland, K. Stears and G. Davies (eds), *Colour in the Ancient Mediterranean World*, Oxford (British Archaeological Reports International Series, no. 1267), pp. 38–43.
29. Ziderman, I.I., Sasaki, Y., Sato, M. and Sasaki, K. (2008). 'Colour change on heating 6-bromoindigotin: key to understanding purple dyeing with banded dye-murex', *Dyes in History and Archaeology* 22, in course of publication.
30. Ziderman *et al.* 2008 (see note 29 above).
31. *Petil Tekhelet* Association, http://www.tekhelet.com, accessed August 2007.
32. Driessen, L.A. (1944) 'Über eine charakterische Reaction des antiken Purpurs auf der Faser', *Melliand Textilberichte* 26, p. 66.
33. Zvi C. Koren, pers. comm., 9 September 2001.

Indigo-reducing Bacteria from the Medieval Woad (*Isatis tinctoria* L.) Vat: Some Aspects of their Interaction with Indigo

Philip John*, Stylianos Arghyros and Sonja Nicholson

Introduction

Indigo is insoluble, and in order for it to be used as a dye it has to be chemically reduced to a soluble form. In the dyeing industry over the past century, alkaline sodium dithionite has been used as the reductant, and continues to be used on a large scale. Before the introduction of chemical methods of reduction, indigo was reduced for dyeing in a fermenting woad vat. This consisted essentially of a controlled anaerobic fermentation of treated (couched) woad leaves, which provided both the indigo and the principal fermentation substrate.

Until our group investigated the scientific basis of the woad vat, the agent responsible for indigo reduction was unknown.[1] We recreated a woad vat according to medieval practice using couched woad as the principal ingredient.[2] A new bacterial species was isolated from the fermenting vat – an indigo-reducing moderate thermophile, which was identified by phenotypic and 16S rRNA gene sequence analysis and named *Clostridium isatidis*.[3] The role of *C. isatidis* in medieval indigo dyeing was confirmed by isolating a virtually identical organism from woad dye-bath waste excavated from 10th-century Anglo-Scandinavian (Viking) deposits in York, UK.[4] The survival of *C. isatidis* spores in a viable state under the waterlogged anoxic conditions in which the material was found was consistent with the proven longevity of bacterial spores.

Modern methods of reducing indigo in the dyeing industry involve either conventional chemistry or electrochemistry.[5] When sodium dithionite is used, it is oxidised to a variety of sulfur compounds, which contaminate the dye-bath waste and rinsing water. As a result, more environmentally sound methods of indigo reduction are being sought, including catalytic hydrogenation, the use of organic reductants and electrochemical methods. However, no entirely satisfactory method is yet available and more novel sustainable biotechnologies could find application. *C. isatidis* (and the closely related *C. carnis*) are the only organisms so far described that have the ability to reduce indigo.[6] Because of regulatory controls over the use of living organisms in biotechnological processes, it is unlikely that *C. isatidis* could now reassume the key role it played in a pre-chemical era and directly replace current chemical and electrochemical technologies for indigo reduction. However, if we understood how *C. isatidis* is able to reduce indigo, then this information could form the basis of a novel indigo reduction biotechnology. For this reason, we have recently been investigating the mechanism of indigo reduction by *C. isatidis*. Specifically, we were interested to determine, firstly, if the bacteria interacted directly with the indigo or reduced it at a distance using chemical intermediates; secondly, what enzymes or other mechanisms might be involved in this interaction. Confocal scanning laser microscopy was employed to view colonies of *C. isatidis* while they were reducing indigo particles *in vitro* and, for reasons explained below, we focused on cellulase as a possible agent involved in the bacteria–indigo interaction.

Experimental

Confocal scanning laser microscopy

Sterile indigo (Kemtex) was attached to sterile microscope slides treated with a layer of nail varnish (Rimmel). The slides were immersed in sterile Reinforced Clostridial Medium, inoculated with cultures of *C. isatidis* (Wv6) and contained in slide troughs. These were incubated in anaerobic jars for 48 h at 47 °C. The slides were removed carefully from the troughs and stained with 5 μL of SYTO 13. Two drops of Vectashield mounting medium H-1000 were then applied and a coverslip added. This was sealed with nail varnish at the edges. The specimen was viewed in a Leica TCS-NT confocal scanning laser microscope.

Particle size analysis and cellulase detection

Particle size distribution was determined using a Malvern Particle Size Analyser (Series 2600c Droplet and Particle Sizer) at room temperature (about 25 °C). *C. isatidis* was from stocks maintained at the University of Reading. All other species were from the Deutsche Sammlung von Mikroorganismen und Zellkulturen, Braunschweig, Germany. Cellulase activity was detected by the addition of bacterial cultures to wells cut into Clostridial Basal Medium, containing 0.5% (w/v) carboxymethylcellulose and 0.1% (w/v) glucose. The control well contained water. Bacteria were cultured overnight anaerobically as previously described,[7] using Reinforced Clostridial Medium. When bran was included as a carbon source, bacteria were grown on Reinforced Clostridial Medium supplemented with 1% (w/v) bran. A recreated woad vat was prepared as previously described.[8] Briefly, 100 g of dried, couched woad balls was added to 500 ml of boiling tap water in a beaker and the resulting woad vat allowed to ferment for eight days at 50 °C. The pH was adjusted to pH 9 by the occasional addition of 2M sodium hydroxide (NaOH) solution. The carboxymethylcellulose plates were incubated under anaerobic conditions for four days at the temperature optima of the bacteria: 37 °C for *Clostridium aurantibutyricum*, *C. celatum*, and *C. papyrosolvens*, 49 °C for *C. isatidis* and 27 °C for *Cellulomonas* sp.; then stained with 1 mg mL^{-1} Congo red for 15 min and destained with 1M sodium chloride (NaCl) solution for 15 min.

Results and discussion

Possible mechanisms by which indigo-reducing bacteria interact with indigo

The insolubility of indigo is not a problem unique to indigo reduction by *C. isatidis*: a variety of bacteria can use insoluble minerals as electron acceptors.[9] For example, for many bacteria in anaerobic environments, Fe(III), which is virtually insoluble at around neutral pH, is an important electron acceptor. What is essential, biochemically, is to establish a link between the reducing centres in the bacteria and the external electron acceptor. In principle, three different mechanisms can be distinguished: (i) a direct interaction between bacteria and indigo; (ii) an indirect interaction via redox shuttle compounds; (iii) dissolution of the indigo to form particles small enough to penetrate the bacterial cell.

Direct interaction between bacteria and indigo

In the case of Fe(III) reduction by *Geobacter metallireducens*, Nevin and Lovley (2000) have argued for a direct mechanism on the grounds that the evidence available does not support alternative mechanisms.[10] Experimental support for a direct interaction in any bacterium is difficult to find. In a Gram positive organism such as *C. isatidis*, the characteristic cell wall structure would militate against such a mechanism. However, our previous work, in which we used carbon electrodes colonised by *C. isatidis*, provided evidence which we interpreted in terms of bacteria coupling electrochemically to the 'insoluble' electrode surface.[11] Thus we concluded that there is 'an electron transfer from the bacterial cell interior to a solid external electron acceptor'.[12] Additional biochemical work is needed to substantiate this conclusion.

Indirect interaction via redox shuttle compounds

There is much evidence in the literature for redox compounds acting as electron carriers between bacteria and their solid electron acceptors. A wide variety of compounds can in principle act as electron shuttles, but most attention has been paid to humic compounds originating from the environment,[13] and to quinones produced by the bacteria.[14] Either type of compound could be important for the indigo-reducing activity of *C. isatidis*. With reference to the latter, it may then be significant that virtually all the published medieval recipes for the woad vat specify the addition of madder, which contains a wide variety of anthroquinones.[15] However, it is known that the traditional woad vat will dye perfectly well without the addition of madder.[16] It has also been observed that madder is not required for pure cultures of *C. isatidis* to reduce indigo.[17] However, members of the genus *Clostridium* are known to produce quinones,[18] despite earlier indications to the contrary.[19] When bacteria produce redox shuttles they form a biofilm in order to localise the shuttle compounds and prevent their loss through diffusion away from the bacterial colony.[20] The biofilm allows for a high cell density, as well as the build-up of the shuttle compounds. We now present direct evidence for biofilm formation in *C. isatidis* in its interaction with indigo (Plate 11). Confocal scanning laser microscopy provides a three-dimensional view of a colony of bacteria which has been allowed to grow on a slide to which particles of indigo were attached. The resulting view (Plate 11) shows the bacterial colony taking the form of a biofilm with bridge-like structures connecting the indigo particles. When *C. isatidis* is grown as a pure culture in liquid media, mucus-like biofilm material is evident. However we do not know whether in the conditions of the undefined, mixed culture of the woad vat, biofilm is produced or is necessary for indigo reduction.

The biofilm shown here would provide an appropriate structure for a shuttle-mediated reduction of indigo by *C. isatidis*. Thus, while conditions are present for redox shuttle compounds to be maintained within the bacterial environment, we still do not have any direct biochemical evidence for their presence.

Dissolution of the indigo to form particles small enough to penetrate the bacterial cell

Currently, the stonewashed look of denim jeans is obtained commercially by treating the jeans with cellulase. The cellulase not only has an action on the cotton fibres, but also interacts with the deposited indigo. Thus, Campos *et al.* (2000) found that 'indigo agglomerates are fractionated into smaller particles when incubated with acid cellulases from *Trichoderma reesei*',[21] while Gusakov *et al.* (2000) reported that 'cellulases ... act as emulsifiers helping indigo float out of cellulose fibres'.[22] Members of the genus *Clostridium* are notable for their production of enzymes like cellulase that hydrolyse complex polysaccharides,[23] thus we were prompted to investigate a possible role of cellulase in indigo reduction. Specifically, we were concerned to answer two initial questions: whether cellulase demonstrably dissagregates indigo; and whether indigo reduction is associated with cellulase.

Figure 1 shows that when indigo particles are incubated with cellulase from *Trichoderma reesei*, particle size decreases. After 30 min there is a significant reduction in abundance of the larger particles seen in the commercial indigo preparation used. Then after a longer incubation period (20 h), during which the untreated particles have aggregated, average particle size

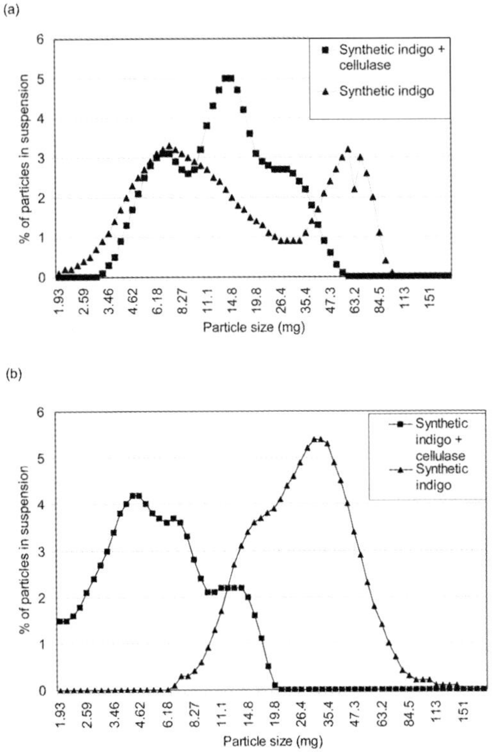

Figure 1 Indigo (0.2 mg mL^{-1}) was incubated at 37 °C with *Trichoderma reesei* cellulase (1 mg mL^{-1}) for (a) 30 min and (b) 20 h, and particle sizes were measured using a Malvern Particle Size Analyser.

is reduced to one-tenth. In this experiment, the indigo particles are still not small enough to diffuse into the bacterial cell. However, the potential ability of enzymes like cellulase to have a profound effect on indigo aggregation is evident.

Having obtained evidence that cellulase could disaggregate indigo, it remained to determine if *C. isatidis* produced cellulase. These determinations were made by adding cultures of *C. isatidis* to wells cut into agar plates containing a soluble form of cellulose, carboxymethylcellulose. After a period of incubation, the carboxymethylcellulose was stained with Congo red. Where the cellulose had been hydrolysed by cellulase activity, a non-stained area appeared around the well. Plate 12 shows that the presence of cellulase could easily be detected in our system from the area clarified by the culture of *Cellulomonas* sp. When we examined *C. isatidis*, we used cultures grown routinely with glucose as the main carbon source, and initially saw no such clear area. This apparent absence of cellulase activity was surprising, since it is known that the fermentation of the woad vat could be maintained for many months if bran was added, and we assume that cellulose, as the major component of the bran, was used by the bacteria for this fermentation. When we examined a recreated woad vat unequivocal evidence was obtained of cellulase activity (Plate 12), and when we grew a pure culture of *C. isatidis* with bran (either in the presence or in the absence of glucose) cellulase activity was readily apparent (Table 1). Thus we conclude that under the conditions of the woad vat, when cellulose is available, *C. isatidis* produces a cellulase. However it is noteworthy that *C. isatidis* cultures will reduce indigo when glucose is supplied as the carbon source, and cellulose is absent.[24]

Our explanation for the disaggregating effect on indigo of *T. reesei* cellulase is that its cellulose binding domain is responsible by disrupting H-bonds that hold together particles of indigo,[25] in the same way as the cellulose binding domain

Table 1 Bacterial indigo reduction and cellulase activity. Cellulase activity was determined by carboxymethylcellulose disappearance; indigo reduction was determined after anaerobic culture in the presence of 0.01% (w/v) indigo.

	Cellulase	Indigo reduction
Cellulomonas sp.	yes	no
Clostridium isatidis (glucose grown)	no	yes
Clostridium isatidis (bran grown)	yes	yes
Clostridium isatidis (bran + glucose grown)	yes	yes
C. aurantibutyricum	yes	no
C. celatum	no	no
C. papyrosolvens	yes	no

disrupts H-bonding in crystalline cellulose.[26] Possibly the cellulose binding domain can be produced in the absence of cellulase for some function we do not yet recognise.

Table 1 shows that a variety of bacteria (some closely related to *C. isatidis*) also possessing cellulase activity was unable to reduce indigo. Thus, even if cellulase activity (or at least a cellulose binding domain) is associated with indigo reduction, and we cannot yet be sure, it is not itself sufficient for indigo reduction.

We are continuing our investigations into the mechanism by which *C. isatidis* reduces indigo. Our understanding remains incomplete, but we note that, while dithionite has been employed industrially for a century, the chemical mechanism by which it reduces indigo is still not completely understood.[27]

Acknowledgements

We are grateful to Paulina Garcia Macias, Hilary Lynch, Nikki Padden and Kalani Seymour for useful discussions; John Edmonds for supplying a recreated medieval woad vat; Moy Robson, Andrew Wilbey and Steven Poutney for expert technical support; and the EC for funding (Spindigo QLK5-CT-2000-30962).

Editor's note

*Author for correspondence: School of Plant Sciences, University of Reading, Reading RG6 6AS, UK (p.john@reading.ac.uk). Paper received 14 March.2003; revised version received 30 July 2004.

Notes and references

1. Padden, A.N., Dillon, V.M., John, P., Edmonds, J., Collins, M.D. and Alvarez, N. (1998) '*Clostridium* used in mediaeval dyeing', *Nature* 396, p. 225.
2. Kokubun, T., Edmonds, J. and John, P. (1998) 'Indoxyl derivatives in woad (*Isatis tinctoria*) in relation to medieval indigo production', *Phytochemistry* 49, pp. 79–87.
3. Padden, A.N., Dillon, V.M., Edmonds, J., Collins, M.D., Alvarez, N. and John, P. (1999) 'An indigo-reducing moderate thermophile from a woad vat, *Clostridium isatidis* sp. nov.', *International Journal of Systematic Bacteriology* 49, pp. 1025–31.
4. Padden, A.N., John, P., Collins, M.D., Hutson, R. and Hall, A.R. (2000) 'Indigo-reducing *Clostridium isatidis* isolated from a variety of sources, including a tenth century Viking woad vat', *Journal of Archaeological Science* 27, pp. 953–6.
5. Roessler, A, Dossenbach, O., Marte, W. and Rys, P. (2002) 'Electrocatalytic hydrogenation of vat dyes', *Dyes and Pigments* 54, pp. 141–6.
6. Padden *et al.* 1999 (see note 3 above).
7. Padden *et al.* 1999 (see note 3 above).
8. Kokubun *et al.* 1998 (see note 2 above).
9. Hernandez, M.E. and Newman, D.K. (2001) 'Extracellular electron transfer' *Cellular and Molecular Life Sciences* 58, pp. 1562–71.
10. Nevis, K.P. and Lovley, D.R. (2000) 'Lack of production of electron-shuttling compounds or solubilization of Fe(III) during reduction of insoluble Fe(III) oxide by *Geobacter metallireducens*', *Applied and Environmental Microbiology* 66, pp. 2248–51.
11. Compton, R.G., Perkin, S.J., Gamblin, D.P., Davis, J., Marken, F., Padden, A.N. and John, P. (2000) '*Clostridium isatidis* colonised carbon electrodes: voltammetric evidence for direct solid state redox processes', *New Journal of Chemistry* 24, pp. 179–81.
12. Compton *et al.* 2000 (see note 11 above).
13. Lovley, D.R., Coates, J.D., Blunt-Harris, E.L., Phillips, E.J.P. and Woodward, J.C. (1996) 'Humic substances as electron acceptors for microbial respiration', *Nature* 382, pp. 445–8.
14. Newman, D.K. and Kolter, R. (2000) 'A role for excreted quinones in extracellular electron transfer', *Nature* 405, pp. 94–7.
15. Gilbert, K.G. and Cooke, D.T. (2001) 'Dyes from plants: past usage, present understanding and potential', *Plant Growth Regulation* 34, pp. 57–69.
16. Kokubun *et al.* 1998 (see note 2 above).
17. Padden *et al.* 1999 (see note 3 above).
18. Yamamoto, K., Murakami, R. and Takamura, Y. (1998) 'Isoprenoid quinone, cellular fatty acid composition and diaminopimelic acid isomers of newly classified thermophilic anaerobic Gram-

positive bacteria', *FEMS Microbiology Letters* 161, pp. 351–8.
19. Collins, M.D. and Jones, D. (1981) 'Distribution of isoprenoid quinone structural types in bacteria and their taxonomic implications', *Microbiological Reviews* 45, pp. 316–54.
20. Hernandez and Newman 2001 (see note 9 above).
21. Campos, R., Cavaco-Paulo, A., Andreaus, J. and Gübitz, G. (2000) 'Indigo-cellulase interactions', *Textile Research Journal* 70, pp. 532–6.
22. Gusakov, A.V., Sinitsyn, A.P., Markov, A.V., Skomoarovsky, O.A., Sinitsyna, A.G., Brlin, A.G. and Ankudimova, N.V. (2000) 'Indigo-binding domains in cellulase molecules', *Biocatalysis-2000: Fundamentals & Applications* 41, pp. 77–80.
23. Matsushita, O. and Okabe, A. (2001) 'Clostridial hydrolytic enzymes degrading extracellular components', *Toxicon* 39, pp. 1769–80.
24. Padden *et al.* 1999 (see note 3 above).
25. Süsse, P., Steins, M. and Kupcik, V. (1988) 'Indigo: crystal structure refinement based on synchrotron data', *Zeitschrift für Kristallographie* 184, pp. 269–73.
26. Xiao, Z., Gao, P., Qu, Y. and Wang, T. (2001) 'Cellulose-disrupting domain of endoglucanase III from *Trichoderma reesei* disrupting the structure of cellulose', *Biotechnology Letters* 23, pp. 711–15.
27. Bond, A.M., Marken, F., Hill, E., Compton, R.G. and Hügel, H. (1997) 'The electrochemical reduction of indigo dissolved in organic solvents and as a solid mechanically attached to a basal plane pyrolytic graphite electrode immersed in aqueous electrolyte solution', *Journal of the Chemical Society Perkin Transactions* 2, pp. 1735–42.

Woad production in Catalonia and Roussillon at the End of the Middle Ages

Anthony Pinto

Around the Mediterranean Sea, the production of woad experienced a considerable growth, connected to the tremendous development of the medieval woollen cloth industry. From the beginning of the 14th century, this increase in production generated substantial revenue. Italian production (in Lombardy and Tuscany) and French production (in Languedoc) were by far the most important.[1] Nevertheless, regions close to Languedoc also developed the production of this dye plant with more or less success. Not far from Languedoc, on the lands of the Crown of Aragon, in Roussillon and in Catalonia, the cultivation of woad emerged on a more or less extensive scale, depending on local conditions.

It is by using an original documentary corpus, thanks to the preservation of notes written by local notaries kept in the archives of the cities of Perpignan and Gerona, that we can retrace the history of this production, its growth, and its decay over two centuries, the 14th and the 15th centuries.

The first mentions of woad cultivation and the role of the inhabitants of Languedoc

It is believed that the earliest attempts at woad cultivation in this north-western part of the Mediterranean basin occurred in Roussillon during the second half of the 13th century.[2] However, the most ancient document giving evidence for this cultivation in Catalonia so far is dated from 1316 and concerns the nearby region of Empordà. It is around the small town of Castelló d'Empúries, which had approximately 6000–6500 inhabitants before the plague of 1348, in the high Empordà countryside that this first attempt to develop the crop occurs. At that date, a Catalan and an inhabitant of Languedoc from Castelnaudary succeeded in producing several hundred kilos of *agranat* woad. We can believe that, in this transaction, the Catalan partner had proposed that his field should be used for the experiment, while the Frenchman had brought the know-how, sown the woad, taken care of the fields and dealt with the leaves. A third partner probably financed the enterprise, expecting benefits from the crop.[3] This first document must be seen as one of the very first attempts to raise the level of production.

Experiments of this type around Castelló d'Empúries did not stop at this stage: woad cultivation continued to develop gradually in the 1330s. In 1335, for example, a local cloth merchant promised a dyer that he would sell him 43 *quintars* of woad, harvested on his lands, which he would deliver later.[4] A few years later, the presence in the region of people from Languedoc cultivating woad is mentioned again. A document concerning the death of a dyer from Carcassonne (caused by the plague of 1348) and his succession provides further proof of the cultivation of the dye plant. However, in spite of these various attempts, no significant increase in woad production occurred in the Empordà countryside. Similarly, in Roussillon, the supply for the local woollen cloth industry continued to depend on imports of woad from Italy and Languedoc.[5]

In the Geronès, the area around the episcopal city of Gerona, the beginning of woad cultivation occurred later: it was first mentioned during the 1340s.[6] The gap in time observed between woad cultivation in this region and in Alt Empordà (Upper Empordà) is best explained by the rather late development of Gerona's woollen cloth trade. Nevertheless, the vigorous growth of this activity, which took place in the 1330s, as well as the close links maintained between the town of Gerona and Languedoc – more precisely Toulouse and its region – had a compensating effect.[7] Once again, the influence of this French province on the spread of woad cultivation appears to be of the utmost importance for northern Catalonia.

Some inhabitants of Languedoc were involved in the early stages of this development of woad cultivation in Catalonia, but even more important is the part played by some local traders: from the beginning, the initiatives of starting woad cultivation were financed by them. In Roussillon and in Empordà, however, in spite of the capital invested and the involvement of some local businessmen, there was no significant increase in woad production. This limited growth could be blamed on a lack of a spirit of enterprise among the local farming community and their inability to break away from their routines.[8] In fact, it may be, rather, that at first the returns were unpredictable and the quality of the harvests unsatisfactory due to a lack of knowledge and experience. An even more likely explanation is that local production could not compete with the massive imports of woad arriving from Tuscany and Languedoc. Any attempt at large-scale cultivation of woad was thus thwarted by the laws of international markets. Some years later, however, in the region around Gerona, woad cultivation did manage to thrive for several decades, thanks to a particularly rich soil.

Original development of the production of woad in Gerona

A favourable situation

Until recently, Catalonia was not recognised as a woad-producing region in the 14th century. Christian Guilleré's work on the history of the city of Gerona and its countryside, however, has brought to light the fact that the crop was originally cultivated in the area. Since its beginning, production of woad in that region benefited from international commercial links and from a favourable local situation. From the point of view of international exchange, it benefited from the decline of woad production in Tuscany from the 1340s due to difficult political circumstances in that region. Some years later, it was in Languedoc that serious trouble occurred: in 1355, the region was devastated by Edward of Woodstock, Prince of Wales, son of the English King Edward III. This brought about a disruption in the export of woad to Catalonia.[9] At this time, only woad from Lombardy could benefit from the truce signed between the Crown of Aragon and Genoa and could thus be produced in great quantity at the end of the 1350s. However, the main part of these imports was increasingly absorbed by big economic centres, such as Barcelona and Valencia – a new state of affairs. Therefore foreign supplies could no longer flood the market of northern Catalonia. But why did this development take place in Gerona and not anywhere else?

More than in any other centre of the woollen cloth trade in northern Catalonia, the growth of this activity in Gerona was due to a good supply of raw materials, and more particularly of dyestuffs. The cloth merchant sector of the city in fact specialised in the finishing of cloths, including fulling and dying. Unlike its competitors, Perpignan and Castelló d'Empúries, the episcopal city of Gerona did not at first appear to be as well supplied by Tuscany and the regions of Narbonne and Foix.[10] Gerona did, however maintain special links with the Toulouse region. It was thus particularly necessary to free the city from the unpredictable nature of dyestuff supply caused by the rise of other cloth trade centres. At that point, the attitude of Gerona's urban merchant elite, as well as the availability of human resources in the surrounding countryside, made the implementation of intensive cultivation possible. At the same time, the outbreak of plague in 1348 and the resultant population decrease contributed to the decline of traditional agriculture, in which grain farming was predominant; this became both less profitable and less vital for local populations. The transformation of the countryside that took place after the great plague outbreak, a major event of Western history, favoured the production of woad.

Production in the countryside

Climatic conditions, the fertility of the land and an expansion of the business of cattle breeding were all favourable to the future success of woad cultivation. Two streams, the Ter and the Onyar, flowing through the rich lands, ensured the necessary irrigation of the plants, while the ever-expanding cattle farming provided manure for the land.[11] The high returns observed in this region provide evidence for the general practice of fertilisation. While in Lauraguais five harvests of woad could be obtained and in Lombardy only four, in Catalonia five to six harvests were the rule.[12] From then on, the development of woad cultivation was instrumental in transforming the countryside, where high profits were sought. The culture of woad was very productive indeed – the first harvest could take place after four months.[13] To achieve this required a large workforce that the village was apparently able to supply because of the development of cattle breeding, which was less labour-intensive than growing grain.

It seems that the technique of purchasing standing crops was rather common in this region, unlike in Languedoc in the 16th century.[14] The crops were sold to the merchants before being harvested.[15] In this type of business deal, the harvester received an advance on the value of the harvest.[16] The buyer could also choose to buy a fixed amount of the crop, as in Lombardy or in Languedoc.[17] The harvest thus belonged to him. In Gerona, however, the production of woad did not depend exclusively on these traders; closely following the example of the merchant class, the farming community quickly learned how to produce and commercialise woad efficiently. They did not hesitate to sell *pastel agranat* (couched woad) directly to the consumers, who were fullers or dyers.[18] These producers included ordinary cloth merchants and also traders who had cultivated woad on their lands around Gerona. The original aspect of this production was that it was, for the main part, in the hands of a small number of entrepreneurs; that it took place within the framework of an intensive economy; and that it was done on small areas of fields.[19]

The *agranat* was the finished product obtained after crushing the leaves, shaping the pulp into balls, drying them, breaking them up and fermenting the pulp before drying it. It was stored by the principal local traders or dyers before being sold locally or at other markets.[20] Woad was generally measured by the cartload or bag, but also by *bramente* or *brunete*, or by *sardinas*, as in Languedoc; it could also be sold in balls (*panibus*).[21] The balls could also be supplied *crudi*, that is, before the second preparation, couching.[22] It is possible that some dyers from Gerona prepared the leaves themselves.[23] All in all, during the second half of the 14th century, about 20 villages in the regions of the Ter and Selva harvested woad.[24] During the 15th century, the number of producing villages remained about the same. It has been demonstrated by Christian Guilleré's work that production moved from Gerona and its environs down towards the plain and the terraces of the Ter and Terri regions, but also towards Selvas.[25] From then on, it may be that the region of the Ter became the main production centre of Catalonia.

If information about production itself is rather scarce, some hints can be found on woad harvesting in the countryside. Of course, cloth merchants, fullers and ordinary merchants trading in the countryside, such as the self-employed horse dealers, could easily bring cartloads of woad to the city. Circulating throughout the countryside, they were the privileged intermediaries between the rural producers and the consumers, and they could evaluate the quality and the quantity of the harvests in order to negotiate the best price.[26] In this way, larger and larger quantities of woad could, at the same time, supply the Geronese market and the other woollen cloth centres of the region.

A regional market

After the first developmental stage, the Geronese woad-producing area began a second regular progression to reach its peak at the beginning of the 15th century. We find that, during this period, large contracts were signed on the local market: one of them amounted to a value of 278 *livres* 5 *cents* (*sous*, s) and 3 *deniers* (d),[27] while in another contract, a fuller sold 145 pounds' worth of woad to a dyer of Gerona.[28] In 1395, a deal amounted to 119 *livres* 10 s 8 d;[29] in 1402, a dyer paid 125 *livres* 7 s 1 d to a trader from Gerona;[30] on another occasion, a company of dyers bought woad for a value of 142 *livres* 15 *cents*, and so on.[31] This demonstrates that sales of woad for values in excess of 100 *livres* were not rare. Initially, the dyers and local textile craftsmen were the main buyers but before long, the episcopal city and its region had enough stocks to supply other textile centres.

The distribution of this excess production took place in a region situated between Céret in Roussillon, Vic to the west of Gerona, and Barcelona more to the south. Within a *c.*100 km circle, a dozen textile centres obtained their woad from this region, including the villages and towns of Besalù, Camprodon, Olot, Sant Feliu de Guíxols, Figueres, Peralada, Monells, Castelló d'Empúries, Bañoles and Torroella de Mongrí.[32] Fullers, dyers and merchants from these villages regularly came to buy fairly substantial cartloads in the episcopal city.[33] The craftsmen of Gerona were buying woad directly, without an intermediary, at very competitive prices, while those craftsmen who were dealing with dyers and fullers from other villages had to pay higher prices. Thus, while in the market of Castelló d'Empúries, where the local producers came to sell woad directly, woad was valued at 97 *cents* a cartload, at the market of Gerona the same cartload cost between 110 and 120 *cents* on average.[34] A merchant from Gerona who bought a standing crop had to pay around 40–50 *cents* for a cartload. It is not possible to state precisely the preparation cost of the harvests, but it is clear that the selling price of the end product was more than double the initial outlay. Thanks to these data, we can now recognise how much financial reward this crop brought to the producers – and even more to the merchants.

The end of a century (1340–1462)

The political problems in Languedoc and Tuscany had an impact on the woad exports in the 1350s and 1360s. At the same time, patricians were keenly interested in their land and farmers were looking for better profits: both these factors favoured the establishment of woad cultivation. It was, however, important for the inhabitants of Gerona to be able to rely on a regular and substantial supply. The first development of production in the region of Gerona occurred in the years 1350–1360. From then on, the local woollen cloth industry could count on a regular supply of dyeing material of good quality at relatively low cost.[35] The following decades saw a continuous increase in the growth of the crops which permitted the conquest of the regional market. This distribution, more developed than that of Roussillon or Barcelona, increased rapidly. The competitive price of the woad from Gerona, due to lower transport costs than those for woad from Languedoc and Italy, and to a far lower number of intermediaries, contributed to its success.

As proof of this success, the steady development that began between the 1360s and the first decades of the 15th century resulted in a total halt in the import of woad from Languedoc and Italy, which had been used until that time in the county of Empúries, especially in the city of Castelló d'Empúries.[36] From the 1420s, the producers of the region of the Ter, together with merchants from Gerona (taking over from those from Castelnaudary), came directly to sell their loads of woad in this centre of woollen cloth production.[37]

This success was short, however: the terrible civil war which broke out in Catalonia in 1462 marked the end of this cycle of production. Although one can still find traces of this cultivation at the end of the 15th century, and even in modern times, it never regained the level it attained during the 14th and 15th centuries.[38] The decline of this production occurred well before the revival of the Languedoc woad trade in the second half of the century, or the import of exotic indigo in modern times.

The brief increase in production in Roussillon in the 15th century

Over a long period of time, the Roussillon woad supply, and more particularly that of Perpignan (which, from studies, appears more and more to be a very important woollen cloth centre), came from Tuscany, then from Languedoc and then from Lombardy, before the Languedoc production experienced a major revival around 1380.[39] Occasional scattered documentary references indicate the permanence of the production of this plant during the 14th century. In the early years of this century, woad fields must have been part of the landscape of the Perpignan countryside, since in 1323 a judgement, in which woad was mentioned, was passed on the rate of tax to be paid on dyeing plants.[40] In the 1380s, there is also mention of a woad mill, with no indication that this was in any way unusual.[41] Information concerning local production is more abundant in the 1430s and the 1440s. This ancient traditional crop was obviously experiencing a revival.

At that same time, several woad mills were working simultaneously in Roussillon. It is known that at least one woad mill was active in Perpignan and there was another a few dozen kilometres from Pia.[42] Two regions of woad cultivation emerge, the more important being the area around Perpignan, in gardens in the suburbs dedicated to extensive and speculative agriculture. The price of the land was accordingly high there. Around Perpignan, the *ortalans* (farmer-gardeners) managed to produce several hundred kilos of woad each, per year, for a profit in excess of 100 *livres*.[43] Other landowners must also have cultivated woad, or leased their plots of land for its cultivation.[44] For example, an inhabitant of Perpignan, living next to the convent of Saint Antoine, sold the harvest he obtained from two fields, one of which was situated by the side of the irrigation canals, to a fuller of Perpignan.[45] More than was the case in Gerona, this production by-passed the intermediaries, who were especially interested in the trade of woad from Languedoc. This may be due to the comparatively small scale of this activity. The second area of production was concentrated in the plain of Salanque, between Perpignan and the coast. Unfortunately we have little information about this area: we only know that the cultivation of woad must have been important enough for a mill to have functioned there before the 1440s, the date at which it was transformed into an oil mill.[46] According to the 19th-century scholar Romain de Saint Malo, woad was also present in the surrounding villages of Pézilla, Néfiach, Millas, Canohès, Le Boulou, Saint-Cyprien and Cabestany, an area corresponding to 10 km around Perpignan.[47]

It should be noted that the fullers and merchants from Perpignan paid for these purchases using the proceeds from selling woollen cloths and, to a lesser extent, from their sales at the fairs of Pézenas and Montagnac. Thus, one characteristic of the woad business in Roussillon was that the most important transactions occurred during these fairs. The debts could be paid in three instalments, the first payment being made at mid-Lent in Montagnac, the second at Whitsun in Pézenas, and the final payment at the same place in September.[48] In another instance, a cartload of woad was paid for on St Hilary's day (in January) in Montagnac.[49] Existing assets in Languedoc, in Montagnac and Pézenas were able to contribute towards and to support the increase in local production. Furthermore, the troubled times in France during the war between Armagnac and Burgundy, the presence of soldiers or swindlers in southern France, and the exceptional situation of the woollen cloth industry in Perpignan at this time all contributed to the development of woad cultivation in Roussillon. The size of capital brought in by the export of woollen cloth and a farming economy that could be considered one of the most progressive in the western Mediterranean world were other important factors. In the 15th century, this region had thus become a *pays de Cocagne*, just like the region of Gerona, but this expansion did not last any longer here than in the Catalan city.

Disrupted by the war, woad cultivation declined in this region from 1462 onwards but without disappearing completely. In the 1480s, the woad mill in Perpignan, which had worked for so many years, was no longer working.[50] It would, however, be a mistake to see the war as the only reason for this decline. It was the exceptional renewal of woad cultivation in Languedoc, and more particularly in Lauraguais, which dominated the Roussillon market during the 1470s and 1480s that made the cultivation of woad in Roussillon less rewarding.

Conclusion

The examples of local woad production that we have been able to discover in the suburbs or in the vicinity of the woollen cloth towns of Catalonia and Roussillon show that there is still a great deal to learn about the production of woad in the countries of the north-west Mediterranean. Well before the middle of the 15th century, various woad-producing regions appear to have emerged and expanded to a greater or lesser extent. Regrettably, even before the triumph of exotic indigo, the region suffered the devastating effects of the Catalan war of 1462, and later the massive import of woad from Languedoc thwarted any new development of woad cultivation around Gerona. However, before this fateful date, the woad cultivation regions of Gerona and Perpignan had been able to thrive in a regional 'supply and demand' relationship, thanks to this activity. This is an aspect of medieval Catalan agriculture unsuspected until now and differs markedly from the previous picture drawn of an underdeveloped and non-speculative industry.

Editor's note

Address for correspondence: 71, rue des Sardes-Grignon, 73200 Albertville, France (anthony.pinto@neuf.fr). Paper received 8 June 2004.

Notes and references

1. See G. Caster's thesis on the Languedoc region: Caster, G. (1962) *Le commerce du pastel et de l'épicerie à Toulouse, 1450–1561*, Toulouse (private publication). For Italy: Borlandi, F. (1949) 'Note per la storia della produzione e del commercio di una materia prima. Il guado nel Medioevo', *Studi in onore di Gino Luzzatto*, 4 vols, Milan, 1949–50, vol. I, pp. 248–97; republished as 'Il commercio del guado nel Medioevo', in C.R. Cipolla (ed.) (1959) *Storia dell'economia Italiana*, Turin, pp. 263–84.
2. Some indications lead us to believe that woad was cultivated in the Roussillon region at the end of the 13th century. Indeed, the early development (between 1250 and 1280) of the woollen cloth industry, combined with considerable efforts to improve woad's agricultural potential, such as irrigation schemes and wetland draining schemes, and work undertaken by religious orders (the Templars, Cistercians and Benedictines) appear to have encouraged its production. It goes without saying that the proximity of Roussillon to the Languedoc region, technically advanced in the development of the woollen cloth industry, must have facilitated this.
3. Arxiu Històric Provincial de Girona (hereafter AHPG), Notariat de Castelló d'Empúries (Not. Cast.), reg. 89, n. f., 8 April 1316.
4. AHPG, reg. 138, f. 10v, 23 November 1335.
5. See for this period Pinto, A. (2001) 'Les sources

notariales, miroir des cycles d'exportation du pastel languedocien en Roussillon et dans le nord-est de la Catalogne (XIVe siècle–premier quart du XVe siècle)', *Annales du Midi* 113, no. 236, pp. 423–55. For Italian imports in Catalonia in the first half of the 14th century see Pinto, A. (2002), 'Commerce et draperie dans le comté d'Empúries (1260–1497)', PhD thesis, Université de Savoie.

6. Guilleré, C. (1995) 'Commerce et production du pastel en Catalogne: l'exemple du diocèse de Gérone au XIVe siècle', in D. Cardon, H. Müllerott, B. Benjelloun *et al.* (eds), *Actes du 2e Congrès international, Pastel, Indigo et autres Teintures naturelles: Passé, Présent, Futur/ Proceedings of the 2nd International Congress on Woad, Indigo and other Natural Dyes: Past, Present, Future*, Arnstadt, pp. 99–104.
7. Guilleré 1995 (see note 6 above).
8. Indeed, other plants used for dyeing and textile production, such as linen and hemp, were produced in the same country areas. It took some time for the production of woad to establish itself, unlike that of madder, which was produced by large numbers of peasants, artisans, sailors and traders and reached amazing proportions in the area around Castelló d'Empúries: Pinto 2002 (see note 5 above), p. 409 ff.
9. Pinto 2001 (see note 5 above), p. 440.
10. Pinto 2001 (see note 5 above), p. 439.
11. Woad production drained the resources of the land, which is why fertilisation with animal manure was so important. In the Languedoc region, cattle grazed on the land for one year in three; the third year the land was left fallow: Caster 1962 (see note 1 above), p. 41.
12. Guilleré 1995 (see note 6 above), p. 101. In the 15th century, production was still important since the same producers sold six harvests a year: see, for example AHPG, Notariat Girona 2 (Not. G.), reg. 154, f. 59v, 28 September 1427.
13. Caster 1962 (see note 1 above), p. 39.
14. Caster, G. (1998) *Les routes de Cocagne: le siècle d'or du pastel 1450–1561*, Toulouse.
15. For example, a woad producer from Sant Martí Vells offered to deliver 10–11 cartloads of woad to a fuller from Gerona: AHPG, Not. G. 2, reg. 112, f. 3v, December 1410.
16. A small producer received 50 *cents* advance for two cartloads of woad worth 90 *cents*: APHG, Not. G. 2, reg. 152, f. 38v, 31 July 1426.
17. Heers, J. (1971) *Gênes au XVe siècle: civilisation méditerranéenne, grand capitalisme et capitalisme populaire*, Paris, p. 316; Caster 1962 (see note 1 above), p. 55.
18. In the 1420s and 1430s, the producers of several villages, such as Bordils, Sant Jordi Vall, Sant Martí Vell, Cerviè and others travelled to Castelló d'Empúries to sell their products to the dyers: AHPG, Not. Cast., reg. 580, f. 92, 25 October 1420; AHPG, Not. Cast., reg. 540, n. f., 3 December 1420; AHPG, Not. Cast., reg. 581, n. f., 27 April 1428; AHPG, Not. Cast., reg. 619, n. f., 18 January 1429.
19. Intensive production was concentrated in small areas; this has also been demonstrated for the north of France: Derville, A. (1990) 'Moulins, cultures industrielles et marchands dans les campagnes artésiennes et flamandes', *Revue du Nord*, LXXII, no. 287, p. 586.
20. It is worth noting that so far there is no indication of a woad mill for this region: exactly how the leaves were processed remains a mystery. However, it seems reasonable to suppose that crushing took place both in the town and in the surrounding countryside.
21. Only one single reference of this type is made. In 1435, a stonemason from Gerona sold four balls of woad: AHPG, Not. G. 8, reg. 36, n. f., 6 September 1436.
22. AHPG, Not. G. 1, reg. 402, f. 1, 23 August 1427.
23. AHPG 1427 (see note 22 above).
24. Guilleré 1995 (see note 6 above), p. 100.
25. For the second half of the 14th century, Christian Guilleré has been able to count more than 20 villages: Guilleré 1995 (see note 6 above), p. 104.
26. An obscure dealer from Gerona, Antoni Miralpex, originally a muleteer from Languedoc who had become a cattle merchant, was very active in the countryside around Gerona. He regularly bought or sold woad: AHPG, G. 2, reg. 112, f. 77v, 1410; AHPG, G. 2, reg. 112bis, f. 137, 23 April 1411. By 1425, his son had taken over from him, and he sold a quality of woad worth 40 *livres* 7 *cents*: AHPG, Not. G. 8, reg. 9, 11 April 1425.
27. AHPG, Not. G. 1, reg. 363, n. f., 14 February 1416.
28. AHPG, Not. G. 5, reg. 462, f. 31, August 1423.
29. AHPG, Not. G. 1, reg. 240, f. 31, 16 October 1395.
30. AHPG, Not. G. 6, reg. 95, n. f., 4 August

1402.

31. AHPG, Not. G. 1, reg. 365, n. f., 1 April 1416.
32. AHPG 1416 (see note 31 above).
33. For instance, 20 sacks were bought by a dyer in the small woollen cloth centre of Besalù, which regularly bought woad: compare AHPG, Not. G. 2, reg. 172, f. 198v, 4 May 1432. The fuller from Olot, Bernat Cassanya, also regularly bought dyes in Gerona: compare AHPG, Not. G. 2, reg. 118, f. 42v, 14 March 1424; AHPG, Not. G. 2, reg. 146, f. 156, March 1421. A merchant from Camprodon, Pere Moxos, also regularly bought cartloads, on one occasion as much as 98 *livres*-worth: AHPG, G. 6, reg. 136, n. f.; AHPG, G. 6, reg. 139, n. f., 9 December 1415.
34. AHPG, Not. Cast., 580, n. f., 27 April 1428.
35. In the 1420s, Gerona woad was worth around 110–120 *cents*, compared with 154 *cents* for woad from Lauraguais.
36. The bankruptcy of a company set up jointly by dyers from Castelló and Raimond Gotviles, a merchant's company of Castelnaudary, can be seen as proof of the decline in woad imports from Languedoc into northern Catalonia: Pinto 2001 (see note 5 above), pp. 423–51. No signs of sales have been traced of woad from Languedoc in Girona between 1380 and 1440.
37. Several people from Bordils, Sant Martí Vells, Cerviè, then later Sant Eugènia and Raset, sold cartloads of woad to craftsmen from Castelló: AHPG, Not. Cast., reg. 543, ff. 74–5, 8 April 1421; AHPG, Not. Cast., reg. 552, n. f., 16 November 1423; AHPG, Not. Cast., reg. 604, n. f., 31 October 1430; AHPG, Not. Cast., reg. 580, f. 92, 25 October 1420; AHPG, Not. Cast., reg. 607, n. f., 21 March 1431; AHPG, Not. Cast., reg. 581, f. 164, 27 April 1426; AHPG, Not. Cast., reg. 637, f. 75v–76, 17 October 1447. The Geronian merchant Pere Madrens, for instance, sold 12 *quintars* of woad: AHPG, Not. Cast., reg. 580, f. 62, 26 June 1420. Another, Joan Angles, sold the same amount to a Peralada merchant: AHPG, Not. G. 2, reg. 140, f. 120, 22 April 1420.
38. Sobresques Vidal, S. (1957) in J. Vicens Vives (ed.) *Historia social y económica de España y America*, Barcelona, vol. 2, pp. 276–7. In 1495, a merchant from Gerona sold two cartloads of woad with an overall weight valued at 10 *livres* in Barcelona: AHPG, Not. G. 3, reg. 124, n. f., 5 May 1496.
39. Pinto 2001 (see note 5 above), p. 442.
40. Archives Départementales des Pyrénées Orientales (hereafter ADPO), Série AA, *Livre vert Majeur*, Perpignan: *Inventaire sommaire*, p. 8.
41. ADPO, Procuration royale, 1B/202, 1299–1415. Another mill operated in the 14th century at Saint-Laurent-de-la-Salanque: Brutails, J.-A. (1841) *Etude sur la condition des populations rurales du Roussillon au Moyen Âge*, Paris (reprinted Geneva, 1974).
42. ADPO, Procuration royale, 1B/375, f. 16.
43. ADPO, Notariat Perpignan (Not. Perp.), reg. 3E3/438, f. 7v, 18 January 1453. That same year, a different amount was paid (55 *livres*) by the same buyer, a dyer from Perpignan, to an *ortalan*: ADPO Not. Perp., reg. 3E3/455, n. f., 28 February 1453. A year before, he acknowledged that he owed another 'gardener' 200 florins of Aragon gold to be paid for woad: ADPO, Not. Perp., reg. 3E3/461, f. [-] 15 December 1452.
44. The widow of a potter sold woad in the name of his heirs; no doubt this came from their fields: ADPO, Not. Perp., 3E3/437, f. 1, 29 December 1431. It is, however, impossible always to know whether the frequent sales of woad by merchants of the Languedoc region correspond to locally produced or imported woad.
45. ADPO, Not. Perp., reg. 3E1/5058, f. 23v, 28 February 1441.
46. ADPO, Procuration royale, 1B/375, f. 16.
47. Saint Malo, R. de (1848) 'Notice sur l'ancienne culture de la garance en Roussilllon', *Bulletin de la Société Agricole, Scientifique et Littéraire des Pyrénées-Orientales* 7, pp. 269–82.
48. ADPO, Not. Perp., reg. 3E3/438, f. 7v, 18 January 1453 and other citations (see note 43 above).
49. ADPO, Not. Perp., reg. 3E1/5058, f. 23v, 28 February 1441.
50. ADPO, Not. Perp., reg. 3E3/703, 1488, Puignau, *Fonds Oms*, f. 214.

The Identification of Purple in Manuscripts

Cheryl Porter

> Skins are dyed with the colour purple, gold melts into letters, the books are clothed with gems – and Christ stands naked outside the door.[1]

The criticisms, by St Jerome and others, of the use of purple parchment and silver and gold letters for luxury manuscripts failed to prevent the production of these magnificent books. In spite of the colour purple being condemned as 'the devil's pomp', and its association with pagans and idolatry, purple manuscripts were produced in western Europe for Charlemagne and other monarchs, and for kings and prelates, throughout the Middle Ages.[2]

The Byzantine emperors mostly used blue- or purple-coloured parchment written in gold and silver for their diplomas, for official documents and for presentation to imperial persons, as well as for commissioned manuscripts. Most (almost all) of them were bibles and gospels. This was a Christian appropriation of imperial symbols. These luxurious manuscripts were imitated in the West, appearing first in Late Antiquity. Subsequently, they were produced by the Anglo-Saxons, the Carolingians, the Ottonians and again during the Renaissance.

It has often been suggested that the purple used to colour and decorate these parchments was shellfish – Tyrian – purple. For example, Diringer (1967) writes, 'The finest purple colour, the Tyrian Purple derived from the murex, was employed for this purpose, especially in Byzantine manuscripts.' Thompson (1956) states, 'Whelks produced the purple of Byzantium, the purple of the great codices written in gold. Parchment was dyed with shellfish purple in France as late, perhaps, as the tenth century' and Laurie (1914), writing of Byzantine and Irish manuscripts, 'I think we are fairly safe in assuming that this brilliant pigment is prepared from the murex.'[3]

During the past seven years, I have been examining purple manuscripts from these periods and, where possible, having them analysed, using the non-destructive technique of X-ray fluorescence (XRF) to look for bromine.[4] The principal constituent of shellfish purple is dibromoindigo, thus the presence of bromine is a reasonable indicator of the presence of shellfish purple on the manuscript, if used as a pigment. No other material generally used as a pigment before the latter part of the 19th century (and the invention of eosin-based pigments) contains bromine. I have never been permitted to take samples for analysis, so where it was not real purple (i.e. where no bromine was detected), the best I have been able to say is that it is organic, since, in every case, the colour used to stain the parchment was organic.[5]

Analysing colour on parchment is fraught with difficulty and is still very much in its infancy.

Questions of transportation (of the manuscript and/or the analytical equipment), insurance and so on have meant that few results have been obtained. A new generation of sensitive and portable tools is beginning to alter this, though the analysis of organic colours used for manuscript illumination is still extremely difficult without the taking of a sample. With the kind permission of the librarians at the British Library and the Vatican Library, I was able to examine (non-invasively) the colours in a number of these manuscripts, in order to understand what colours were used for their decoration and, in particular, to look for evidence of the use of real purple.

From the British Library two purple fragments were examined to determine the nature of the purple. Additional MS 40107, the Codex Palatinus, a Latin manuscript of the 5th–6th century, is a single leaf coloured purple on both sides, with some degree of fading. Surprisingly, the grey-black writing was shown not to be silver, but the rather lowly metal, tin. Since neither bromine nor any other mineral element was detected on the purple, it can be seen that the colouring was from some sort of organic material and not the real purple.

Another manuscript, Cotton MS Titus C 15 in the British Library, is a Greek manuscript from the 6th century. The British Library allowed the examination of four leaves (folios 2–5) from the manuscript. The parent manuscript is Codex Purpureus Petropolitanus, in St Petersburg Public Library, Gr. 537 (182 leaves). There are also 32 leaves in Patmos, two in Vienna and six in the Vatican Library. Again, analysis showed that the purple parchment was not real purple, but, rather, some organic red. The writing was indeed silver and some mercury was also detected, suggesting the presence of the mineral cinnabar (natural vermilion, red mercuric sulfide).

Some six months later, the six leaves of this same manuscript in the Vatican Library, MS Lat. 3785, were examined. The results of this analysis concurred with the previous analysis of the other leaves from this manuscript in the British Library, in that the red colour of the very thin calf parchment was an organic colorant, and no bromine was detected. As with the related leaves in the British Library, the writing was confirmed as silver ink and here too, the purple colour of the leaves has faded with exposure to the light. In fact, fading is not unusual with these full-page purple leaves. On a number of fragments, the only evidence of the colour is to be seen in the veins of the (calf) parchment, signifying that it is not real purple, which tends to be most resistant to fading.

From analysis thus far, it would seem that these full-page purple manuscripts were not coloured with the real purple. I decided then to look at purple manuscript decoration from the relevant periods to see if real purple had been used in that way. There are a number of references to purple-dyed cloth used to cover couches in Roman times and one such purple-coloured couch is depicted in the Codex Virgilius Romanus (Vatican, Biblioteca Apostolica, Cod. Vat. Lat. 3867, f. 100v), dating from the late 5th to the early 6th century AD. When the fragment was analysed, it was clear that, once again, another organic colour and not real purple had been used, not only for the couch, but also to depict the red robes of the figures. Other colours used for the painting of this fragment were red lead (minium, lead tetroxide) for the border of the image and for the robe of the central figure; the green curtain behind the figure seems to be painted with an organic blue mixed with green earth, with lead, calcium, iron and copper also detected. The red sash along the curtain is painted with cinnabar (mercuric sulfide).

Folio XLr, Pict 26 of the Virgilius Vaticanus (Vatican, Biblioteca Apostolica, Cod. Vat. Lat. 3225) is a painting of the death of Dido, dating from the 5th or 6th century. She is depicted on a purple couch in a porphyry-clad chamber. No bromine was detected in either of these areas, although the colours are organic. Other pigments used in this fragment were identified as lead white mixed with an organic pink for Dido's robe; lead and iron were found in the depiction of the porphyry wall, suggesting the presence of a red ochre (with a lead-containing pigment). The red border contained high levels of mercury (and some iron), suggesting the presence of cinnabar.

Much influenced by Byzantine art, the sumptuous Ottonian manuscripts, with their magnificent display of purple and gold leaf, might have used real purple to enhance their magnificence. Analysis of the calendar page of the Vatican manuscript Gospels of Henry II (Vatican, Biblioteca Apostolica, Cod. Ottob. Lat. 74, f. 9r) showed that here at least, they did not. This Gospel book, made for Henry II at Regensburg in about AD 1020, used a high quality gold leaf (average 84% with little copper). Red lead was used and the greens seem to have been an earth colour, with iron and some lead detected. The beautiful blue could be indigotin or ultramarine, since no copper was detected.[6]

In Codex Reg. Lat.124 (Vatican, Biblioteca Apostolica), made in about AD 820 in Fulda, north Germany, again no evidence was found for the presence of bromine in the purple background, which was of an organic nature. Silver (69.46% as a percentage of the silver alloy) was much used for its decoration, as was gold (52.87%). Copper (20.52%) was a large component of the gold, which probably accounts for the greenish colour showing through on the verso of the leaf.

Finally, MS Barb. Lat. 570 (Vatican, Biblioteca Apostolica), an Insular Gospel book probably made in the second half of the 8th century AD, was examined. Three of the four Evangelist portraits used purple in the decoration, but in no case was bromine detected. The pigments found in the portrait of St Matthew (f. 11v) included red lead for the border and for the finely painted lines over the green tunic of the upper body, and orpiment (arsenic sulfide, As_2S_3) for the halo and as a component of the saint's green robe. All evidence points to the use of indigotin (derived from woad or from another indigo-yielding plant species) as the blue component of this green, since no copper was detected and it is unlikely – visually, historically and economically – that ultramarine would have been used mixed with orpiment.[7] The pale blue is undoubtedly indigo mixed with a calcium salt and the pink used for the saint's lap is an organic red lake mixed with a calcium salt; iron, lead and potassium were also detected. An organic yellow was used in the border.

In almost every one of the analyses performed, calcium and lead were identified. To account for the calcium was not difficult, since chalk was often used in the making of parchment, applied to the surface to 'key' the inks and the pigments, and also to degrease the skin. The presence of lead proved more problematic. At first, it was thought that lead white may have been applied to the parchment surface, perhaps to whiten it. When the surface of the parchment was tested, however, lead was found to be absent. It may be that many of the pigments were mixed in a lead container, and perhaps this could account for its ubiquitous presence.

The ink used to outline the drawings and for the text is iron gall ink. In folios 12r and 50v (showing the portraits of SS Mark and Luke respectively), red lead was identified, but all the other colours – pink, red and green-blue – were organic.

On folio 125r bromine was finally found in the pink-red background of the script at the opening of the gospel of St John. Of all the manuscripts examined, this was probably the least likely to have contained real purple. In fact, so unlikely was the identification of bromine that I did not publish the finding for some years, and then only after Professor Giacomo Chiari (then at the University of Turin, now at the Getty Conservation Institute) had re-analysed the manuscript (Fig. 1).[8] In some ways, the discovery raised as many questions as it answered, the first of which was: why was bromine identified on only one area when other areas, which seem to be the same colour, tested negative?[9] The next question was: does this tell us anything about the manuscript? There is no indication of date, where it was made or its later history and, since so few books remain from this era, there is not enough information to group what we do have into centres or schools. What seems to be agreed is that it is probably from the second half of the 8th century; that the scribe was English-trained; that the parchment looks insular (that is, from somewhere in Britain or Ireland) and that there are Eastern influences in the iconography (for example, the drapes on the Evangelists' thrones).

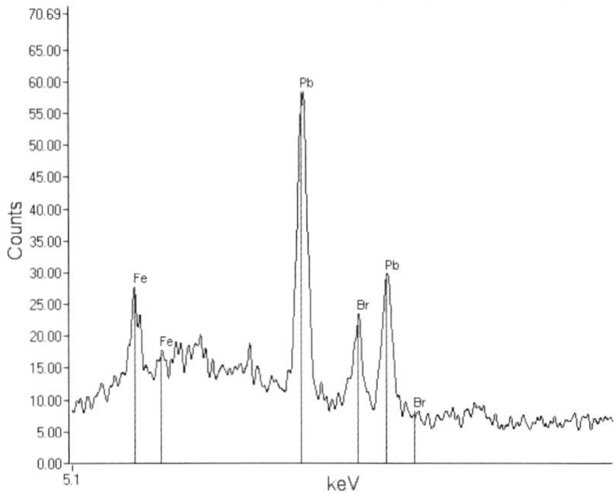

Figure 1 MS Barb. Lat. 570 (Vatican, Biblioteca Apostolica), folio 125r. XRF analysis of pink-red background of script at the beginning of the Gospel of St John, showing the presence of bromine.

The problem is that an English-trained scribe does not imply that the manuscript was written in England. The parchment seems insular, but parchment travels. The iconography looks Eastern (Byzantine/southern Italian), but scribes travelled too and could be working in Britain, or Ireland, or anywhere else. Real purple was available in the Mediterranean and off the coast of Britain and Ireland, and so we need to be able to say: 'This purple is from Britain or Ireland, and not from the Mediterranean', before this information could be used for suggesting a location where the manuscript was produced.

While much study has been done on the purple molluscs of the Mediterranean region, comparatively little has been carried out on *Nucella lapillus* L., the dog whelk of Britain and Ireland. A number of archaeological sites in Britain and on the west coast of Ireland have deposits of dog whelk middens. Recent excavations of a hill fort in north Brittany, occupied by an abbey between AD 500 and 1000, also found heaps of dog whelk shells, virtually all broken to extract the animal intact.[10] The use of purple is recorded in Bede's *Historia Ecclesiastica*, but there is no evidence of textile dyeing with this purple in Britain or Ireland,[11] and it is unlikely to have been so used, since the amount of juice extracted from these whelks is minimal, compared to the Mediterranean molluscs. It would, however, be sufficient to decorate a manuscript.

In manuscript MS Barb. Lat. 570 (Vatican, Biblioteca Apostolica), the Barberini Gospels, the colour is used as a pigment, applied with a brush or a pen and with a binder. Most interestingly, it is not used to colour the garments of the Evangelists, but only to decorate the letters of the opening of the Gospel, indicating the importance of the 'Word' to these early Christians. For Christians, from at least the 8th century, sacred texts were considered talismanic. The Gospel of St John was considered particularly efficacious and it is claimed that St Cuthbert cured a man by laying a copy of St John's Gospel upon him. During the Middle Ages, its *incipit* 'In the beginning was the Word, and the Word was with God, and the Word was God' was often worn around the neck as a protective charm. St Augustine even records the belief that these words could cure headaches.

Analysis of these manuscripts, though only a small sample from a wide geographical area and over a great range in time, has raised some interesting points. Of course, the identification of the use of real purple has been the most profound discovery, but of great interest also has been the enormous use of organic colours in these books – reds, blues, greens and yellows used to paint in books from the earliest examples that we have of manuscript painting, from the 5th century until the invention of printing and beyond.

Finally, and by way of conclusion, I would like to make some comments on the significance of this research. The identification by scientific means of pigments and organic colours in the context of manuscripts has a relatively short history. Conservators quickly recognised its importance, for if they were to treat damaged pigments, they must be sure of the chemistry, lest treatments change not only the refractive index (and thus the appearance of the colours), but other properties also, quite apart from the damage caused incidentally. Since the technologies have been applied, historians are gradually becoming aware of the data. Until recently, manuscript scholars and art historians have been heavily dependent upon observations by eye. In discussing the Barberini Gospels recently with a very distinguished calligrapher and manuscript scholar, the observation was made that descriptions of this book noted the greasy character of its parchment and the above average, but not magnificent, scribal hand. These indicated that the manuscript was not of the highest quality. However, the use of real purple, a difficult and expensive pigment to use, may indicate that the book was more important than has been thought. Recently, I have been working on another manuscript of 15th-century English origin. Here, ultramarine and kermes dyestuff (in a red lake pigment) have been identified in a completely unexpected context. The data of colour has much to contribute to modern scholarship, by indicating the prestigious status of such books. The new information raises questions, but in doing so immensely expands our understanding of craftsmanship through history, and in the context of the later period, it casts yet more light on the patron who paid for these rare and precious colours. The discovery of purple in this manuscript adds to the understanding of the book, and by recognising its limited use in a specific context, it has demonstrated the significance of the colour and the importance of the 'Word'.

Acknowledgements

I owe thanks to Giacomo Chiari and Andrea Cavallo for the analysis of purple, as well as to Vincent Daniels (British Museum, Conservation Department) and Sebastiano Schuiti; to the curators and librarians of the British Library and the Vatican Library, for access to their manuscripts and to Hero Granger-Taylor, Mark Clarke, Dominique Cardon, Chris Cooksey and Jan Wouters for their help.

Editor's note

Address for correspondence: Thesaurus Islamicus Foundation, 21 Misr Helwan al-Ziraaa'i Street, al Ma'adi, Cairo, Egypt (chezzaporter@yahoo.com). Paper received 26 September 2003; revised version received 23 February 2004.

Notes and references

1. Jerome [Eusebius Hieronymus Stridonensis], *Saint* (1933) *Select Letters*, tr. F.A. Wright (Loeb Classical Library), London, Letter XXII to Eustochium.
2. See Tertullian [Quintus Septimus Florens Tertullianus] (1987) *De idololatria* [On idolatry]*: Critical Text, Translation and Commentary*, eds J.H. Waszink and J.C.M. van Winden, Leiden and New York, ch. 18; Roberts, A. and Donaldson, J. (eds) (1979–86) *The Ante-Nicene Fathers: Translations of the Writings of the Fathers down to AD 325*, 10 vols, Grand Rapids (reprint of edition published Edinburgh 1867–94), vol. IV, ch. viii.
3. Diringer, D. (1967) *The Illuminated Book: Its History and Production*, new edn, London; Thompson, D.V. (1956) *The Materials and Techniques of Medieval Painting*, New York (Dover), p. 156; Laurie, A.P. (1914) *The Pigments and Mediums of the Old Masters*, London, p. 63.
4. The presence of bromine can also indicate that the object has been buried under the earth, or contaminated by seawater. In manuscripts, this is an unlikely source of contamination.
5. As far as I know, only one purple manuscript has been analysed to identify the organic colorant. Codex Sinopensis, made in Syria or Mesopotamia in the 6th century, now at the Bibliothèque Nationale in Paris, was analysed at the CNRS using gas chromatography–mass spectrometry (GC–MS). The colour was identified as *folium*, *Chrosophora tinctoria* A. Juss. (The results of this analysis must be doubted, however, since this method is not suitable for identifying *folium*.)
6. Lighter elements (below sulfur, atomic number 16) are not easily determined by XRF, thus ultramarine (a sulfur-containing sodium aluminium silicate) is often undetectable.
7. There are many recipes for this combination in documentary sources, sometimes known as a type of 'vergaut'.
8. XRF analysis was performed by Andrea Cavallo of Assing SpA, Rome. An ASSING XRF portable instrument was used for the measurements; this was equipped with a Peltier cooler (rhodium) X-Ray tube, operated at 30kV, 0.1–0.5 mA. The detector resolution ranges from 5.9 keV to 189 eV, thus allowing for the detection of elements heavier than sulfur (analytical range: ^{16}S – ^{92}U). The distance of the detector from the sample can

vary from 1 to 200 mm, and the exact spot of the analysis can be detected by the convergence of two laser pointers. The dimensions of the spot vary from 5 to 10 mm^2, depending on the distance of the detector. The instrument could be easily adjusted into the various positions required for analysis of the manuscripts by mounting it on a sturdy camera tripod. Manipulation of the manuscripts themselves was thus minimal.

9. There is the question of sensitivity of the instrument and it would seem that much has to do also with luck, and the fact that sometimes elements can be detected within a relatively small sample, while at other times, what seems to be a 'good' sample gives no result.

10. Gregory Campbell, pers. comm. For further references see: Murray, E.V. (2002) *Early Evidence for Coastal Exploitation in Ireland*, PhD Thesis, Queens University Belfast; Jackson, J.W. (1917) *The Geographical Distribution of the Shell-Purple Industry*, London, pp. 1–29; Cocagne, J.Y. (1997) 'Le purpura (*Nucella lapillus*) et son utilisation comme tincture en Amorique', *Annales de Bretagne et des Pays De L'Ouest* 104, 4, pp. 7–22.

11. There is only evidence of linen markings: see Cole, W. (1685) 'A letter from Mr William Cole of Bristol, to the Phil. Society of Oxford; containing his observations on the Purple Fish', *Philosophical Transactions of the Royal Society* 15, pp. 1278–86.

Missal Blue: Anthocyanins in Nature and Art

Maria João Melo

Abstract Anthocyanins are the water-soluble colorants, ubiquitous in flowers and fruits, which are responsible for their impressive red and blue colours. This paper discusses some of the old recipes used to produce a blue colour, such as missal blue, from the cornflower (*Centaurea cyanus* L.). The 'capture' of a blue from a flower or berry is not an easy task, due not only to the complexity of the natural supramolecular structure of the colorant, but also to the various structural transformations that it undergoes in aqueous solutions driven by pH changes. Therefore a 'correct' recipe can be an important piece of evidence testifying to the use of these colorants in the past. Our present knowledge of the molecular properties of anthocyanins places us in a good position to analyse these old recipes.

Introduction

The colours of flowers and fruits have attracted the attention of humankind throughout history: flowers as a fundamental source of artistic inspiration and beauty and fruits as an important part of the human diet. Anthocyanins are the ubiquitous water-soluble colorants found in flowers and fruits that are responsible for their impressive red and blue colours (Plate 13, Fig. 1).[1] Several references on the use of these compounds as pigments in decoration, namely, as substitutes for inorganic pigments when these are not available, or in ancient illustrations to give special lighting effects, have been reported. Descriptions of this type of use range from the Roman empire, by the famous architect Vitruvius,[2] to the paintings of the Maya civilisation.[3] Their use as 'watercolours' was frequently described in treatises or recipe books on illumination such as the Strasbourg (or Strassburg) manuscript,[4] an early Portuguese work on manuscript illumination, 'The book on how one makes colours of all shades', and the Italian *De arte illuminandi*.[5] Anthocyanins could be used to produce clothlets (*pezzuole*) as described by Cennino Cennini in the 15th century.[6]

Due to the pioneering work of Richard Willstätter,[7] who received the Nobel Prize for Chemistry in 1915 for his research into plant

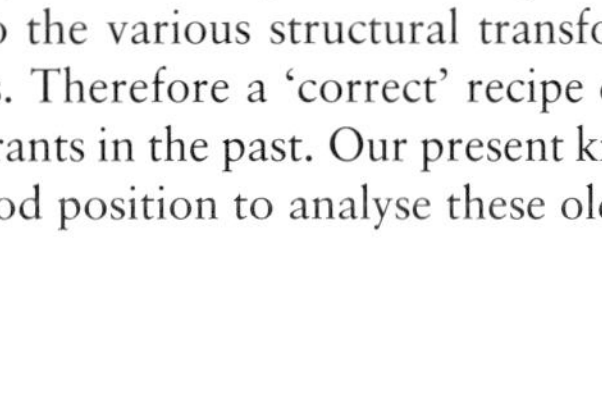

Figure 1 The basic chemical structure of anthocyanins: a hydroxyl group is present in positions 4′ and 7, and a sugar in position 3 (monoglycosides) or 3 and 5 (diglycosides).

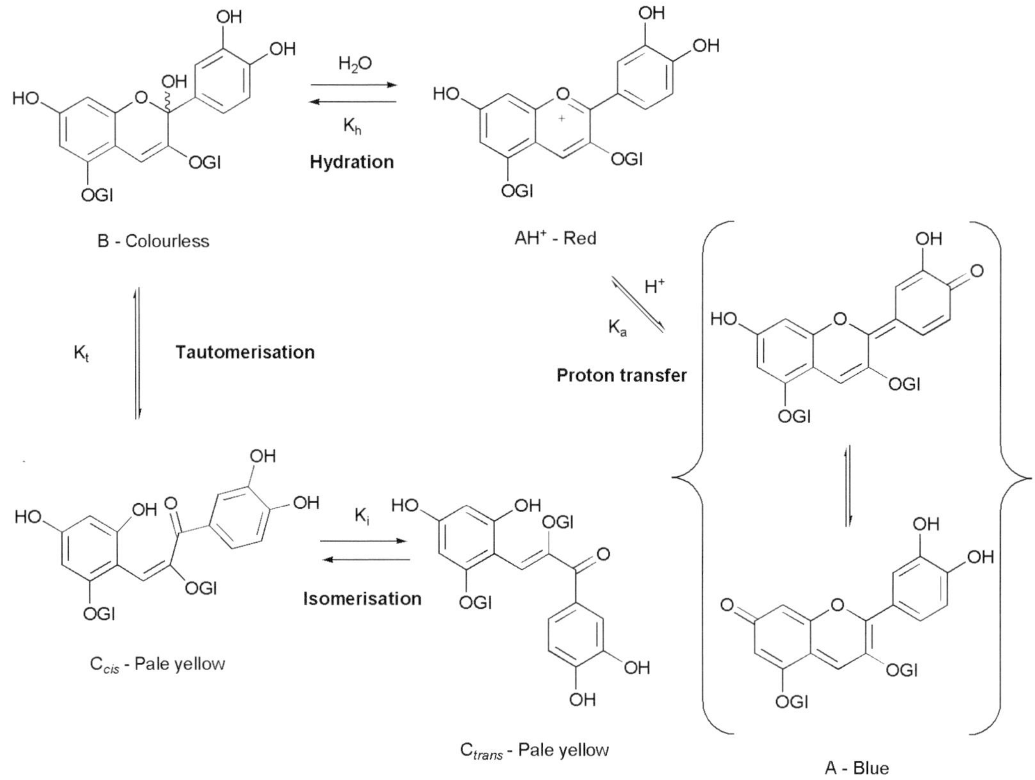

Scheme 1 Equilibria in aqueous solution for cyanin: AH^+ flavylium cation; A quinoidal base; B hemiacetal; C_{cis} *cis* chalcone; C_{trans} *trans* chalcone.

pigments, much information has been accumulated regarding the role played by anthocyanins and the colours they produce.[8] Isolation of the most common anthocyanins has shown that they follow a general pattern, summarised in Scheme 1 for cyanidin 3,5-diglucoside.[9] The red flavylium cation (AH^+) is the dominant species in very acidic solutions, but with increasing pH a series of more or less reversible chemical reactions take place: (i) proton transfer leading to the blue quinoidal base (A); (ii) hydration of the flavylium cation, giving rise to the colourless hemiacetal (B), (iii) a tautomerisation reaction responsible for ring opening, to give the pale yellow *cis*-chalcone form (C_{cis}), and finally, (iv) *cis-trans* isomerisation to form the pale yellow *trans*-chalcone (C_{trans}).

A useful way to illustrate the pH domain of the red and blue colours is to represent the pH dependence of the mole fraction distribution of the species (Plate 14a), when the system has reached the thermodynamic equilibrium. In this illustration, the key parameter is pK_{ap}, defined as the pH value for which the mole fraction distribution of the red flavylium cation is reduced by one half,[10] and as a consequence colour is also reduced by one half.

Anthocyanins in nature

Anthocyanins are located inside vacuoles, where the pH can assume a large range of values, depending on the plant species.[11] However, as can be visualised in the band at the top of Plate 14a, the colouring power of cyanidin 3,5-diglucoside is very low, even for the most favourable pH values that can appear in the vacuoles of certain plants. There is no doubt that cyanidin 3,5-diglucoside, for example, at a pH of about 3.1, does not exhibit any substantial colouring power, neither red nor blue. At this pH value neither the blue quinoidal base nor the red flavylium cation are thermodynamically stable species, and therefore the system will naturally evolve to the colourless species, the hemiacetal and the chalcones. For the most common natural anthocyanins, the major species, at neutral pH, is the colourless hemiacetal.

This is the reason why, when extracting an anthocyanin at neutral pH the colour disappears. This reaction can be reversible, if the only process taking place is proton transfer. In this case the anthocyanin will behave as an acid-base indicator and colour can be restored by changing the pH.

For the formation of a blue colour some interaction between the blue quinoidal base and other compounds is required (Plate 14b). This phenomenon is called co-pigmentation and can involve other flavonoid compounds and complexation with metals.[12] The mole fraction distribution of the several cyanidin 3,5-diglucoside forms obtained upon complexation can be simulated for a simplified model system, consisting of a single 1:1 equilibrium with association constant K'_{cp} involving the species A and the co-pigment.[13] The results are shown in Plate 14b, and indicate that a blue colour can now be obtained at moderately acidic pH values.

Colour in nature, as far as anthocyanins are concerned, depends also on the substitution pattern of the benzopyrilium ring (Fig. 2). Most of the anthocyanins present in flowers and fruits have a sugar substitution in position 3 (monoglucosides) (Fig. 3). A sugar can also be present in position 5 (di-glucosides) or sometimes in position 7. The ring B can also be substituted with hydroxyl or methoxy groups.

Luteolinidin is a prehistoric anthocyanin colorant, more precisely a 3-deoxyanthocyanidin, present in mosses and ferns (Fig. 2). Luteolinidin gives wonderful sunset colours, ranging from yellow to violet, without forget-

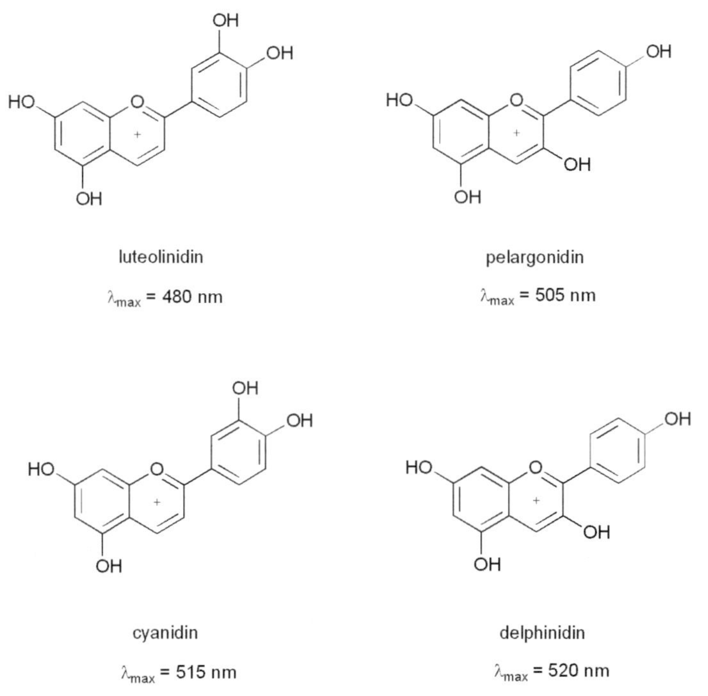

Figure 2 Molecular structures, common name and λ_{max} of absorption in aqueous acidic solutions for four anthocyanidins.

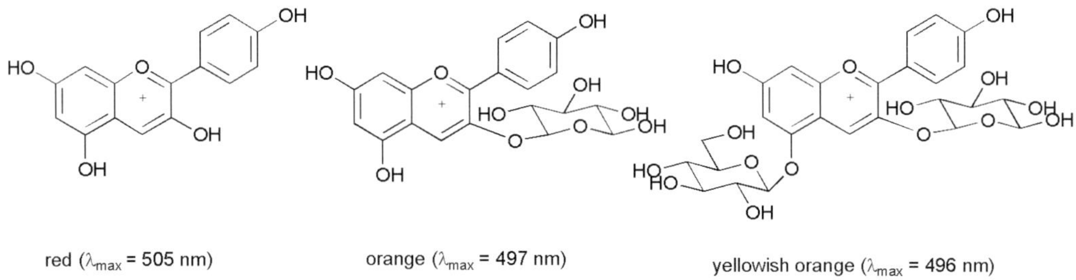

Figure 3 λ_{max} in aqueous solution at pH = 1 for pelargonidin, pelargonidin-3-glucoside and pelargonidin-3,5-diglucoside.

ting the splendid orange colours – almost as beautiful as a Turner sunset (Plate 13) – but no red and no blue. As can be observed there is no substitution in position 3. With a hydroxyl (OH) group in this position, the wavelength at maximum absorption (λ_{max}) will increase by about 25 nm in the visible region, and more importantly, the possibility of creating a red flavylium and a blue quinoidal base will also increase. On the other hand, with a hydroxyl group in this position the quinoidal base will decompose. However, if this OH group is not free – for example, if a glucosyl bond is formed – the base will not degrade. The presence of this sugar causes a reduction in λ_{max} of 5–10 nm in the visible region, but with no particular consequences for the observed colour; the net result is still very positive: + 25 nm (OH substitution) – 5 nm (OGl substitution) = + 20 nm in the visible region; and, more importantly, the quinoidal base can be blue! Another sugar in position 5 has almost no influence on λ_{max} (Fig. 3), but can reduce the pH domain of the red colour (that is, of AH^+).[14]

With these compounds and depending on the strategy used, nature is able to construct the exuberant red displayed by poppies, the delicate blue of a cornflower, and also all the imperial purples that lie in-between. Many examples of the strategies used by nature to obtain colour in plants are based on the formation of a supramolecular structure, through molecular recognition and self-assembling of the components in solution.[15] In effect, many of the structures that have been found to be responsible for the blue colour are supramolecular complexes shaped by metallic cations and anthocyanins (compare the values for cyanidin and its glycosides in Table 1 and the spectra in Fig. 4). The presence of other flavonoids is also necessary for the build-up of the blue.[16] *In vivo*, the metal centre organises a hydrophobic core, where the sugars are directed to the outside, interfacing with the more polar, aqueous media. Several examples of intermolecular, intramolecular or self association co-

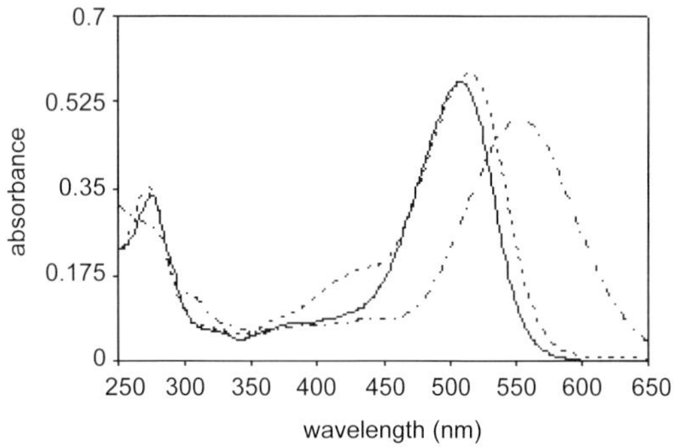

Figure 4 UV-visible spectra in aqueous solutions at pH = 1 for cyanidin (—), cyanidin-3,5-diglucoside (- -) and for a 1:1 complex between cyanidin-3,5-diglucoside and Al^{3+}(- · · -).

Table 1 Absorption max ima (λ_{max}) for cyanidin, cyanidin-3-glucoside and cyanidin-3,5-diglucoside in aqueous solutions at pH = 1.

	λ_{max} AH^+
Cy-3,5-OGl	508 nm
Cy-3-OGl	509 nm
cyanidin	515 nm

pigmentation have been reported,[17] in particular the fascinating structures unveiled by Goto and Kondo (1991) and Kondo *et al.* (1992).[18]

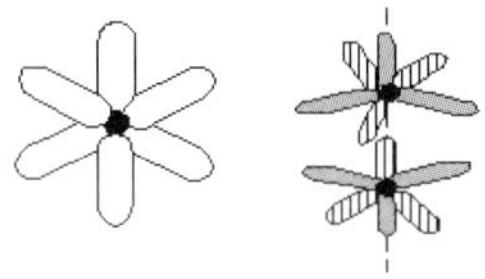

Figure 5 Schematic representation of the blue supramolecular complex involving anthocyanins. The 2 metals (black spheres) form 2 planes under a ternary axes; a top view of one of this planes is represented on the left. Each metal coordinates three anthocyanin chromophores (striped); three flavones (light grey) are intercalated with the anthocyanins, filling the void spaces.

Anthocyanins in art

Blue clothlets and missal blue

As reported in the introduction, there are several references dating from antiquity to the use of watercolours for painting, particularly for illuminations. These watercolours could be used as described by Cennino Cennini.[19]

For a saffron yellow watercolour, a carotenoid, the procedure is rather easy: the colour is extracted in warm water,[20] then a piece of cloth is dyed with this solution and allowed to dry in the dark. The operation is repeated until a saturated cloth is obtained (that is to say, it has taken up enough of the colorant). Finally, a gum arabic coating is applied. The colour is then kept in a wooden box, carefully protected from light and humidity (the cloths were thoroughly dried first). For use as a watercolour, a little piece is cut into a convenient receptacle (mussel or scallop shell, or horn), some gum arabic is added and the colour is ready to be applied.

Extracting and capturing an anthocyanin blue is not as easy, because the supramolecular structure must be conserved, otherwise the colour is lost. For the cornflower (*Centaurea cyanus* L.), this implies that the concentration of the supramolecular elements must be maintained, and also that pH must be controlled. This is the case for all the recipes described in the Strasbourg Manuscript (see the Appendix).[21]

The structure of the self-assembled supramolecular pigment from the blue cornflower was elucidated by Kondo *et al.* (1994);[22] the supramolecular components are succinylcyanin (Sucy, a cyanidin chromophore with a sugar in position 3), malonylflavone (Mafl, a flavone with a sugar in position 3) and magnesium and iron(III) ions, the exact composition being ($Sucy_6Mafl_6Fe^{3+}Mg^{2+}$). These authors proposed that the supramolecular complex should be similar to that of commelinin (Fig. 5). The metal centre organises the geometry of the complex, which is held together by weak hydrophobic interactions, as the π-π stacking of the aromatic rings and hydrogen bonding through the sugar moieties.[23] The anthocyanin molecules bond to the metal through the two OH groups present in ring B (catechol unit). The inner parts of the molecule are hydrophobic and the surface of the supramolecule is hydrophilic. This not only enables the formation of a blue complex, but also can protect against water attack in the 2 position and consequent formation of the hemiacetal.

For cornflower blue as well as for commelin the blue colour is only stable in concentrated solutions. When diluted the solution quickly becomes colourless due to decomposition of the supramolecular complex.

The very first step in the recipes used to obtain a blue clothlet from cornflower is to 'break only the petals, crush them in a mortar and wring the colour through a straining cloth' (the process is shown schematically in Plate 15).[24] In this case, the first requirement for the stabilisation of the supramolecular structure is observed; that is to say, by squeezing the flower juice the concentration of the supramolecular components is maintained and therefore they will be in the stoichiometric ratio necessary to self-assemble. The pH of the cornflower juice is about 4.7; this value must also be maintained around pH = 5 ± 0.5, to profit from all the blue colour. This is precisely what is recommended as the second step: 'take a *settit* of sal-ammoniac and put it in the coloured juice and it will dissolve at once'. The sal-ammoniac is our modern ammonium chloride (NH_4Cl), which, used as described, will buffer the pH of the solution to a value around 5: perfect for the blue colour!

The next day the procedure is repeated, but the sal-ammoniac is substituted by alum (aluminium potassium sulfate, $AlK(SO_4)_2 \cdot 12H_2O$) and some gum arabic is also added to the juice; the gum arabic will form a protective coating and will also be the binding medium. The presence of alum shifts the pH of the medium to lower values and also to a more violet shade. This violet shade can be due to the lower pH value, but also to substitution of the natural coordinated metal ions, Fe^{3+} or Mg^{2+}, by Al^{3+}.

Finally, it is also worth mentioning that it is stated that the blue obtained with these recipes is a durable colour, quoting for missal blue: 'it is a good and valuable colour'.[25]

Dyeing parchment

In the 'Book on how one makes colours of all shades',[26] Abraão B. Judah Ibn Hayyim uses the term 'cor de mora'; in Portuguese we still have the verb 'morar', which means to give the colour of a blackberry, as the Portuguese common name for blackberry is 'amora'. Dyeing with anthocyanins follows the same general concept as dyeing with anthraquinones such as alizarin, or flavonoids such as quercetin. Dyeing is accomplished through the very densely charged aluminium ion, Al^{3+} that will bond to the fibre as well as to the colorant. In cyanidin the bonding is performed through the *ortho*-dihydroxy groups located in ring B, the catechol unit.

In blackberry, as in raspberry, the basic chromophore is cyanidin exhibiting 3-substituted sugars;[27] surprisingly, the anthocyanin concentration in blackberry is roughly half that present in raspberry. The strong blue, almost black, colour of blackberry, when compared with the charming raspberry red, is an indication that the capture of colour in these two berries has followed different strategies. In the raspberry, colour is due to the colorant power of the single anthocyanin, which is present in large concentration;[28] for blackberry, a more complex structure must be involved.

The colours obtained when dyeing parchment can range from a reddish-violet to a bluish-violet or bluish-grey colour, depending on the berry, but also, as expected, on the pH and on the Al^{3+} concentration (Plate 16).

Conclusion

Anthocyanins are the water-soluble colorants responsible for the marvellous diversity of red, purple and blue colours found in flowers and fruits. The exuberant red displayed by poppies and the delicate blue that tints a cornflower have a common chemical origin: the cyanidin chromophore.

The use of anthocyanins as pigments was already reported by Vitruvius. These molecules were used in manuscript illuminations as water-colours, where their light effects and softness were irreplaceable.[29] In common with other pigments already in use in antiquity, such as the anthraquinones alizarin, present in madder dye, and kermesic and carminic acids in scale insect dyes, anthocyanins share the ability to complex metal ions such as Al^{3+} (present in alum); an important difference is that anthocyanins form water-soluble supramolecular structures and the anthraquinones insoluble lakes. The supramolecular structures of anthocyanins, if obtained in the past using the previously described procedures, could be easier to identify than the anthraquinone lakes,[30] not only because they might be present in higher concentrations, but also because they are water-soluble compounds that can easily be extracted. A century-old missal blue is still waiting to be discovered, to prove that the delicate anthocyanins can defy time!

Appendix

The recipes transcribed are from the Strasbourg (or Strassburg) Manuscript, dating from the beginning of the 15th century; the raw materials are flowers, namely, cornflowers, thus the colorants obtained and used could be anthocyanins. In one of the recipes for a violet colour the flowers used are not identified, but are described as 'large red flowers': could these be poppies?[31]

> If you want to make a bright rag colour, proceed as follows. During the first week after Whitsun pick a quantity of fully opened cornflowers – these should be gathered during the morning, that is, before midday. Break off the flowers at the top of the stem, put them in a clean vessel and pound them in a clean mortar very thoroughly till they are reduced to a mash. After this, put them in a clean straining-cloth and wring out the juice into a glazed pan. Then take a *settit* of sal-ammoniac and put it in the coloured juice and it will dissolve at once. After this take a well-washed piece of material, such as a piece of fine old veiling or of an old tablecloth and put this, cut in small pieces, into the colour so that the 'rags' soak up all the coloured liquid and become neither too wet nor too dry, but just soak up the colour and are evenly tinted. After this one should hang up the rags on a line in the garden, letting them dry well in the open air.
>
> The next day, pick the same amount of fresh flowers as before and repeat the process of breaking off the heads of the flowers, of pounding them as before and of wringing the juice through a straining-cloth into the pan. Then take very clear gum arabic that has been previously softened. This gum must first be well stirred round with the finger and then be mixed with the juice of the flowers and stirred with a wooden stick. Then take a *settit* of alum crystals ground to a fine powder and put this into the above-mentioned juice and stir it well until the alum has melted. Then take the already coloured rags (which have been drying overnight) and dip them once again in the juice and leave them steeping in the colour till they have soaked up enough to dye them thoroughly. After this, hang up the rags again in the open air and let them dry thoroughly. Lastly, wrap them up in clean paper and keep them in a clean wooden box somewhere high up and well ventilated, in order that they may not get damp.[32]

This is followed by a similar recipe for a violet rag colour using pink cornflowers, gathered at the same time of the year, using 'one *settit* of alum crystals; if you have a lot of flowers, ... but you should note that if the flowers amount to a quarter pound the correct amount of alum would be one *settit*, but if the amount of petals were less, put less alum'.

> When you want to make a violet rag colour which is useful for many things, you must in the first place, manage one morning before midday to gather some of those beautiful large red flowers – as many as you require – you must break off only the petals or the colour will be spoilt; when they have all been pulled off you must crush them into a clean stone vessel or in a mortar and wring the colour thoroughly through a straining cloth and collect it into a glazed earthenware dish. Then take a piece of alum the size of a hazelnut, ground to powder on the slab and sprinkle it into the colour and stir well together until the alum is quite dissolved. Then take well-washed linen rags and soak them in the colour so that they are thoroughly saturated and yet not wet to drip and then hang these rags in the open air and leave them to dry. Now soak them again in the colour and once again leave them hanging up to dry. Next take some wine of a good red colour, at least a *mos* and put in it ½ oz of pure gum arabic and let it steep in the red wine for half a day to soften; then stir it thoroughly in the wine and lastly soak the coloured rags in the red wine so that it is all sopped up. Keep turning them over in the air until they are quite dry and then store them in a drawer to keep them quite clean until you need them – and this is how you make a really good violet colour.[33]

On the same page,[34] instructions are given for 'a lovely bright blue rag colour, according to the London practice, with which you can dye yarn and which is also of use for many things such as letters in blue and serves for a transparent glaze over silver.' Finally, 'the blue rags must be wrapped in paper and be laid in a clean wooden box and must be kept on some high shelf where it smells sweet and clean and where the colour will not be affected by mould. If the rags are kept in this way, they will remain fresh and lovely for as long as twenty years without fading (*wol behalten frisch und schon*). In Paris and in London this colour is called blue for missals, but here in this country rag blue colour and it is a good and valuable colour.'

Editor's note

Address for correspondence: Departamento de Conservação e Restauro e REQUIMTE/CQFB, Universidade Nova de Lisboa, Portugal (mjm@dq.fct.unl.pt). Paper received 11 April 2003; revised version received 14 September 2003.

Notes and references

1. Strack, D. and Wray, V. (1994) *The Flavonoids, Advances in Research*, ed. J. B. Harborne, London, ch. 1. Anthocyanins, in nature, are glucosyl compounds; when they are free from sugar substitution, they are denominated anthocyanidins.
2. Picon, A. (1995) *Les dix livres d'architecture de Vitruve*, Paris, Book VII, ch. XIV.
3. Colombo, L. (1995) *I colori degli Antichi*, Florence, pp. 171–3.
4. *The Strasburg Manuscript: A Medieval Painter's Handbook* (1966), tr. V. and R. Borradaile, New York (Transatlantic Arts).
5. Blondheim, D.S. (1928) 'An old Portuguese work on manuscript illumination', *Jewish Quarterly Review* NS, XIX, pp. 97–135; Brunello, F. (1992) *De arte illuminandi, e altri trattati sulla tecnica della miniatura medievale*, Vicenza.
6. Cennini, C. (1997) *Il libro dell' arte*, ed. F. Brunello, Vicenza.
7. Willstätter, R. and Everest, A.E. (1913) 'Untersuchungen über die Anthocyane. I. Über den Farbstoff der Kornblume', *Justus Liebigs Annalen der Chemie* 401, pp. 189–232; Willstätter, R. and Mallison, H. (1915) 'X. Über Variationen der Blütenfarben', *Justus Liebigs Annalen der Chemie* 408, pp. 147–62 (this is the final paper of a series of 10 published in the same volume, beginning on p. 1).
8. Shibata, K., Shibata, Y. and Kasiwagi, I. (1919) 'Studies on anthocyanins: color variation in anthocyanins', *Journal of the American Chemical Society* 41, pp. 208–20; Pratt, D.D. and Robinson, R. (1922) 'CLXXXVIII. A synthesis of pyrylium salts of anthocyanidin type', *Journal of the Chemical Society, Transactions* 121, pp. 1577–85; Robinson, R.and Robinson, G. (1932) 'Developments in the chemistry of the anthocyanins', *Nature* 130, p. 21; Robinson, R. and Robinson, G.M. (1939) 'The colloid chemistry of leaf and flower pigments and the precursors of the anthocyanins', *Journal of the American Chemical Society* 61, pp. 1605–7.
9. Brouillard, R. and Dubois, J.E. (1977) 'Mechanism of structural transformations of anthocyanins in acidic media', *Journal of the American Chemical Society* 99, pp. 1359–64; Santos, H., Turner, T.L., Lima, J.C., Figueiredo, P., Pina, F. and Maçanita, A.L. (1993) 'Elucidation of the multiple equilibria of malvin in aqueous solution by one- and two-dimensional NMR', *Phytochemistry* 33, pp. 1227–32.
10. Pina, F. (1998) 'Thermodynamics and kinetics of flavylium salts – Malvin revisited', *Journal of the Chemical Society Faraday Transactions* 94, pp. 2109–16; Pina, F., Melo, M.J., Maestri, M., Ballardini, R. and Balzani, V. (1997) 'Photochromism of 4'-methoxyflavylium perchlorate. A *"write-lock-read-unlock-erase"* molecular switching system', *Journal of the American Chemical Society* 119, pp. 5556–61; Pina, F., Maestri, M. and Balzani, V. (1999) 'Photochromic flavylium compounds as multistate/multifunction molecular-level systems', *Journal of the Chemical Society Chemical Communications*, pp. 107–14.
11. Steward, R.N., Norris, K.H. and Asen, S. (1975) 'Microspectrophotometric measurement of pH and pH effect on color of petal epidermal-cells', *Phytochemistry* 14, pp. 937–42; Yoshida, K., Kondo, T., Okazaki, Y. and Katou, K. (1995) 'Cause of blue petal colour', *Nature* 373, p. 291.
12. Shibata *et al.* 1919 (see note 8 above); Cai, Y., Lilley, T.H. and Haslam, E. (1990) 'Polyphenol anthocyanin copigmentation', *Journal of the Chemical Society Chemical Communications*,

p. 380; Dangles, O., Saito, N. and Brouillard, R. (1993) 'Kinetic and thermodynamic control of flavylium hydration in the pelargonidin cinnamic acid complexation – origin of the extraordinary flower color diversity of Pharbitis-nil', *Journal of the American Chemical Society* 115, pp. 3125–32; Goto, T. and Kondo, T. (1991) 'Structure and molecular stacking of anthocyanins – flower color variation', *Angewandte Chemie International Edition* 30, pp. 17–33; Kondo, T., Yoshida, K., Nakagawa, A., Kawai, T., Tamura, H. and Goto, T. (1992) 'Structural basis of a blue-colour development in flower petals from *Commelina communis*', *Nature* 358, pp. 515–18; Pina, F. (1998) 'Caffeine interaction with synthetic flavylium salts. A flash photolysis study for the adduct involving 4′,7-dihydroxyflavylium', *Journal of Photochemical Photobiology A* 117, pp. 51–9.

13. Melo, M. J., Moncada, M. and Pina, F. (2000) 'On the red colour of raspberry (*Rubus idaeus*)', *Tetrahedron Letters* 41, pp. 1987–91.
14. Melo *et al.* 2000 (see note 13 above).
15. Supramolecular chemistry is a novel field of science, with a novel terminology. As Jean-Marie Lehn, winner of the 1987 Nobel Prize in Chemistry (together with D. J. Cram and C. J. Pederson), wrote, 'As a novel field of science emerges, grows and matures, it evolves basic concepts and generates novel terminology to name the concepts that define it and to describe the objects that constitute it. Such conceptualizing and naming play a very important role, not only for shaping the field but also by the ground they offer to creative imagination.' This is the main reason for the transcription of some important definitions: see Lehn, J.-M. and Ball, P. (2000) *The New Chemistry*, ed. N. Hall, Cambridge, pp. 301–4. The authors present the ideas, concepts and also the lexicon of this novel area in science, with some wonderful analogies. Supramolecular chemistry is also about molecular architecture or as Ball and Lehn stated, 'It is as if the brick-makers have suddenly realized that their products need not be an end in themselves but provide a means for them to become architects.' The structures that nature developed to capture the blue colour are wonderful molecular architectures, where the components self-assemble through molecular recognition.

    'Civilization combats entropy through a network of information exchange. ... When molecules need to get organized, they adopt analogous strategies. This is why the key concepts of supramolecular chemistry embrace not just those of traditional molecular chemistry – structure and energy – but also a third, information ... The components of supramolecular chemistry communicate, they form associations, they have preferences and aversions, they follow instructions and pass on information. Central to these exchanges is the idea of molecular recognition, whereby one molecule is able to distinguish another by its shape or properties ... Supramolecular systems are groups of two or more discrete molecules that interact in specific ways to form an organized assembly ... The forces that hold together supramolecular systems are not strong covalent or ionic bonds, but interactions that are weaker, commonly by an order of magnitude in bonding energy. Hydrogen bonding, donor-acceptor interactions, metal-ion coordination, hydrophobic interactions – these are the glues that stick supermolecules together. This subtle attraction means that molecules can communicate without becoming locked into barren unions ... Self-assembly is the spontaneous association of several (more than two) molecular components into a discrete, non-covalently bound aggregate with a well-defined structure. This will generally involve more than one kinetically distinct step. Self-assembly involves molecular-recognition processes – binding events, but not "mere" binding. Rather, one may say that recognition is binding with a purpose.'
16. Cai *et al.* 1990; Dangles *et al.* 1993; Goto and Kondo 1991; Kondo *et al.* 1992; Pina 1998 (see note 12 above for all).
17. Picon 1995 (see note 2 above).
18. Goto and Kondo 1991 (see note 12 above); Kondo *et al.* 1992 (see note 12 above).
19. Cennini 1997 (see note 6 above), pp. 168–9, ch. CLXI.
20. Cardon, D. and du Chatelet, G. (1990) *Guide des teintures naturelles*, Paris-Lausanne.
21. *The Strasburg Manuscript* 1966 (see note 4 above).
22. Kondo, T., Ueda, M., Tamura, H., Yoshida, K., Isobe, M. and Goto, T. (1994) 'Composition of protocyanin, a self-assembled supramolecular pigment from the blue cornflower, *Centaurea cyanus*', *Angewandte Chemie International Edition* 33, pp. 978–9; Kondo, T., Ueda, M., Isobe, M. and Goto, T. (1998) 'A new molec-

ular mechanism of blue color development with protocyanin, a supramolecular pigment from cornflower, *Centaurea cyanus*', *Tetrahedron Letters* 39, pp. 8307–10.
23. Kondo, T., Oyama, K.-i and Yoshida, K. (2001) 'Chiral molecular recognition on formation of a metalloanthocyanin: a supramolecular metal complex pigment from blue flowers of *Salvia patens*', *Angewandte Chemie International Edition* 40, pp. 894–7.
24. *The Strasburg Manuscript* 1966 (see note 4 above).
25. *The Strasburg Manuscript* 1966 (see note 4 above).
26. Blondheim 1928 (see note 5 above).
27. Melo *et al.* 2000 (see note 13 above).
28. Melo *et al.* 2000 (see note 13 above).
29. Wallert, A. (1993) 'Natural organic colorants on mediaeval parchment', in J. Bridgland (ed.), *ICOM Committee for Conservation, 10th Triennial Meeting, Washington, DC, USA, 22–27 August 1993: Preprints*, 2 vols, Paris, pp. 516–23.
30. Strack and Wray 1994 (see note 1 above); Kondo *et al.* 1994 (see note 15 above); Kondo *et al.* 1998 (see note 15 above); Kondo *et al.* 2001 (see note 23 above).
31. *The Strasburg Manuscript* 1966 (see note 4 above), pp. 29 and 37.
32. *The Strasburg Manuscript* 1966 (see note 4 above), p. 29.
33. *The Strasburg Manuscript* 1966 (see note 4 above), p. 37.
34. *The Strasburg Manuscript* 1966 (see note 4 above), p. 37.

Sanguis draconis, Dragon's Blood: The Use and Analysis of a Manuscript Illuminators' Colorant

Arie Wallert* and Maarten R. van Bommel

Introduction

Dragon's blood is one of the more mysterious colours in the history of painting. It is the name for a red, resinous colorant. The resin received its name from a description in Pliny's *Historia Naturalis*. Pliny describes in Book VIII, chapter xii the continuous warfare between dragons and elephants:

> Elephants are very cold blooded, and consequently in very hot weather are sought after by the dragons; and it is for this reason they submerge themselves in rivers and lie in wait for the elephants when drinking, and rising up coil round the trunk and imprint a bite inside the ear, because that place only cannot be protected by the trunk; and that the dragons are so large that they can hold the whole of the elephants blood, and so they drink the elephants dry, and these when drained collapse in a heap and the dragons being intoxicated are crushed by them and die with them (Fig. 1).[1]

Later, Pliny gives an account of the advantage that could be taken of this feud: '..."Indian cinnabar", for that is the name the Greeks gave to the gore of a dragon crushed by the weight of dying elephants, when the blood of each animal gets mixed together, as we have said; (VIII. 34) and there is no other colour that properly represents blood in a picture. That kind of cinnabar is extremely useful for antidotes and medicaments'; and: 'In old times dragon's blood cinnabar was used for painting the pictures that are still called monochromes.' Pliny described dragon's blood as one of the more expensive pigments, for which the patron had to pay separately (Book XXXV, ch. XII): 'The brilliant colours, which the patron supplies at his own expense are cinnabar (natural vermilion, red mercury(II) sulfide), Armenium, dragon's blood.'[2]

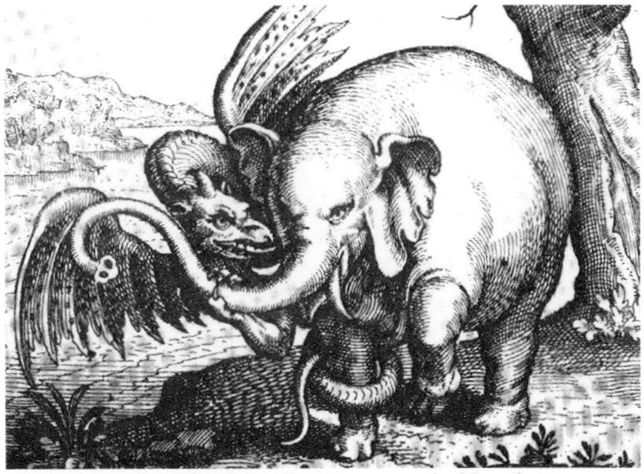

Figure 1 *The Elephant being strangled by the Dragon*, from Petrus Bonus, *Pretiosa Margarita Novella de thesauro ac pretiosissimo philosophorum lapide ... compositus 1330*, Venice 1546.

Dragon's blood was used primarily as a stain for colouring varnishes, most of which was applied over gold, silver or other metals: it gives shiny, deep red, translucent glazes. Its application is described in the first part of the *Mappae clavicula*, dating from about the 9th century, and also in a later part of this compilation; in Book III of Heraclius's treatise dating from the 13th century; and in Jehan Le Begue's 15th-century compilation.[3] For use in manuscript illumination dragon's blood was prepared with glair, beaten egg white. The 12th-century treatise *De Clarea* states: 'With this glair, vermilion, minium (red lead), saffron and dragon's blood are tempered.'[4] The use of dragon's blood as a colorant for manuscript illuminations is confirmed by the remark in the late 14th-century *Liber de coloribus* that: 'clear [transparent] and thick [opaque] colours on parchment: these are ... dragon's blood', which is taken from the introduction of the Philipps-Corning manuscript of the earlier *Mappae clavicula*.[5] In the concordance of colours in the *Liber de coloribus*, advice is given that shadows on dragon's blood should be made with black and highlights with white. For an orange colour dragon's blood should be mixed with orpiment (yellow arsenic sulfide) to give an orange colour, on which the shadows were laid with pure dragon's blood, the highlights again with white. According to this treatise dragon's blood was also used in miniature painting to make shadows on areas in the figures painted with vermilion.[6] The 14th–15th-century *Liber diversarum arcium* recognises dragon's blood as a plant product, rather than one of animal origin. It also gives an indication of the geographical origin of the product: 'Some say dragon's blood is the juice of a herb, which is worthless; it is, rather, the gum of a tree growing in Persia and India.'[7] Dragon's blood was mentioned briefly in the table of synonyms in Jehan Le Begue's compendium: 'Dragon's blood is a murrey (purple) or dark red colour.'[8]

By about the 14th century the use of dragon's blood had declined, possibly because there were so many other, more reliable or more powerful organic reds available. By that time, apparently, it was used almost exclusively for the painting of miniatures. Cennini dismissed it as a suitable painters' pigment, with a rather brief description: 'A colour known as dragon's blood is red. This colour is used occasionally on parchment, for illuminating. But leave it alone, and do not have too much respect for it; for it is not of a constitution to do you much credit.'[9] It was mentioned, however, in the later 15th-century Bolognese Manuscript.[10] In the 17th-century Paduan Manuscript it features among the colours that were typically used in miniature painting: ground with sal ammoniac (ammonium chloride) and gum it was added to lead white and minium.[11] Thus almost all the major historical technical sources on the art of painting show some awareness of the existence of dragon's blood; its actual application hardly ever finds description, however.

Three sources of the colorant

There are at least three different major botanical sources of the substance we know as dragon's blood.

- *Dracaena draco* L. (Dracaenaceae), indigenous to the Canary Islands, is now almost extinct (Plate 17). The related *D. cinnabari* Balf. f. is the source for so-called Socotran dragon's blood. *D. ombet* Kotschy & Peyr. occurs mainly in the countries around the Red Sea. *D. boerhavii* Ten. is found primarily in Mediterranean countries and may well have been the major source for dragon's blood in classical painting. The colorant is usually obtained from spots on the bark of the tree.
- *Daemonorops draco* Blume (Arecaceae; formerly *Calamus draco* Willd.) grows primarily in the Indonesian islands of Sumatra and Borneo, and on the Moluccan archipelago. The resin exudes naturally from the husks of the ripe fruits.
- *Croton draco* Schltdl. (Euphorbiaceae) is widespread in the tropical rainforests of Central and South America.[12] When the

bark of the tree is cut the resin immediately flows freely from the fresh wounds.

The geographical origin, colouring principles, chemical structures, painting properties and history of use of all these dragon's blood resins are different in nature.

History of the colorants

Dracaena draco

The dragon's blood we find described in Cennini's treatise and in those of the medieval alchemists was the substance that must have come from *Dracaena draco*. This name is a corruption of the Greek word χινναβαρισ (*cinnabaris*, the name Pliny used); that is, 'female dragon'. In medieval western Europe, the dragon tree must have been the epitome of exoticism. It served to give the desired exotic touch to the background of Martin Schongauer's engraving of *The Flight into Egypt*, of 1470–75. The tree is also illustrated in a woodcut in Hartmann Schedel's *Liber chronicarum* (Nuremberg 1493) and by Albrecht Dürer in his woodcut of *The Flight into Egypt*, which appeared in 1502.[13] The dragon tree is also shown in the left-hand panel, *The Earthly Paradise* (*Garden of Eden*), of Hieronymus Bosch's *The Garden of Earthly Delights* (Madrid, Museo Nacional del Prado), of about 1504.

This dragon tree species is the pride of the Canary Islands. In other areas, however, related species occur, similarly yielding dragon's blood. *Dracaena cinnabari* Balf. f. is found on the island of Socotra, off the Yemeni coast, *D. schizantha* Baker in the coastal areas of Somalia, and other species on Mauritius and the mountains of Cameroun.[14] A *Dracaena* species in Yunnan, China (*D. terminalis* L., or Chinese Colli) is also known to yield red resin;[15] Chinese dragon's blood was used for the famous red Chinese varnish. Middle American species such as *D. americana* Donn.Sm. and *D. cubensis* Marie-Vict., and the Hawaiian *D. aurea* H.Mann, can also be expected to yield the coloured resin.[16]

Because of their massive size and alleged age, the dragon trees of Tenerife aroused great interest among 18th-century naturalists. The dragon tree shared with the baobab tree the distinction of being the oldest known living representative of the vegetable kingdom. Alexander von Humbold estimated in 1799 that the big dragon tree of Orotava would already have been growing around 1400. Of course, Charles Darwin was very eager to see this tree and wrote on July 27, 1831, that: 'I am sure nothing will prevent us seeing the great Dragon tree.'[17] In Europe most of the botanical knowledge on the dragon tree came from the famous 16th-century botanist, Karel van der Sluis or Charles de l'Écluse, better known as Carolus Clusius (Atrecht/Arras 1526 – Leiden 1609). The illustration in his *Rariorum aliquot stirpium per Hispaniae observatum historia* (Antwerp 1576) of the dragon tree he saw in Portugal in 1565 served as the model for many of the illustrations in the herbals that were to follow. The same illustration is found in editions of the herbals by Matthias de L'Obel (Lobelius) and Rembert Dodoens (Dodoneus), as well as in the nine editions – from 1585 to 1718 – of Castore Durante's *Herbario* (first edition Rome 1585). The same tree is still depicted in the 1731 edition of the *Kraüterbuch* of Jacob Theodorus (Tabernaemontanus). Jean Félix Watin described dragon's blood as the exudate from unspecified trees, but does so with reference to Linnaeus.[18] The most extensive botanical study of the dragon tree was published in 1864 by Rauwenhoff.[19] A more recent study on the physiology, biochemistry and general taxonomy of the plant is that by Symon (1974).[20] Although the tree seems to be almost extinct in its original habitat, it grows relatively easily in all Mediterranean climates. It can be found in subtropical sections in botanical gardens and occasionally even in private gardens.[21]

Contrary to common belief, the colorant does not flow freely from cuts in the bark of the tree (Plate 18). It is, in fact, very difficult to collect the colorant from the tree: fresh cuts usually result in colourless liquid exuding in very small amounts, if at all.[22] The production of the red colorant seems to be triggered, or increased, through the action of various microorganisms.[23]

Daemonorops draco

After the Middle Ages, dragon's blood became synonymous with the exudate from a rattan palm tree, *Daemonorops draco* Blume. The exudate was used as an astringent for medicinal purposes, as an ingredient in incense mixtures and as a sexual stimulant.[24] Sheikh Nefzani mentioned it in his treatise as a component in erotic recipes.[25]

There are several types of rattan palm, only a few of which actually produce the colouring matter. The most important habitat with trees yielding the highest quality dragon's blood is Sumatra (*rotan djerenang*). Some of the dragon's blood may have reached Europe through the Dutch East India Company only after the middle of the 17th century. Before that time, some may have reached the West through Arab traders.

The leaves of the genus *Daemonorops* have prickly stalks which may grow into long tails. The bark is provided with flattened spines. The coloured resin is obtained from the fruit, pointed berries about the size of a cherry. When ripe, they are covered with a reddish, resinous substance which may be gathered in several ways. The simplest is by shaking or rubbing the berries in a coarse canvas bag. The hard, brittle resin easily breaks off. Alternatively, the fruits could be boiled to separate the resin; the hot, molten resin would be skimmed off the surface of the water and pressed into beads, or packed in leaves, or strips of cane. The description of dragon's blood in the 1742 Dutch edition of the treatise of Filippo Buonanni (or Bonanni) on varnishes (first published Rome 1720) strongly suggests that this product is a *Daemonorops* exudate.[26] By the 19th century the rattan product was the only substance produced commercially under the name dragon's blood. A very elaborate and accurate description of its processing and production appeared in a 1839 handbook on laquers.[27]

The constituents of the exudate have been described as compounds of benzoic and bezoyl-acetic acids, dracoresinotannol esters, dracoalban, dracoresene, calcium oxalate and calcium phosphate. Most of the material is soluble in alcohol, with a deep red colour. Tschirch and Stock (1933–36) describe the purification of the exudate by extraction with petroleum ether, which is then evaporated to dryness and the residue taken up in ethanol. Dracoresene will go into solution, leaving the insoluble dracoalban. The dracoresene solution can then be further purified from undesirable components by fractional distillation.[28]

The colouring principle of this variety of dragon's blood is described as containing dracorubin and dracorodin. Efforts have been made to characterise it by Raman spectroscopy.[29]

Croton draco

The use of dragon's blood from *Croton draco* in a European context can, of course, only have occurred in the 16th century after the conquest of South and Central America, when the export of colonial wares to Europe had begun. Therefore 'Mexican' dragon's blood probably only appeared on the European market around the 1570s. The red juice of *Croton draco* is used in Mexico as a vulnerary and astringent.

The *Croton* colouring matter may have been confused with that of another substance: a strong purple colorant called *quauhayohuachtli* was obtained from *Jatropha curcas* L. This shrub, with three- or five-lobed, long petiolate leaves and greenish-yellow flowers is widely distributed in tropical America. Because of the strong purgative properties of its seeds, which are enclosed in large, three-celled drupaceous capsules, this so-called *piñon de Indias* or *sangregado* was often cultivated. On cold pressing, the seeds produce up to 40% of oil, making it a plant of such commercial value that it was also grown in the Dutch East Indies.[30] Sometimes its colour was also called *sangre de drago* – dragon's blood – thus causing some confusion with the red resinous colouring matter from *Daemonorops draco*. Several plants are described as *sangre de drago* in Francisco Hernández's *Historia natural de Nueva España* (Madrid 1570), who also first described its use as a source of colorant.[31] *Sangre de drago* can similarly be obtained from *Croton hibiscifolius* Kunth, *C. gossypiifolius* Vahl (Venezuela), *C. sanguifluus* H.B. & K. and other species.

In many other cultures, different indigenous red plant exudates are known as 'dragon's blood'. Good examples are the gums from *Eucalyptus resinifera* Sm. and *Eucalyptus terminalis* F.Muell. in Australia.

Chemical constitution

Some of the colouring matter in the red resin of *Dracaena draco* was identified by Masaoud *et al.* (1995), mainly on the basis of nuclear magnetic resonance (NMR) spectroscopy.[32] This substance could be obtained by successive extractions with hexane, chloroform and methanol, followed by a series of separations over Sephadex and silica gel. By means of the NMR data they demonstrated the connectivity of the aromatic rings, as well as the 4- and 4″- positions of methoxy groups in the newly found structure. They identified this as cinnabarone ($C_{32}H_{32}O_7$), a biflavonoid composed of dihydrochalcone linked by a carbon–carbon bond to a deoxotetrahydrochalcone. In addition to cinnabarone, the colouring components of *Dracaena draco* include a number of other flavonoids, one example being a triflavonoid, damalachawin. Himmelreich *et al.* (1995) isolated a small fraction (0.04%) of this component by the previously described successive extractions and separations.[33] They established a structure composed of a flavan

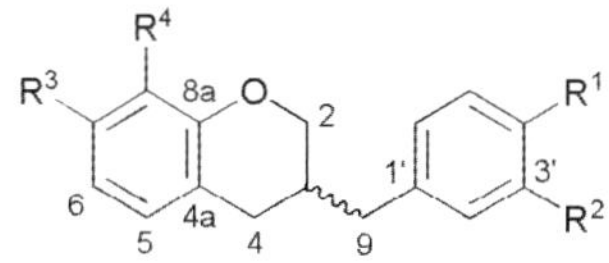

	R1	R2	R3	R4
1	OMe	OH	OH	H
2	OH	H	OH	OMe
3	OH	H	O-CH_2-O	
4	OH	H	OH	H

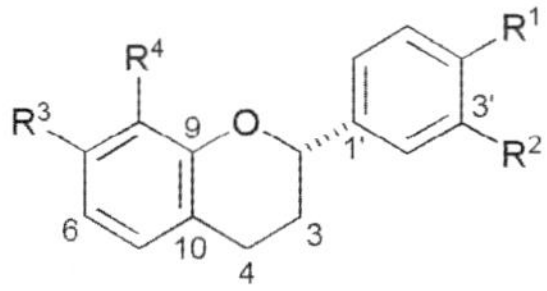

	R1	R2	R3
5	OH	OMe	OH
5a	OAc	OMe	OAc
6	OMe	OH	OH
6a	OMe	OAc	OAc
7	H	H	OH

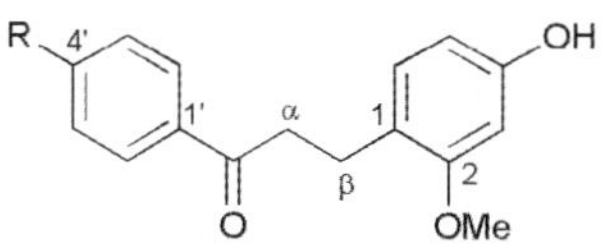

8 R = H
9 R = OH

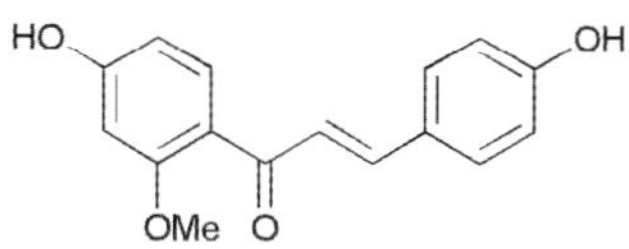

10

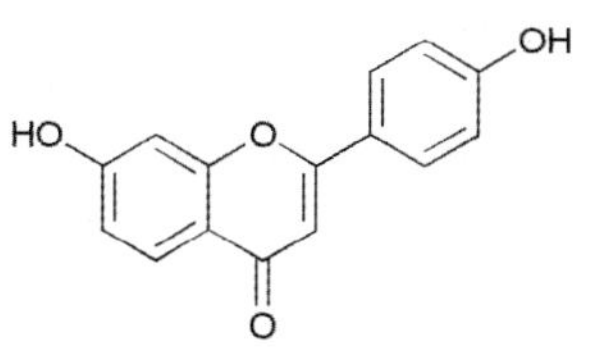

11

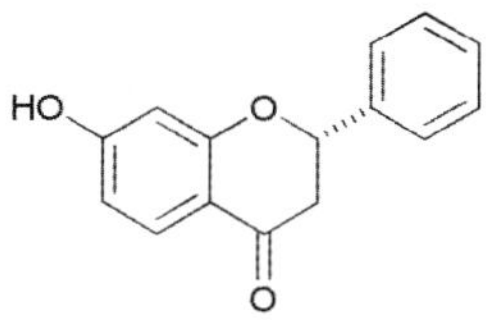

12

Figure 2 Flavan constituents in *Dracaena* exudates (redrawn after Masaoud *et al.* 1995; see note 32).

Table 1 Constituents of dragon's blood resins.

Dracaena draco L.	*Daemonorops draco* Blume
cinnabarone	dracorhodin
damalachawin	dracorubin
various flavonoids (chalcones, flavans, flavonols)	various flavonoids (chalcones, flavans, anthocyanidins)
	carotenoids
	terpenoid resin
	triglycerides

and two deoxotetrahydrochalcone moieties, differing from cinnabarone by replacement of the keto group by a 7-hydroxyflavan-6-yl group and hydrogen.

Dracaena exudates also contain a number of homoisoflavans with seemingly the same absolute configurations (Fig. 2) These can be isolated with the same series of extractions and separations. One of these (1) was established as $C_{17}H_{18}O_4$. The elemental composition of (8) is $C_{16}H_{16}O_3$. The other homoisoflavans differ only in the position of hydroxy- and methoxy groups. Compound 5, for instance, possesses the same substitution pattern as compound 6 and must differ only in the position of the O-methyl group.

The colouring components of *Daemonorops draco* are generally described as the two quinone methides, dracorhodin and dracorubin. However, the resin has been shown to contain a large number of other substances that contribute to the colour. Flavans, biflavonoids, anthocyanidins, triflavonoids and chalcones have been found. Merlini and Nasini (1976) have isolated a secobiflavonoid by elution of dragon's blood resin with hexane and ether over a silica gel column, followed by preparative thin-layer chromatography (benzene:ether 4:1, Rf 0.43 in ether and 0.15 in benzene:ether (9:1) on Bakerflex IB-F plates; detection with cerium(IV) sulfate in sulfuric acid).[34] Edwards *et al.* (1997) have even found indications for the presence of carotenoid substances.[35]

The constituents present in these two varieties of dragon's blood are summarised in Table 1.

Adulteration, substitution

Because the colorant was so rare and costly, adulterations and substitutions often occurred. This practice was already described in antiquity. Pliny (XXXIII, 7) stated that: 'it is adulterated with goat's blood or crushed service berries'.[36] On folio 122r of a Venetian manuscript in the British Library (MS Sloane 416, London, British Library), dating from the second half of the 15th century, we find a recipe for an imitation of dragon's blood.[37] The recipe calls for extracts of sticklac, kermes and red sandalwood precipitated with alum, to be mixed with varnish, gum arabic and Armenian bole. The resinous mixture is melted and then cooled into a cake that would look quite similar to the real dragon's blood. Another falsification is described in Matthioli's commentary on the *Materia medica* of Dioscorides: 'It is brought in loaves counterfeited with goat's blood, ground brick and dried sorb apple.'[38] Tchirch mentions substitution with iron oxides, Armenian bole and colophony, and mixtures with resin sandalwood and plant gums.

There is some evidence to suggest that the dragon's blood of antiquity, imported by the Egyptians from the land of Punt (believed to have been located in the region of northern Ethiopia/Eritrea), actually came from Socotra. Periplos rightly observed that the χιvvαβαριϭ (*cinnabaris*) of his time was collected there as exudates from trees, which – appropriately – must have been *Dracaena cinnabari*. Not surprisingly, the Arab dragon's blood came from the same source. It seems that exudates of *Dracaena draco* from the Canary Islands appeared in western Europe only in the later Middle Ages. It was in the late 16th and early 17th centuries that dragon's blood from *Daemonorops* spp. appeared in European apothecaries' lists. Pomet (1717) elaborates on the properties of the dragon's blood from the Canary Islands and Socotra and refers to the description by Clusius. He also refers to the so-called Indian dragon's blood, indicating a *Daemonorops* product.[39] Simon Eikelenberg (*c.*1690–1720), who wrote about painting, reveals that it was often substituted with a mixture of gum arabic and brasilwood red; in this as in other instances he is quoting Buonanni. He states that the falsification can easily be detected as the false product would dissolve and give off colour in warm water, which the real dragon's blood does not.[40]

By the end of the 19th century, only dragon's blood from Sumatra (*Daemonorops* sp.) was still available on the market as a commercially viable product.

Identification

Different analytical procedures have been suggested to distinguish between the different dragon's blood species. To evaluate the suitability of these procedures, the exudates from *Daemonorops* sp. and *Dracaena draco* were examined by all the methods tested; the solubility test and high-performance liquid chromatography (HPLC) were not tried on the *Croton* sp. exudate.

Solubility tests

Dragon's blood from *Daemonorops* sp. can be distinguished from that produced by the various *Dracaena* spp. with the help of a simple solubility test (Dieterich'sche Dracoalbanprobe). An extract of the unknown substance in ether is poured into the same amount of alcohol and the mixture is allowed to stand for one hour. The formation of a white flocculent precipitate indicates the presence of a rattan palm dragon's blood; the absence of the precipitate suggests a *Dracaena* dragon's blood. According to Tschirch, *Daemonorops* dragon's blood dissolves in benzene and carbon disulfide (CS_2), whereas *Dracaena* exudates do not.[41]

UV-visible absorption spectrometry

The three colorants dissolved in HPLC grade methanol may be distinguished by the differences in the shapes and absorption maxima of their UV-visible spectra.

The *Dracaena* solution showed a shoulder at 574 nm, a rather strong absorption around 457 nm, minor peaks at 346 nm and 328 nm, and a very strong absorption at 289 nm. The solution of *Daemonorops* showed an important, but rather broad maximum at 475 nm, a shoulder around 367 nm, a slight shoulder at 310 nm, and a very strong maximum at 285 nm. *Croton* exudates have a shoulder at 570 nm, with the maximum for the colorant at 465 nm, a slight shoulder around 325 nm, and the strongest absorption at 287 nm.

All three types of dragon's blood have their strongest absorption in the same ultraviolet wavelength range. In the visible light range there are some marked differences that may help to distinguish between *Daemonorops* (475 nm), *Dracaena* (457 nm), and *Croton* (465 nm) exudates. The most important maxima, however, are too broad to allow an unambiguous identification.

Fluorescence spectrometry

Both the *Dracaena* and the *Daemonorops* exudates contain highly fluorescent substances and can therefore be identified by fluorescence spectroscopy in very small quantities. The *Croton* colorant, however, has some fluorescent components, but also contains substances that tend to quench the fluorescence to a considerable extent. Therefore the identification of this colour poses more problems and the sample quantities required are larger.

A solution of *Dracaena draco* exudate in spectroscopic grade methanol showed two distinct fluorescent maxima. An emission spectrum with the excitation wavelength set at 336 nm showed a peak at 413 nm and shoulders around 360 nm and 530 nm. The excitation spectrum, acquired with an emission wavelength of 413 nm, showed a strong peak at 336nm, and a

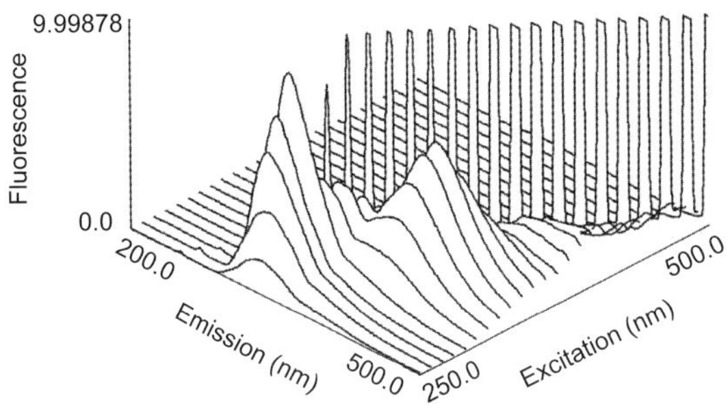

Figure 3 Three-dimensional fluorescence spectrum of *Dracaena* dragon's blood in methanol produced using the excitation and emission wavelengths described in the text.

rather small one at 267 nm. The other major maximum of the *Dracaena* extract is located in a lower wavelength region: it can be found at λ_{ex} max. 290 nm and λ_{em} max. 318 nm, with the excitation spectrum showing a rather broad shoulder around 265nm (Fig. 3).

As one of the major constituents in the *Dracaena* resin has a flavonoid stucture, the detection limits by fluorescence spectrometry may be brought down to the pico/nanogram range by using various complexing reagents.[42]

Thin-layer chromatography

It is fairly easy to distinguish between the three types of dragon's blood by thin-layer chromatography (TLC).

Excellent separations of *Daemonorops* resin were obtained on Merck HTPLC Kieselgel 60 F254 plates, with a benzene:methanol (95:5, v/v) mixture as eluent. The chromatogram is developed in a saturated chamber for 15 minutes, taken out and dried in a stream of hot air. The same plate is then run twice again under the same conditions.

The *Dracaena* exudate, however, showed only a little transport on the plate and the separations were not satisfactory, while the *Croton* red showed no reaction at all. This resin showed a better migration in a mixture of methyl ethyl ketone:methanol:formic acid (65:30:5, v/v), but the other two colorants showed incomplete separation with much tailing.

Although there is no optimal technique that can be used to achieve good results for all three colorants within one system, TLC on the silica gel plates with a toluene:ethyl formate:formic acid (50:40:10, v/v) mixture appears to give the best results.[43] Here again, chromatography has to be performed at least three times for 15 minutes each time, in a saturated chamber, to obtain a reasonable separation. The *Croton* exudate then separates into three non-fluorescent bands and one other fraction that has the same Rf as one of the major bands of the *Daemonorops* red. This band also shows the same fluorescence under UV illumination (254 nm). The *Daemonorops* red itself separates into a large number of yellow, orange and red fractions, all of which have a strong fluorescence in UV illumination. The *Dracaena* resin also gives a good result, with many purplish-red, carmine-red, orange and yellow fractions. The order of the colours, their number and their intensity, in both ambient and UV light, makes it quite easy to distinguish the chromatographic pattern of the *Dracaena* resin from those of the other two.

High-performance liquid chromatography

The application of HPLC can give good results in the analysis of the *Dracaena* exudate. An isocratic system can be used with a mobile phase of a 7:3 methanol:water mixture to which 1% acetic acid is added. When the sample, dissolved in HPLC grade methanol, is delivered at a flow rate of 0.8 mL min^{-1} with an average pressure of 1200 psi, through a Nucleosil C_{18} column with a 5 μm packing, the most significant components of *Dracaena* can be detected within 15 minutes. Detection was carried out using a variable wavelength detector set at 260 nm.The *Dracaena* exudate separated into ten major fractions; this includes fractions that absorb in the UV range, but do not contribute to the actual colour.

The colouring components of *Dracaena* and *Daemonorops* extracts are separated better with the dragon's blood resins dissolved in dimethyl formamide. Chromatography is performed on a Luna C_{18} HPLC column (100 × 2 mm i.d., supplied by Phenomenex, Torrance, CA, USA)

Table 2 HPLC gradient (all changes are linear).

Time (min)	%A (10% methanol/water, v/v)	%B (100% methanol)	%C (1% formic acid, v/v)
0	74	16	10
15	0	90	10
20	0	90	10
23	0	100	0
27	0	100	0
30	74	16	10
45	74	16	10

with the use of a gradient of water, methanol and 1% formic acid with a flow rate of 0.2 mL min^{-1}. A Luna C_{18} guard column (4 × 2 mm i.d.) protects the HPLC column. The gradient, with a total analysis time of 45 min, is given in Table 2.

The HPLC eluent was supplied by a Waters 616 LC pump, controlled by a 600 S controller (Waters, Etten-Leur, the Netherlands). A Waters in-line degasser degassed all eluents used. Samples were injected using a Waters 717 autosampler. Detection was performed with a Waters 996 PDA detector equipped with a 10 μL detector cell, scanning from 200 to 700 nm at a scanning rate of 1 scan/sec and a resolution of 4.8 nm.

The chromatogram of the *Dracaena* dye is shown in Figure 4. Five red compounds and one yellow compound were found. The spectra obtained are shown in Figure 5A–F. The main compound is the yellow compound, eluting at 19.78 min (Fig. 5B). The two main red compounds elute at 21.06 and 21.42 min respectively.

The *Daemonorops* dye consists of more orange-reddish compounds, compared with the *Dracaena* dye, thus the two dyes are easily distinguished. The chromatogram of the *Daemonorops* dye is shown in Figure 6; six compounds were found. Spectra of these compounds are shown in Figure 7A–F. The main compound, eluting at 20.36 min, shows a strong absorption at 443 nm (Fig. 7C).

When the analysis of an organic pigment from an art object is required, extraction with hydrochloric acid is usually performed to break up any bonds between the colorant and the mordant or substrate and to dissolve the colorants. Since before analysis it is unknown if a sample contains dragon's blood or a red lake pigment, it is necessary to investigate the effect of the hydrolysis on the pigments. Therefore, both the *Dracaena* and the *Daemonorops* dye were hydrolysed at 100 °C in a solution of water: methanol:37% hydrochloric acid (1:1:2, v/v/v) for 10 min. Next, the solutions were evaporated to dryness under a nitrogen flow and the extracts were dissolved in dimethyl formamide. Prior to

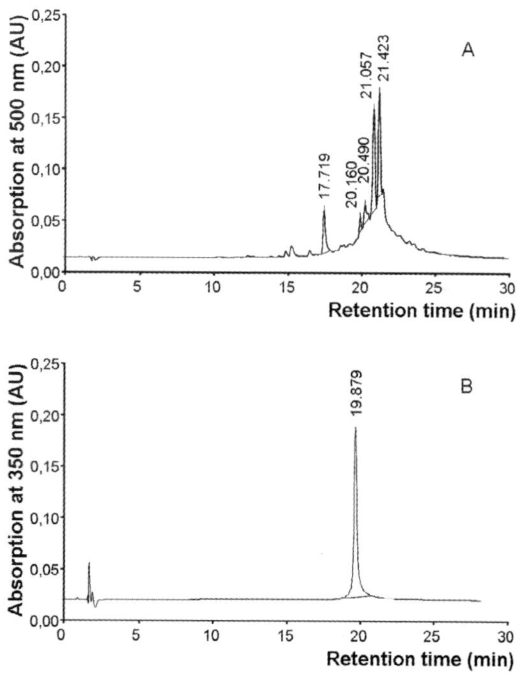

Figure 4 HPLC chromatogram of *Dracaena* dragon's blood; five red compounds and one yellow compound are separated. Wavelength set at 500 and 350 nm.

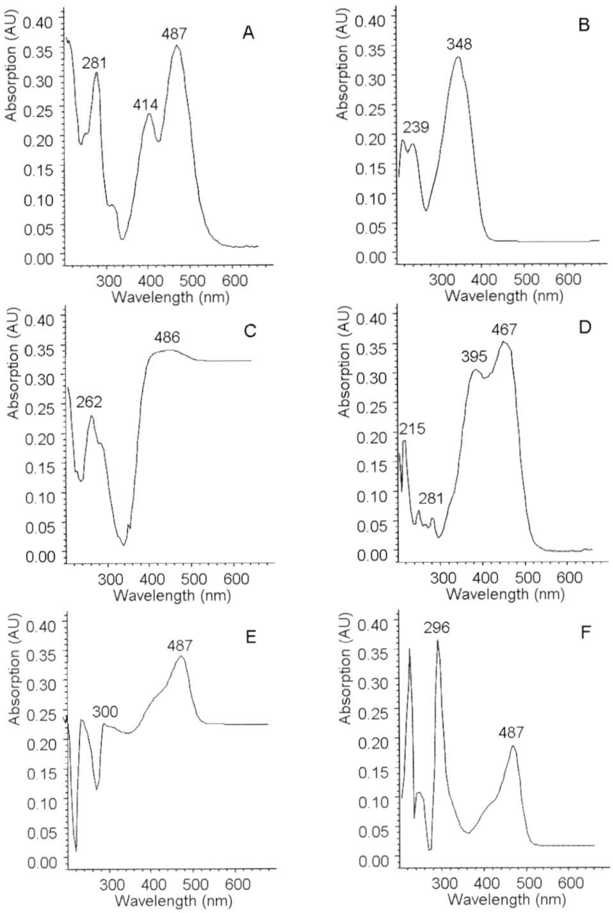

Figure 5 Absorption spectra of the different fractions of *Dracaena* dragon's blood. T_r (retention time) A = 17.72 min, T_r B = 19.88 min, T_r C = 20.16 min, T_r D = 20.49 min, T_r E = 21.06 min, T_r F = 21.42 min. Spectra not to scale.

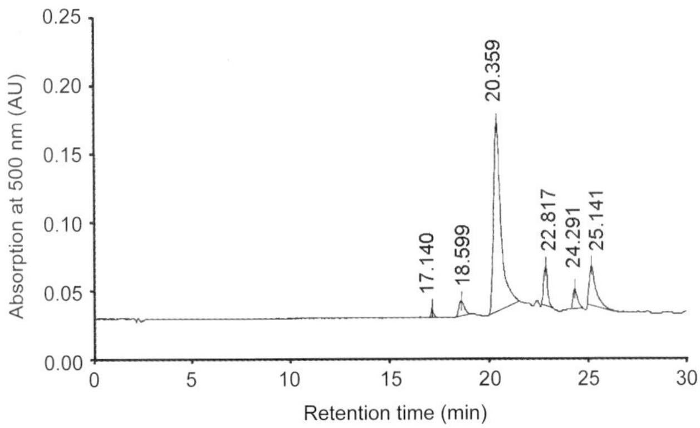

Figure 6 HPLC chromatogram of *Daemonorops* dragon's blood; six red compounds are separated. Wavelength set at 500 nm.

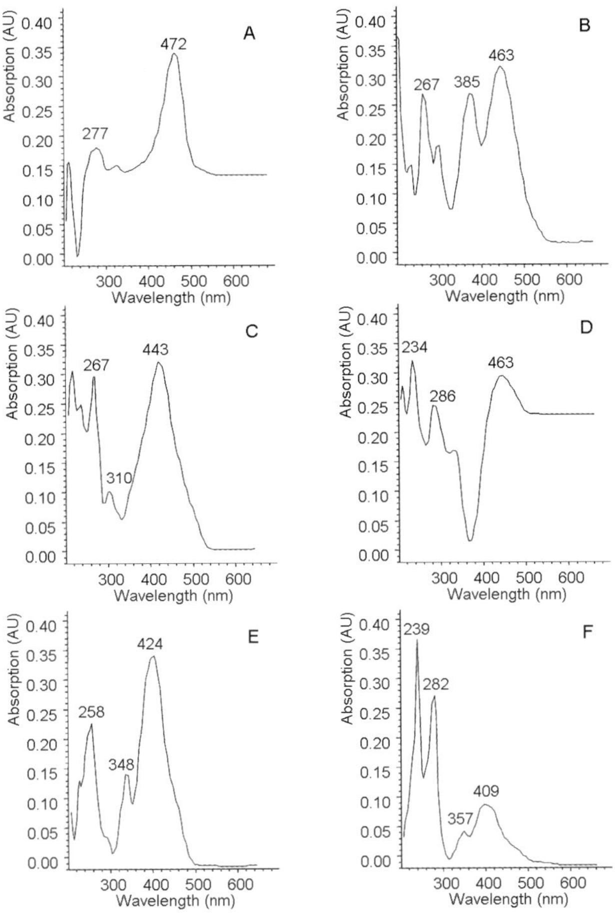

Figure 7 Absorption spectra of the different fractions of *Daemonorops* dragon's blood. T_r A = 17.14 min, T_r B = 18.59 min, T_r C = 20.36 min, T_r D = 22.83 min, T_r E = 24.29 min, T_r F = 25.14 min. Spectra not to scale.

HPLC analysis, the samples were centrifuged at 2000 rpm for 2 min.

The hydrolysis of the dyestuffs did not affect the results of HPLC analysis of the *Daemonorops* dye. The ratio of the colorants from the *Dracaena* dye was altered a little, but the two main red components and the yellow component were not affected by the hydrolysis.

The results of TLC and HPLC analysis clearly show that the dragon's blood dyes consist of a mixture of compounds. Therefore, separation of the compounds is preferred before mass spectrometric (MS) analysis. HPLC–PDA (photodiode array detection) separation was performed according to the procedure described above, followed by fraction collection of each single compound. To determine the result of the fraction collection and to determine if the fractions were pure, they were re-analysed by HPLC–PDA. Although the purification was successful, all fractions still contained other compounds in addition to the main compound. However, the ratio of the other compounds compared to the compound of interest was much smaller. A good example is given by the *Daemonorops* extract, which separated into six red components. The main fraction came off at 20.16 min and had its strongest absorption at 443 nm. However, this fraction also had a significant absorption in the UV range (272 nm), suggesting that small amounts of other non-colouring components may be present in this fraction. This affects the results of subsequent mass spectrometric analyses considerably. These results were therefore not included in this paper.

Conclusion

Although the substances we know as dragon's blood are of nearly identical colour and are soluble in similar solvents, they come from botanically very different plant species and are of distinctly different chemical composition. Mass spectrometry on dragon's blood samples from works of art may be problematic as some of the most important fatty acids in binding media, such as palmitic acid (C16:0, *m/z* 256) and stearic acid (C18:0, *m/z* 284), can give similar values to some of the flavonoids ocurring in dragon's blood.

For the identification of dragon's blood pigments on art objects, HPLC with diode array detection seems to be the method of choice, especially when detection limits can be brought down by means of fluorescence detection. For structural characterisation the separate collection of the individual fractions has its problems. We foresee a further characterisation of the various dragon's blood components with the use of chemical ionisation mass spectrometry (CI–MS), HPLC coupled with electrospray ionisation–mass spectrometry (HPLC–ESI–MS), and ESI–MS–MS techniques.

Acknowledgements

The authors are beholden to several persons and institutions who provided them with suggestions, comments, advice and botanically certified reference materials. Jane Frimodig, curator of El Alisal – The Lummis Home and Garden in Pasadena, gave permission to take material from the *Dracaena* tree in the Lummis garden. Dr J.J. Bos of the Department of Plant Taxonomy of the Agricultural University in Wageningen, the Netherlands, provided us with valuable information and botanical specimens.

Editor's note

*Author for correspondence: Rijksmuseum Amsterdam, PO Box 74888, 1070 DN, Amsteram, The Netherlands (A.Wallert@rijksmuseum.nl). Paper received 13 July 2003; revised version received 31 March 2005.

Notes and references

1. Pliny [Gaius Plinius Secundus] (1938–62) [*Naturalis Historia*] *Natural History*, tr. H. Rackham (and others) (Loeb Classical Library), 10 vols, Cambridge, Mass., and London, vol. III (repr. 1984), VIII.xii.34.
2. Pliny 1938–62 (see note 1 above), vol. IX (repr. 1984), XXXIII.xxxviii.116; XXXV.xii.30.
3. Smith, C.S. and Hawthorne, J.G. (1974) 'Mappae clavicula: A little key to the world of medieval techniques', *Transactions of the American Philosophical Society*, N.S., 64, 4, pp. 3–128, ch. 41, p. 34; Merrifield, M.P. (1849) *Original Treatises dating from the XIIth to XVIIIth Centuries on the Arts of Painting*, 2 vols, London (reprinted New York, Dover 1967), vol. I, Manuscripts of Jehan Le Begue, Heraclius III, pp. 52, 56 and 58.
4. Straub, R.E. (1965) 'Der Traktat de Clarea in der Burgerbibliothek Bern', *Schweizerisches Institut für Kunstwissenschaft, Jahresbericht 1964*, pp. 89–114; p. 94: 'De ista clarea temperatur vermiculum, minium, safranum et sanguis draconis.'
5. Thompson, D.V. (1926) 'Liber de coloribus illuminatorum siue pictorum, from Sloane MS no. 1754', *Speculum* I, pp. 280–307; pp. 288–91: 'Colores in percameno clari et spissi: hii sunt... sanguis draconis ...'; Smith and Hawthorne 1974 (see note 3 above), no. 8, p. 27.
6. Thompson 1926 (see note 5 above), pp. 288–91.
7. '*Liber diversarum arcium*' [Montpellier, Ecole de Médecine, MS 277, 14th–15th century, text no. 17, ff. 81v–100v] (1849) *Catalogue général*

des manuscrits des bibliothèques publiques des départements, vol. I, Paris, pp. 739–811; p. 756: 'Sanguis draconis dicunt quidam quod sucus herbe sit, quod frivolum est; est autem gummi arboris in Persia et India nascentis.'

8. Merrifield 1849 (see note 3 above), vol. I, Manuscripts of Jehan Le Begue, *Tabula de vocabulis sinonimis*, p. 25: 'Drachonis sanguis est color morellus seu rubeus obscurus.'
9. Cennini, C. (1954) *The Craftsman's Handbook: The Italian 'Il libro dell' arte'*, tr. D.V. Thompson Jr, New York (Dover Publications), p. 26.
10. Merrifield 1849 (see note 3 above), vol. II, *Segreti per colori* (the Bolognese Manuscript), pp. 448–9 (at the end of a recipe to make a lake from sticklac).
11. Merrifield 1849 (see note 3 above), vol. II, *Ricette per far ogni sorte di colore* (the Paduan Manuscript), no. 131, pp. 706–7.
12. Mell, C.D. (1930) 'Dragon's blood of Mexico', *The Textile Colorist*, 52, p. 397.
13. Tavernier, R. (1989) 'De Drakenboom: enkele sprokkelingen over een ongewone plant', *Ex Officina*, 2, 6, pp. 76–85.
14. Tavernier 1989 (see note 13 above), p. 83.
15. Shan, Y. (1974) 'A new plant resource in Yunnan – the "dragon blood" (*Dracaena cambodiana*)', *Acta Botanica Sinensis* 15, pp. 126–7 (trans. from Shan, Y. (1973) in *Chih-wu Hsueh-pao* 15, pp. 145–6).
16. Dr J.J. Bos, Department of Plant Taxonomy of the Agricultural University Wageningen, the Netherlands, personal communication, 21 September 1995.
17. Darwin, C. (1958) *The Autobiography of Charles Darwin and Selected Letters*, ed. F. Darwin, New York (Dover Publications), p. 121 (first published London, John Murray, 1892).
18. Watin, J.F. (1774) *Der Staffirmaler, oder die Kunst anzustreichen, zu vergolden und zu lackiren ...*, trans. from 2nd French edn, Leipzig, p. 192.
19. Rauwenhoff, N. (1864) 'Bijdrage tot de kennis van *Dracaena draco* L.', *Verhandelingen der Koninklijke Akademie van Wetenschappen* 10 (Amsterdam), 54 pp.
20. Symon, D.E. (1974) 'The growth of *Dracaena draco*, dragons-blood tree', *Journal of the Arnold Arboretum of Harvard University* 55, pp. 51–8.
21. Connelly, K. (1995) 'El Alisal: ten (or 98) years old and still growing', *Southern California Gardener* 5, pp. 2–4.
22. Several attempts to collect a red exudate from fresh cuts in the *Dracaena* tree in the Lummis Garden, Pasadena, California and from trees in San Diego, California, gave very poor results. Similar experiences were had in botanical gardens in Leiden and Delft. The resin seen in Plate 18 had collected on the edge of an old damage to the bark.
23. Jiang, D.F., Ma, P., Wang, X.H., Zhang, L.Q., Li, Q.D., Wang, J.L., Cheng, Z.Y. and Yang, C.R. (1995) 'The study of fungal population and relationships between fungi and forming of dragon's blood resin in *Dracaena cochinchinensis*', *Acta Botanica Yunnanica* 17, pp. 79–82 (from Biosis abstracts no. 98335155).
24. Spek, T.M. van der (2002) 'Medicijnen door de eeuwen heen', *Medicijnen van vroeger; Bijsluiter bij de tentoonstelling Drakebloed & Bevergeil*, exhibition catalogue, Museum Boerhaave, Leiden, pp. 19–31, esp. 28. Dragon's blood from *Dracaena* spp. was used as an astringent in cases of diarrhoea and dysentery, as well as an antiseptic, haemostatic and anti-ulcer remedy. Some of the homoisoflavans in *Dracaena* dragon's blood have antioxidative properties: see Badib, A.S. (1991), *The Medicinal Plants of Yemen*, Sanaa (Maktabah Alirsad), p. 15.
25. Rätsch, C. (1990) *Pflanzen der Liebe*, Bern-Stuttgart.
26. Buonanni, F. (1742) *Verhandeling over de vernissen, waar in de wyze opgegeven wordt, om 'er een toe te stellen dat 't Chinees vernis volmaakt gelykt ...*, Leyden, p. 48 (first published Rome 1720).
27. *Leerboek der Verlakkunst, of praktische handleiding om alle soorten van Lakken en Vernissen te bereiden* (1839) tr. W. Thomson, rev. edn by T.G. Entrup Bavink, Groningen, p. 267.
28. Tschirch, A. and Stock, E. (1933–36) 'Palmdrachenblut', in *Die Harze, Die botanischen und chemischen Grundlagen unserer Kenntnisse über die Bildung, die Entwicklung und die Zusammensetzung der pflanzlichen Exkrete*, 3rd edn of A. Tschirch, *Die Harze und die Harzbehälter*, by W.O.A. Tschirch, 2 vols in 4 parts, Berlin, vol. II, part 1 (1935), pp. 140–55, esp. p. 146. See also Howes, F.N. (1949) *Vegetable Gums and Resins*, Waltham, Mass. (Chronica Botanica), pp. 139–40.
29. Edwards, H.G., Farnell, P.W. and Quye, A. (1997) 'Dragon's blood: characterization of an ancient resin using Fourier transform Raman spectroscopy', *Journal of Raman Spectroscopy*

28, pp. 243–9; Edwards, H.G., Oliveira L.F. de and Quye, A. (2001) 'Raman spectroscopy of coloured resins used in antiquity – dragon's blood and related substances', *Spectrochimica Acta A: Molecular and Biomolecular Spectroscopy* 57, 14, pp. 2831–42; Edwards, H.G., Oliveira L.F. de and Prendergast, H.D. (2004)'Raman spectroscopic analysis of dragon's blood resins – basis for distinguishing between *Dracaena* (Convallariaceae), *Daemonorops* (Palmae) and *Croton* (Euphorbiaceae),' *Analyst* 129, pp. 134–8.

30. Greshoff, M. (1900) '*Jatropha curcas* L.', *Indische Mercuur* (Amsterdam) 23, 19, pp. 315–16, 320.
31. Hernández, F. (1959) *Historia natural de Nueva España* (Madrid 1570), *Obras completas*, vols II and III, Mexico (Universidad Nacional de Mexico), vol. II, *Historia de las plantas de Nueva España*, I, 213, CLVI; for the use of dragon's blood as a colorant see Hernández, F. (1651) *Rerum medicarum Novae Hispaniae thesaurus, seu plantarum, animalium, mineralium mexicanorum historia*, Rome, no. 8788.
32. Masaoud, M., Ripperger, H., Himmelreich, U. and Adam, G. (1995) 'Cinnabarone, a biflavonoid from dragon's blood of *Dracaena cinnabari*', *Phytochemistry* 38, 3, pp. 751–3.
33. Himmelreich, U., Masaoud, M., Adam, G. and Ripperger, H. (1995) 'Damalachawin, a triflavonoid of a new structurural type from dragon's blood of *Dracaena cinnabari*', *Phytochemistry*, 39, 4, pp. 949–51.
34. Merlini, L. and Nasini, G. (1976) 'Constituents of dragon's blood. Part II. Structure and oxidative conversion of a novel secobiflavonoid', *Journal of the Chemical Society, Perkin Transactions I* (1976), pp. 1570–76.
35. Edwards *et al.* 1997 (see note 29 above), pp. 246 and 248.
36. Pliny 1938–62 (see note 1 above), vol. IX (repr. 1984), XXXIII.xxxix.117: 'aduteratur sanguinine caprino aut sorbis tritis'.
37. MS Sloane 416 (London, British Library, 15th century): see Tosatti, B.S. (1991) *Il Manoscritto Veneziano; un manuale di pittura e altre arti ... del quattrocento*, Milan, p. 221.
38. Matthioli, P.A. (1565) *Commentarii in libros sex Pedacii Dioscoridis Anazarbei de medica materia*, Venice: 'ci si porta contrafatto in pani co' sangue di becco, mattone pesto, e sorba secche'.
39. Pomet, P. (1717) *Der aufrichtige Materialist und Specerey-Händler*, Leipzig (reprinted Weinheim, 1987), pp. 389–92, figs 288 and 289.
40. Eikelenberg, S. (*c.*1690–1720) 'Aantekeningen over schilderkunst' (the Eikelenberg Manuscript), Regionaal Archief Alkmaar, collectie aanwinsten, MS nr 390–394, f. 847; see also ff. 63 and 667.
41. Tschirch and Stock 1935 (see note 28 above), p. 144.
42. Wallert, A. (1995) 'Identification of flavonoid type yellows in small paint samples by solution absortion and fluorescence spectrophotometry', *Dyes in History and Archaeology* 13, pp. 46–58.
43. Schweppe, H. (1993) *Handbuch der Naturfarbstoffe: Vorkommen, Verwendung, Nachweis*, Landsberg, p. 642.

Some Aspects of Medieval and Renaissance Lake Pigment Technology

Jo Kirby

Abstract Red lake pigments prepared from the dyestuffs extracted from lac, kermes and madder were widely used for easel painting in northern Europe and Italy during the period 1200–1600. Recipes for lac lakes suggest that dyestuff was extracted directly from the resin-like raw material using alkali and pigment analysis has provided evidence to support this. In contrast, written sources indicate that textile shearings were used as a source of kermes dyestuff. Experimental evidence suggests that this may be one explanation for the small proportion of flavokermesic acid often observed in 15th- and 16th-century kermes lakes. An understanding of the production of madder lakes is complicated by the lack of documentary evidence and the variable composition of the extracted madder dye.

Introduction

The red lake pigments, prepared from natural dyestuffs by precipitating them onto a suitable substrate insoluble in the paint medium, have clear links with the dyeing industry. Over a long period of time, the same range of dyestuffs was used for both pigments and dyed materials: lac, kermes, the Old and New World cochineals, madder and brazilwood. The most valuable sources of information on the technology of lake pigment making are recipes, the number and variety of which differ for each dyestuff. Broadly speaking, throughout the period 1200–1600 and across western Europe, there are more recipes for brazilwood pigments than any other. The only other dyestuff to feature in recipes from most parts of Europe throughout this time is lac, although it is mentioned less frequently in later sources. On the other hand, madder is very rarely mentioned at all.

The ways in which the dyestuffs are treated also differ. Brazilwood and sticklac were used directly as the source of dyestuff for pigment-making throughout the period, but, after Roman and early medieval times, reference is rarely made to kermes and madder as direct sources of dye until the early 17th century.[1] In the case of kermes in particular, a study of the recipes suggests that, from the late 14th or early 15th century, shearings or clippings of dyed textile were used as the source of dyestuff, rather than the scale insect itself: a clear link between the dyeing industry and the pigment makers and an economical recycling of expensive dyestuff. In Italian sources, shearings of fabric dyed with *grana* (grain: kermes, *Kermes vermilio* Planchon) are often stipulated and the resulting pigment is the *lacca di cimatura di grana*. In many recipes, however, particularly those from northern Europe, the dyestuff used on the original textile is not specified.

An understanding of the technology of lake pigment production from a study of the recipes can be supplemented by analysis of the pigments themselves. As samples obtained from easel paintings are usually small, the information that can be gained is limited: often only the major components of the dyestuff (or those that give signals strong enough to be monitored by the instrument detector) can be detected. However it may still be possible to learn something about the way in which the pigment was made: for example, the dyestuff constituents present may be different, quantitatively or qualitatively, in dye that has been extracted directly from the raw material and dye extracted from textile matter. Three dyestuffs commonly identified in red lake pigments used during the period under discussion are those extracted from the scale insects lac (*Kerria lacca* Kerr) and kermes (*Kermes vermilio* Planchon), and from the madder root (*Rubia tinctorum* L.). Examination of samples from medieval and Renaissance paintings by high-performance liquid chromatography (HPLC) has revealed features indicative of the methods used to make pigments from these dyestuffs. Although recipes for pigments containing the dyestuff extracted from species of brazilwood (*Caesalpinia* spp.) are common, these lakes have so far been identified rather infrequently in easel paintings, perhaps because the dyestuff is so impermanent, perhaps because the lakes may have been more regularly used in other types of painting. These pigments will therefore not be discussed here.

A common feature of most recipes for making red lakes during this period is that the dyestuff was extracted in alkali; alum (aluminium potassium sulfate, $AlK(SO_4)_2{\cdot}12H_2O$) was then added to precipitate the pigment, the substrate being a form of amorphous hydrated alumina. The strength of the alkali used is an important factor to which attention is often drawn in the recipes; it will be seen that it may influence the chemical constitution of the final pigment. Medieval alkalis were derived from a range of materials, including urine (often stale), wood ash, calcined wine lees and lime, or a combination of any of these, soaked in water. The alkaline salt in wood ash, for example, is essentially potassium carbonate (K_2CO_3); if a solution prepared from wood ash is left to stand over quick lime (CaO), a stronger alkali containing potassium hydroxide (KOH) is the result.[2] The actual concentration of medieval or Renaissance alkalis in present-day terms is more difficult to assess. A widely used contemporary test consisted of dipping a feather into the solution: if the plume of the feather could be stripped off its central quill after exposure to the alkali, the solution was deemed sufficiently strong.[3] Alkalis made in the National Gallery laboratory using twigs from oak, beech and other English woodland deciduous trees had a pH of about 11 to 12 (equivalent to that of an aqueous solution of commercial potassium carbonate) before treatment with quicklime and 14 (equivalent to potassium hydroxide solution) afterwards. Their strengths of about 0.07–0.1M were strong enough to cause individual barbs of the feather to begin to melt together, but not so strong that they could be stripped from the quill, unless the treatment was prolonged.

Lac lakes

Recipes for pigments based on lac dyestuff – together with kermes and madder – appear in European recipe collections dating from around the late 8th century and increase in number and sophistication from this time onwards.[4] The number of recipes for the pigment and its frequent identification in paintings of the 13th–16th centuries indicate its importance as a source of dyestuff for dark red pigments. It was also used to dye leather and textiles.[5]

The water-soluble constituents of lac dyestuff are the laccaic acids, the principal components being laccaic acids A and B. Laccaic acid D, also known as flavokermesic acid, is the simplest structurally speaking and is also a constituent of the kermes dye and that of the Old World cochineals. Other anthraquinones, similar in structure to laccaic acid D but lacking the carboxylic acid group, are also present in sticklac, the raw material: erythrolaccin, isoerythrolaccin and deoxyerythrolaccin.[6] These are soluble in alkaline solutions. The resin-like material secreted by the lac insect as a protective coating is also alkali-soluble. This became a valuable commodity in its own right: cleaned and with the dyestuff removed it is known as shellac, but its use for varnishes and lacquers seems to have been unknown in Europe before the 17th century.

As discussed above, according to the recipes, the dyestuff was extracted from the raw material with alkali and precipitated by the addition of alum to make the lake. This seems to have been general practice all over Europe, from the earliest recipes onwards. There is no documentary evidence for the use of the indirect method, by means of shearings of dyed fabric, although the essentially anonymous nature of this method means it cannot be ruled out for any dyestuff. The significant variations in the recipes are concerned with the strength of the alkali, the length of time for which the alkali-sensitive sticklac is exposed to it, and the degree of heating. This can be illustrated by comparing two recipes from well-known 15th-century recipe collections: one is Italian, from the so-called Bolognese Manuscript,[7] the other German, from a collection known as the Strassburg Manuscript.[8]

In the Bolognese Manuscript recipe, hot alkali made from stale urine (a source of ammonia) and ashes (potassium carbonate) was poured over finely ground sticklac (gum lac) in a glazed pot and stirred. The purple solution was then poured off. Powdered roche alum was dissolved in water and added to the hot alkaline dyestuff solution until effervescence occurred, precipitating the lake. In the Strassburg Manuscript recipe, for a pigment named 'Paris red', a wood-ash alkali was also used and the gum lac was similarly ground to a fine powder, which was sprinkled into the hot alkali, stirred and left to soak overnight. The following morning it was boiled briefly 'for half as long as one would boil fish', with stirring, before the completion of the recipe by the addition of a little alum, filtering the coloured solution and precipitating the lake with alum. If lac is boiled in water, it forms a lump which impedes the release of the dyestuff. The ground material left to stand in alkali tends to coalesce; some dissolution of the shellac constituents may also occur. If this mixture is boiled the gum lac dissolves. When the lake is precipitated, the dissolved shellac is also brought down as an integral part of the pigment. In this case, the presence of alkali-soluble anthraquinones would be expected.

Analysis of samples

Lakes were prepared in the laboratory following both methods (see the Appendix). For comparison, a lake was also made by treating the ground sticklac with hot distilled water to dissolve the dyestuff, adding potassium carbonate to the dyestuff solution and precipitating the lake with alum. To extract the dyestuff, 5–10 μL 4% boron trifluoride/methanol reagent (BF_3/MeOH) was added to a small sample of each dry powdered lake.[9] The samples were then left overnight. The following morning the dyestuff solutions were diluted with a few microlitres of methanol if necessary, centrifuged and analysed by high-performance liquid chromatography (HPLC) using Hewlett Packard HP1100 equipment as described in the Appendix.

The chromatograms given by the three lakes are illustrated in Figure 1. In all three, the principal constituent is laccaic acid A, seen as the free acid and as its two methylated derivatives; a similar series is present for laccaic acid B. Some laccaic acid D – flavokermesic acid – can be detected in all three lakes; its methylated derivative elutes some 30 minutes later, at about 146 minutes under the present conditions. In the lake prepared according to the Strassburg Manuscript recipe (2c), a substantial, poorly resolved feature is present in this region, comprising at least two constituents. Methylated laccaic acid D is present in the first, smaller, region. The second part of this feature (retention time *c.*148 minutes) has spectral characteristics close to those published for erythrolaccin (Fig. 2).[10] This constituent is detectable (and more cleanly resolved) in the lake prepared according to the gentle alkaline extraction described in the Bolognese Manuscript recipe (L2), but there is markedly less of it.[11] It is absent in the lake prepared by aqueous extraction (L6). Deoxyerythrolaccin and isoerythrolaccin may also elute in this region, perhaps with retention times similar to that of methylated laccaic acid D, but unfortunately no samples of these compounds were available for comparison. Figure 3 illustrates the chromatograms given by lac lakes used in painting the *Virgin and Child* from Dal, dated about 1250–75 (Oslo, University Museum of Cultural Heritage; Plate 19), and *The Story of Papirius*, painted by the Sienese painter, Domenico Beccafumi, around the mid-1520s (London, The National Gallery; Plate 20). These are very similar to those given by the experimental samples.

Analysis of red lac lake pigments from easel paintings and polychromed sculpture of the 13th–16th centuries, from northern Europe and Italy, indicates that erythrolaccin is frequently present, often in quite substantial amounts, confirming the use of alkaline extraction of the dyestuff in the preparation of the lakes. Table 1 illustrates this in the form of the ratio between the estimated amount of alkali-soluble constituents present, together with laccaic acid D (to allow for the poor separation between these constituents and methylated laccaic acid D), and the estimated amounts of laccaic acids A and B. In many of the early pigments, the erytholaccin content is substantial, even larger than that observed in the pigment prepared following the Strassburg Manuscript recipe: in the case of the Tingelstad frontal, the amount of alkali-soluble material (and laccaic acid D) present is actually greater than that of laccaic acids A and B. While it should be remembered that some fading of the dyestuffs may have taken place and that conceivably A and B are less light-stable than erythrolaccin, this also suggests that the treatment of the sticklac was quite intense: it may have been subjected to prolonged soaking and probably boiling, resulting in the incorporation of alkali-soluble dyestuff constituents and some shellac in the final pigment. As erythrolaccin is orange, this might be expected to have an effect on the final colour, giving a less purple-red. Even the careful treatment recorded in the Bolognese Manuscript recipe results in the incorporation of some alkali-soluble material, but the amount present as the consequence of a very gentle extraction may not be detectable in a small paint sample. The amount present in the pigment used for the Dal Madonna is perhaps unusually low by comparison with that in the other Norwegian samples, which are similar in date. In this case, the binding medium had been extracted from the paint sample using TMTFTH (3-(trifluoromethyl)phenyltrimethylammonium hydroxide) before dyestuff analysis was carried out and it is possible that this treatment may have affected the dyestuff content to some extent.

Lakes from shearings

Alkali that can damage a feather is also strong enough to affect shearings of a proteinaceous textile, such as wool or silk. This property was used in the preparation of recycled dyestuff to be used for dyeing lower quality textiles such as blankets or aprons: the *bourre* mentioned in certain French sources,[12] and the Dutch *volle greijn*.[13] In these cases the alkali used was strong enough to reduce the textile matter to a jelly-like consistency. In recipes for pigments using dyestuff extracted from textile material, the alkali used was not always of this strength. In another 15th-century recipe from the Bolognese Manuscript, for *lacca di cimatura* from a woollen cloth material, the textile material was boiled gently in the alkali for a length of time measured as '2 *paternosters*' – so several minutes at least – and the alkaline dyestuff solution was strained off. It was boiled, added again to the shearings on the strainer, then filtered off by pressing the shearings hard with the hand. The lake was precipitated by the addition of powdered roche alum to the coloured alkaline solution.[14] If the alkali had been strong enough to reduce the shearings to jelly, pressing them would have been impossible. Alkali strong enough to strip the plume of a feather from its quill is, however, used in another German recipe – again for so-called

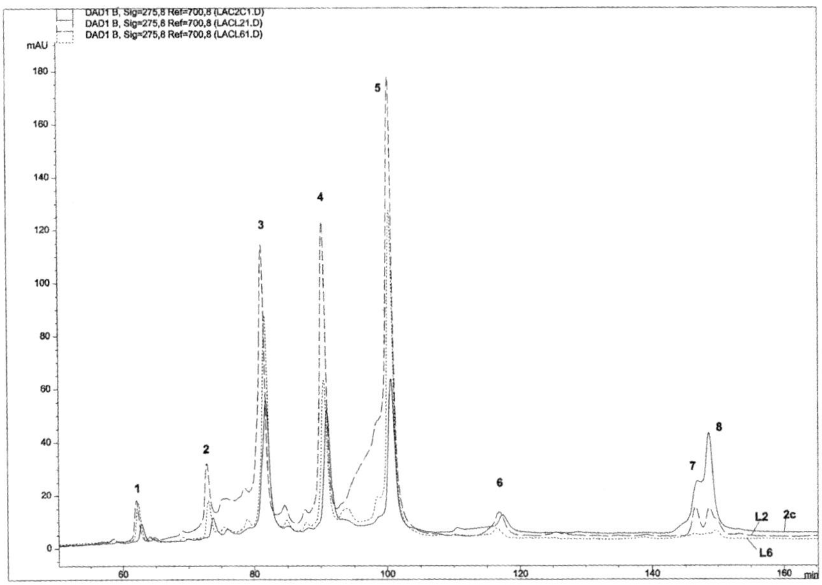

Figure 1 Chromatograms of lac lakes made using dyestuff extracted by boiling sticklac in alkali (2c), pouring hot alkali over sticklac (L2) and aqueous extraction (L6), recorded at 275 nm; detail of region from 60 to 160 min. 1: laccaic acid B; 2: B + 1 methyl group; 3: laccaic acid A, here co-eluting with B + 2 methyl groups; 4: A + 1 methyl group; 5: A + 2 methyl groups; 6: laccaic acid D (flavokermesic acid); 7: D + 1 methyl group; 8: erythrolaccin.

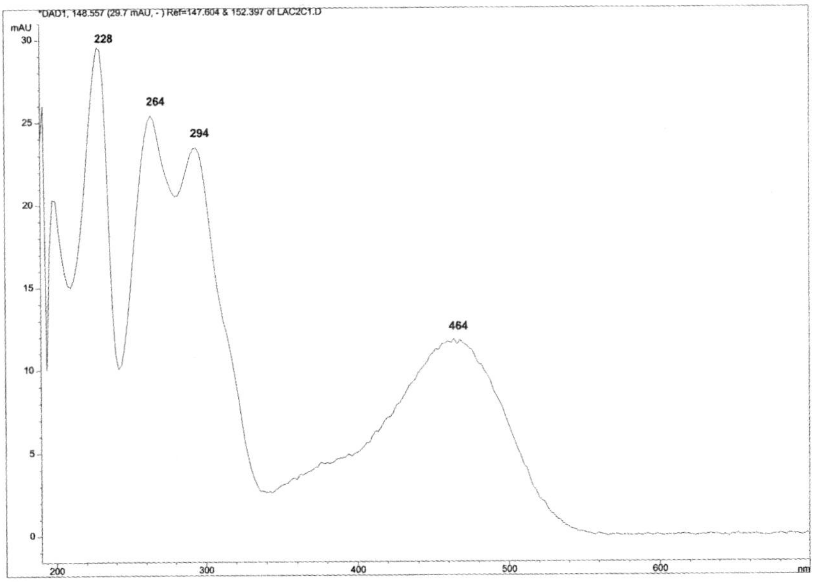

Figure 2 UV-visible spectrum of constituent t_R *c*.148 min, erythrolaccin (λ_{max} 228, 264, 294, 464; published values λ_{max} (EtOH) 228, 266, 295, 466 nm [log ε 4.26, 4.18, 4.13, 3.74]).

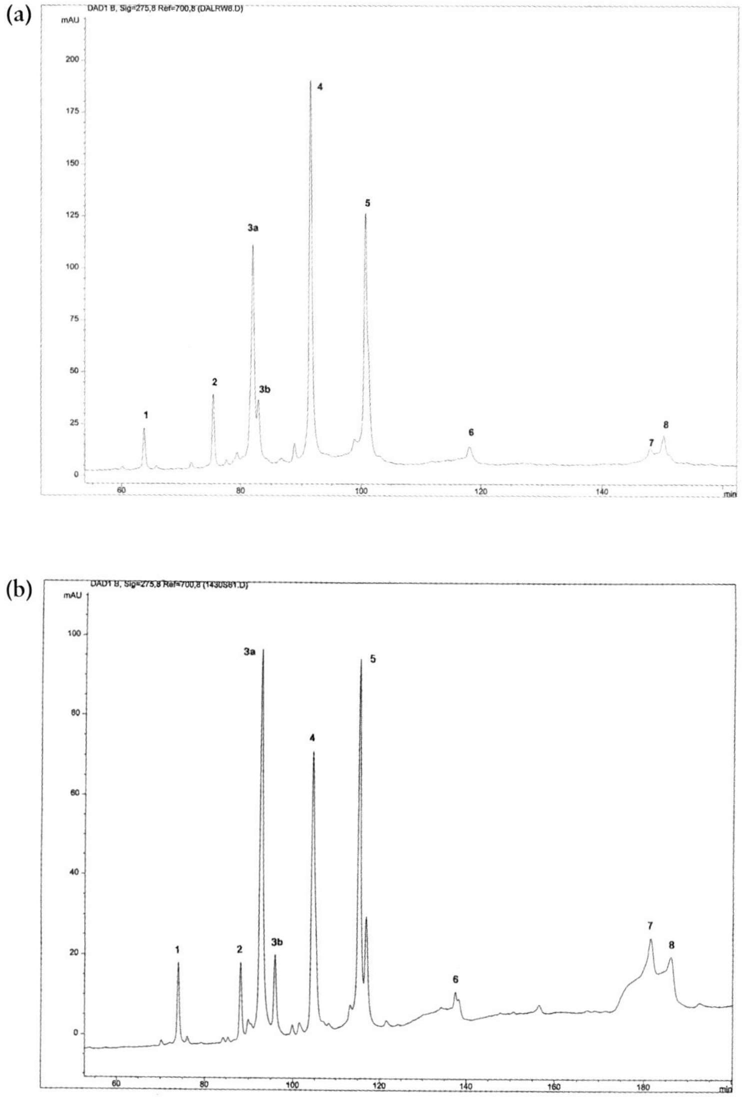

Figure 3 (a) *Virgin and Child*, Dal (Oslo, University Museum of Cultural Heritage, Inv. no. C 1410), *c.*1250–75; chromatogram of lake used for jewel in crown, after extraction of paint medium with TMTFTH; (b) Domenico Beccafumi, *The Story of Papirius* (London, The National Gallery, NG 1430), *c.*1540–50; chromatogram of lake used for crimson drapery of woman in background: both recorded at 275 nm, detail of region from 60 to 180 min (note: longer retention times for this sample). 1: laccaic acid B; 2: B + 1 methyl group; 3a: laccaic acid A; 3b: B + 2 methyl groups; 4: A + 1 methyl group; 5: A + 2 methyl groups; 6: laccaic acid D (flavokermesic acid); 7: D + 1 methyl group; 8: erythrolaccin.

Table 1 Examples of lac lakes containing alkali-soluble dyestuff constituents.

Artist and school	Title and date	Area examined	(D + alkali-soluble constituents)/ A + B) [a]
English School	Westminster retable, *c.*1260–80 (London, Westminster Abbey)	scene IV: dark red shadow on drapery of left-hand figure	0.155
English School	Thornham Parva retable, *The Crucifixion with Saints*, *c.*1330 (Thornham Parva, Suffolk, via the Hamilton Kerr Institute, University of Cambridge)[1]	St Peter's robe	0.925
English School	*The Virgin*, *c.*1380 (Colchester Castle Museum)	drapery of angel above Virgin's head	0.552
Norwegian School	*Virgin and Child*, *c.*1250 (Hedalen parish church, Oppland, formerly on loan to University Museum of Cultural Heritage, Oslo, Inv. no. C 32476)[b]	red glaze of modelling on throne	0.473
Norwegian School[2]	Heddal frontal, *Christ in Majesty*, 1250–74 (Oslo, University Museum of Cultural Heritage, Inv. no. C 34746)	red mantle of figure on far left, upper row	0.869[c]
Norwegian School	*Virgin and Child* from Dal,[3] 1250–75 (Oslo, University Museum of Cultural Heritage, Inv. no. C 1410)[b]	red glaze on imitation precious stone[d]	0.076
Norwegian School[2]	Tingelstad frontal (I), *Virgin and Child, with Scenes from the Life of the Virgin*, 1275–1300 (Oslo, University Museum of Cultural Heritage, Inv. no. C 5040)	central scene: red glaze from area below Virgin's throne	1.619[c]
Domenico Beccafumi (Sienese School)	*The Story of Papirius*, mid-1520s (London, National Gallery, NG 1430)	crimson drapery of woman in background	0.414
Experimental lakes	Strassburg Ms recipe (2c): sticklac boiled with alkali	–	0.448
	Bolognese Ms recipe (L2): hot alkaline extraction from sticklac	–	0.093
	Lake made using hot aqueous extraction from sticklac (L6)	–	0.029

Notes

a: A, B, D = laccaic acids A, B, D. Dyestuff extraction from paint samples using boron trifluoride/methanol gives the free acids and their methylated derivatives, inconvenient for quantitative analyses, particularly in the case of tiny samples, thus this ratio can only be approximate.

b: Polychromed sculpture.

c: No laccaic acid B constituents detectable.

d: Paint medium (linseed oil) extracted with TMTFTH before dyestuff analysis.

References

1. Bucklow, S. and Kempski, M. (2003) 'The paint layers', in *The Thornham Parva Retable: Technique, Conservation and Context of an English Medieval Painting*, ed. A. Massing, London (Harvey Miller), p. 56.
2. Plahter, U., Morgan, N., Hohler, E. and Wichstrøm, A. (2004), *Norwegian Altar Frontals 1250–1350*, London (Archetype); vol. 2, Plahter, U. *Materials and Technique*, pp. 148–9, 269–72, 224–8.
3. Plahter, U. and Wijk, S.A. (1970) 'The Virgin and Child from the church of Dal (Norway) – examination and restoration', *Studies in Conservation* 15, pp. 304–15.

Paris red – from the 15th-century Nuremburg Kunstbuch.[15] The red wool was boiled in the alkali 'until you see nothing but a red water' and no hairs were visible. It was filtered through a thick cloth, washed well, then ground with alum (which would neutralise the alkali), boiled in water and filtered.

Analysis of samples

In order to test the recipes, samples of wool and silk yarns were dyed with madder and kermes respectively, using an alum mordant (see the Appendix). The recipe used for the silk dyeing was based on one in Gioanventura Rosetti's

Plictho de larte de tentori (Venice 1548) and dyeing was carried out at about 60 °C.[16] A second dyeing on wool was carried out, but this was not used for pigment preparation. Ground '*krap*' grade madder was used for the madder dyeing, carried out at about 60 °C as before, following a modern recipe.[17] Two further dyeings were carried out, but these wools were not used for pigment preparation. Portions of the dyed yarns were then clipped very finely with scissors and lake pigments were made following the two recipes. Following the Bolognese Manuscript recipe, 0.1M potassium carbonate solution (about pH 11) was used for the silk clippings. The wool clippings were divided into two portions, one of which was treated with 0.1M potassium carbonate solution, the other with 1M potassium hydroxide solution (pH 14). This may be slightly more concentrated than that described in the German recipe, as, when tested, the feather began to dissolve.

The 0.1M potassium carbonate solution extracted the kermes dyestuff from the dyed silk almost completely, also causing the clippings to felt together a little. When the same concentration of alkali was used on madder-dyed wool a very incomplete extraction was achieved and the wool barely changed colour, although the clippings did felt together a little as before. The madder lake prepared by the addition of alum was, however, of surprisingly intense colour. Extraction of the dyestuff with the 1M alkali caused the wool to dissolve with an unpleasant sulfurous smell. The extraction was halted before complete dissolution had taken place; the red alkaline solution was filtered and the lake precipitated by the addition of alum as before. When this lake was examined by Fourier transform infrared microspectroscopy (FTIR microscopy) it was possible to assign certain bands to protein constituents – in other words, dissolved wool protein – within the pigment particles.[18] Claims have been made for the presence of fibres in lake pigments that would certainly be possible if a recipe of this type were used, although there are other reasons why textile fibres might be present by chance in a paint sample.[19]

Kermes lakes

The limited evidence available from the analysis of pigments suggests that lac lakes were used earlier than the *lacca di cimatura di grana* (see, for example, Tables 1 and 2).[20] The use of kermes lakes was at its peak in the 15th and 16th centuries and was not confined to Italy, as Table 2 shows, but it is unknown whether the pigments used by painters from northern Europe were imported from Italy, made in the north (kermes was widely imported for dyeing and medicinal purposes), or both.

The most important constituents of the dyestuff extracted from the adult female *Kermes vermilio* insect are the red kermesic acid and the yellow flavokermesic acid – laccaic acid D, discussed above. The lesser constituents include precursors of these compounds, which, depending on any pre-treatment the insects may have received and perhaps on their original source, could contribute a proportion of the colouring matter on the textile.[21] The aqueous extract from the Algerian kermes insects used for the dyeing experiments described contained small amounts of minor yellow and red dyestuff constituents, which subsequently appeared on yarns dyed in the bath and in the kermes lake prepared from the silk. Similarly small amounts of these precursors have been observed when using dyestuff extracted from other batches of insects.[22] Minor dyestuff constituents (possibly in methylated form) are also present in some of the pigments listed in Table 2, but it is not necessarily possible to correlate these with any observed in the test samples, partly because of the restrictions imposed by sample size, partly because the kermes used for the tabulated pigments may have received some pre-treatment. Kermes dyestuff also contains small amounts of non-dyeing constituents, some of which appear in lakes prepared directly from the insects, but not in lakes prepared from cloth shearings. In the paintings listed in Table 2, none of the samples examined contained such non-dyeing constituents, suggesting that the lakes were not prepared directly from the insects.

It has been found that, after acid hydrolysis, the percentage of flavokermesic acid in the extracts from adult female kermes insects varies from 15 to 27%. In the dyestuff found on textiles, however, the proportion of flavokermesic acid is variable, but considerably lower: in one published study of Florentine silk textiles the proportion varies from 6 to 12%.[23] Similarly, it seems that many of the kermes lakes found in paintings contain little or no detectable flavokermesic acid, as indicated in Table 2. An example is Domenico Ghirlandaio's *Portrait of a Young Man in Red*, painted in about 1480–90 (London, The National Gallery; Plate 21): the chromatogram of the dyestuff extracted from a sample from the sitter's red coat is illustrated in Figure 4. The explanation for this probably lies in the method of preparation from textile shearings, particularly, it seems, when these are from dyed silk.

Figure 5 shows HPLC chromatograms for the dyestuff extracted from kermes insects as used in the dye-bath described above, the dyestuff extracted from the lake pigment prepared from the dyed silk (both after treatment with 4% BF_3/MeOH) and the dyestuff remaining in the bath after this dyeing. To show the proportions of kermesic and flavokermesic acid present more clearly, two chromatograms, recorded or calculated at a wavelength at which each absorbs strongly – 491 nm (recorded) for kermesic acid and 430 nm (calculated from spectral data) for flavokermesic acid – are shown in each case. Comparing the chromatograms calculated at 430 nm, flavokermesic acid is clearly visible in the dyestuff itself as a small peak following closely behind the kermesic acid peak. However, the lake pigment extract contains largely kermesic acid; flavokermesic acid is only detectable on the trailing edge of the peak. A closely similar dyestuff profile is given by an extract of the dyed silk in 4% BF_3/MeOH. The residual bath after dyeing contains more flavokermesic acid; kermesic acid is barely detectable on the leading edge of the peak calculated for 430 nm and it can be seen from that recorded at 491 nm that

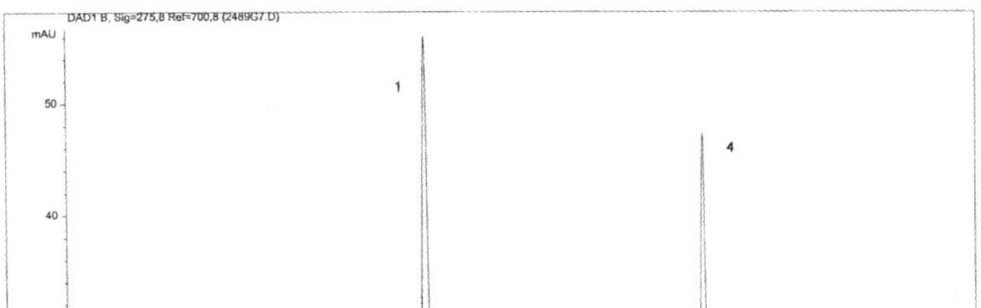

Figure 4 Domenico Ghirlandaio, *Portrait of a Young Man in Red* (London, The National Gallery, NG 2489), probably *c.*1480–90; chromatogram of lake used for red coat after extraction of paint medium with TMTFTH, recorded at 275 nm; detail of region from 80 to 180 min. 1: kermesic acid (no flavokermesic acid detectable); 2: minor kermes constituents; 3: residual traces of TMTFTH; 4: methylated kermesic acid.

Table 2 Kermes lakes in 15th- and 16th-century paintings.

Artist and school	Title and date	Area examined	Flavokermesic acid content
French School	*Battle of Pavia*, after 1525 (Oxford, Ashmolean Museum)	area of red paint protected by frame	trace only
German Schools			
Albrecht Altdorfer	*Christ Taking Leave of his Mother*, probably 1520 (London, National Gallery, NG 6463)	red lake on sleeve of woman, left-hand side	none detectable
Albrecht Dürer	*Suicide of Lucretia*, 1518 (Munich, Bayerische Staatsgemäldesammlungen, No. 705)[1]	red tablecloth	trace only
Italian Schools			
Domenico Beccafumi	*Marcia*, probably *c.*1520–25 (London, National Gallery, NG 6369)	(i) purple skirt, shadow on fold[a]	some present
		(ii) purple skirt, deepest shadow lower down on same fold[b]	none detectable
Sandro Botticelli	*The Virgin Adoring the Sleeping Christ Child*, *c.*1481–84 (Edinburgh, National Gallery of Scotland)	shadow on Virgin's red tunic	none detectable
Workshop of Botticelli	*The Virgin and Child with a Pomegranate*, probably *c.*1480–1500 (London, National Gallery, NG 2906)	red dress	none detectable
Giovanni Battista Cima da Conegliano	*Saint Catherine of Alexandria*, *c.*1502 (London, Wallace Collection)[2]	saint's dress	none detectable
Lorenzo Costa	*A Concert*, *c.*1485–95 (London, National Gallery, NG 2486)	sleeve of man on right-hand side	trace only
Florentine School	*Portrait of a Lady in Red*, probably 1460–70 (London, National Gallery, NG 585)	red dress	none detectable
Garofalo (Benvenuto Tisi)[3]	*The Virgin and Child with Saints Dominic and Catherine of Siena*, *c.*1500 (London, National Gallery, NG 3102)	red canopy of throne[c]	none detectable
	The Madonna and Child with Saints William of Aquitaine, Clare, Anthony of Padua and Francis, 1517 (London, National Gallery, NG 671)	Virgin's red robe	trace only
	Two Couples with Cupid, *c.* 535–45 (London, National Gallery, NG 1362)	red drapery around figures on right-hand side[d]	none detectable
Domenico Ghirlandaio	*Portrait of a Young Man in Red*, *c.*1480–90 (London, National Gallery, NG 2489)	red garment	none detectable
Giovanni del Ponte	*The Ascension of Saint John the Evangelist, with Saints, c.*1410–20 (London, National Gallery, NG 580)	glaze on fold of St John's robe	some present
Attributed to Lorenzo Marini da Cattaro	*The Annunciation*, mid-15th century (via Hamilton Kerr Institute, University of Cambridge)	red glaze over gold, angel's drapery	none detectable

Perugino	*The Archangel Raphael with Tobias*, *c*.1496–1500 (London, National Gallery, NG 288.3)	Raphael's red drapery	none detectable
	The Virgin and Child with Saints Jerome and Francis, *c*. 1507–15 (London, National Gallery, NG 1075)	St Jerome's red robe	none detectable
Raphael	*The Crucified Christ with the Virgin Mary, Saints and Angels* (*The Mond Crucifixion*), *c*.1503 (London, National Gallery, NG 3943)	St John's red cloak[c]	trace only
	The Madonna and Child with Saint John the Baptist and Saint Nicholas of Bari (*The Ansidei Madonna*), 1505 (London, National Gallery, NG 1171)	(i) St John's red drapery (ii) purplish shadow under throne canopy	none detectable none detectable
	Saint Catherine of Alexandria, *c*.1507–08 (London, National Gallery, NG 168)	red drapery[c]	trace only
Netherlandish Schools			
Hugo van der Goes	*The Adoration of the Magi* (*The Monforte Altarpiece*), *c*.1470 (Berlin, Staatlichen Museen Preussischen Kulturbesitz, Gemäldegalerie, No. 1718)	sleeve of king's garment, right-hand side	none detectable
Follower of Joos van Cleve	*The Crucifixion* (Triptych), probably 1525–35 (London, National Gallery, NG 1088.1–3)	central panel: glaze on St John's robe	trace only
Workshop of Marinus van Reymerswaele	*Two Tax Collectors*, *c*.1540 (London, National Gallery, NG 944)	(i) red sleeve of man, right-hand side[c] (ii) purple sleeve of man, left-hand side[e]	some present some present

Notes

a: Lac lake also present, perhaps in the underpaint mixed with lead white.

b: Kermes lake mixed with azurite; same pigments in underpaint mixed with lead white.

c: Madder lake in underpaint.

d: Kermes lake mixed with madder lake.

e: Kermes lake apparently applied over madder lake in deepest shadow, over azurite with lead white.

References

1. Goldberg, G., Heimberg, B. and Schawe, M. (1998) *Albrecht Dürer: Die Gemälde der Alten Pinakothek*, Munich, pp. 440–61.

2. Dunkerton, J. (2000) 'The restoration of two panels by Cima da Conegliano from the Wallace Collection', *National Gallery Technical Bulletin* 21, pp. 58–69, esp. p. 64.

3. Dunkerton, J., Penny, N. and Spring, M. (2002) 'The technique of Garofalo's paintings at the National Gallery', *National Gallery Technical Bulletin* 23, pp. 20–41, esp. pp. 34, 40–41.

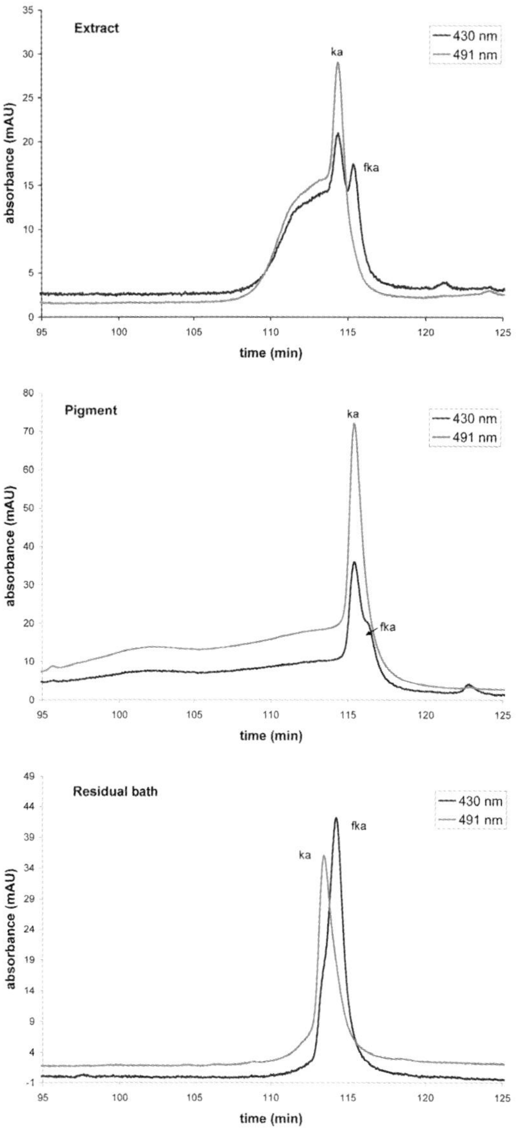

Figure 5 Chromatograms of dyestuff extracted from kermes insects, dyestuff extracted from the lake pigment prepared from kermes-dyed silk (both after treatment with 4% BF_3/MeOH for comparison) and dyestuff remaining in the bath after this dyeing, recorded at 491 nm and calculated for 430 nm using HP Chemstation software; detail of kermesic acid/ flavokermesic acid region. ka: kermesic acid; fka: flavokermesic acid.

kermesic acid is a less important constituent. A second dyeing from the dye-bath (on wool) was a warmer crimson than the silk and contained more flavokermesic acid. A lake made from the residual bath after this dyeing was brownish in colour and contained almost equal amounts of kermesic and flavokermesic acids. It is noticeable that kermes lakes made by way of cloth shearings (silk in particular, but also wool) are a more vibrant and slightly bluer crimson than those made directly from the scale insects themselves, which tend to be rather dull or brownish. This may have been as good a reason for using textile shearings, rather than the insects themselves, as any economic motivation.[24]

This implies that extracting the dyestuff from shearings of a textile from the first dyeing from the kermes bath gives a lake with a high kermesic acid content. However, the experimental study and earlier work carried out at the National Gallery suggest that the situation is rather less straightforward than this. In wool-dyeing experiments where wool was the first fibre dyed, rather than the second as described above, both kermesic and flavokermesic acids were taken up: wool takes up more dyestuff than silk in any case and, in particular, it appears that, from the same batch of insects, more flavokermesic acid is taken up by the wool than the silk. This is the case regardless of which sample of kermes was used and whether the wool was mordanted with alum alone, or with alum and cream of tartar (potassium hydrogen tartrate, the mono potassium salt of (+)-2,3,-dihydroxybutanedioic acid). The dyestuff profile of lake pigments prepared from dyed wool clippings is closely similar to that of the wools themselves. However, analysis of kermes dyestuff extracted from a different sample of silk, in this case dyed using an alum and cream of tartar mordant, gave a similar result to the silk yarn described above: the dyestuff consisted largely of kermesic acid and very little flavokermesic acid was present – the same pattern as that observed in the samples from paintings.

This, admittedly limited, study suggests that one explanation for the low proportion of flavokermesic acid found on silk textiles could be that initially kermesic acid is taken up preferentially. Clearly other factors may come into play: the dyestuff content and quality of insects used; pre-treatment of the insects or the silk; the mordants used; the temperature of dyeing and other conditions of the bath. The same pattern of constituents is then found in lakes prepared from the textiles. It has also been suggested that flavokermesic acid may be more vulnerable to ageing effects so has survived less well on the textiles.[25] However, some preliminary observations of faded lake pigments in laboratory samples and on paintings suggest that, on the contrary, as far as light-ageing is concerned, flavokermesic acid may be less vulnerable than kermesic acid.[26]

Occasionally kermes lakes showing a more typical kermesic acid/flavokermesic acid ratio are identified in paintings. In Domenico Beccafumi's *Marcia* (London, The National Gallery; Plate 22 and Fig. 6), painted in about 1520–25, an extraordinarily deep red kermes lake, containing a marked amount of flavokermesic acid, has been used for some, but not all, areas of shadow on the subject's purple dress (the deepest, bluest shadows are painted using a kermes lake low in flavokermesic acid, mixed with the blue pigment azurite). It is hard to say what relationship this lake has to the dye-bath. Possibly its dyestuff was extracted from a textile from a late dyeing. There is no evidence that it was prepared directly from the scale insects. A rather similar dyestuff pattern is seen in the kermes lake used in a painting of *Two Tax Collectors*, from the workshop of Marinus van Reymerswaele and dated to the 1540s.[27] These lakes, which may have appeared very slightly warmer in colour than usual, may have been cheaper or dearer: either factor may have influenced the artist's

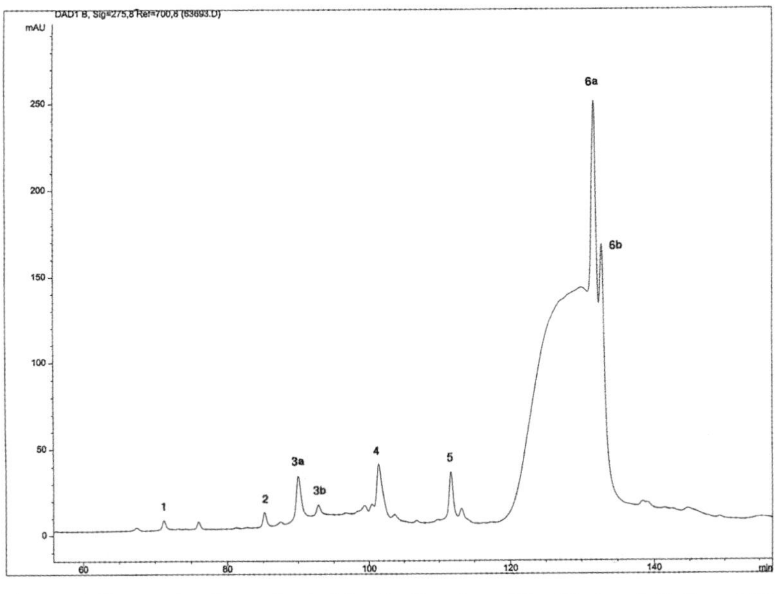

Figure 6 Domenico Beccafumi, *Marcia* (London, The National Gallery, NG 6369), *c.*1520–25; chromatogram of kermes lake used for reddish-purple shadow on purple dress with a lac lake probably in the underpaint, recorded at 275 nm; detail of region from 60 to 150 min (note: longer retention times for this sample). 1: laccaic acid B; 2: B + 1 methyl group; 3a: laccaic acid A; 3b: B + 2 methyl groups; 4: A + 1 methyl group; 5: A + 2 methyl groups; 6a: kermesic acid; 6b: flavokermesic acid (laccaic acid D). The contribution made by the small amount of lac dyestuff to the flavokermesic acid peak is negligible; any alkali-soluble lac constituents present are masked by the large methylated kermesic acid/flavokermesic acid peak (not shown).

choice. Beccafumi seems to have chosen subtle, unusual colour schemes for his paintings and his selection of this intensely coloured pigment may well have been on aesthetic grounds.

Clear evidence for the relationship between the dye-bath and the painter's pigment is not restricted to kermes dyestuff. The dyestuffs from both Polish cochineal (*Porphyrophora polonica* L.) and, later, New World cochineal (*Dactylopius coccus* Costa) have also been identified in lake pigments in late 15th- and 16th-century paintings. In several cases, a little ellagic acid is also present, probably derived from galls or another source of tannin used to weight the silk or to assist in the mordanting process.[28] One example is the pigment used by the Antwerp painter Joachim Beuckelaer in *The Four Elements: Air. A Poultry Market, with the Prodigal Son in the Background* (NG 6587, dated 1569–70), which contains New World cochineal dyestuff. There is no other reason for ellagic acid to be present in the pigment. Several examples of a single pigment containing a mixture of scale insect dyestuffs have also been found. These include a lake used in *The Adoration of the Kings* (NG 3098, about 1490), formerly attributed to the Venetian painter Vittore Carpaccio, but now thought to be by Giovanni Bellini, which contains a mixture of probably Polish cochineal and kermes dyestuffs. In a dyeing context, a mixture of this sort in the dye-bath would not be unexpected, although it could also derive from the use of a mixed bag of shearings.

Discussion: the case of madder lakes

Madder lakes were quite widely used in northern Europe and Italy throughout the 15th and 16th centuries, alone or beneath a layer of another, more crimson, lake (Table 3). It is thus curious that so very little is known of their manufacture. Madder lakes of this period tend to contain rather scant dyestuff, thus it is usually only possible to see the principal constituents: generally pseudopurpurin, sometimes with a little purpurin; alizarin is frequently undetectable. However, even when the sample from the painting is generous, the pattern of constituents present is often little different: alizarin is present in trace amounts or is apparently absent.[29] This is unlike the results obtained from contemporary madder-dyed textiles: in a study of Bruges textiles from the 16th and 17th centuries, for example, the proportion of alizarin detected varied between 15 and 95%, but was typically 30–45% (the other major constituent being purpurin), results typical of the dyestuff from madder (*Rubia tinctorum* L.).[30] The dyestuff content of some lake pigments has a pattern closer to that of wild madder (*R. peregrina* L.),[31] yet it seems improbable that this species should have been used on a large scale for pigment production while *R. tinctorum* was used for dyeing and was traded on a large scale all over Europe. While *R. peregrina* may have been used occasionally, another explanation for the pigment results may lie in the technology of the dye-bath and the particular characteristics of the madder dyestuff itself.

The principal anthraquinones in *R. tinctorum* are present as glucosides in the root itself: alizarin, for example, is found as alizarin-2-β-primeveroside (ruberythric acid), and pseudopurpurin as the 1-β-primeveroside (galiosin). Purpurin is largely generated by the gradual decarboxylation of pseudopurpurin (purpurin-3-carboxylic acid); little is present in the root. Ruberythric acid does not dye mordanted fibres;[32] for dyeing it is necessary to liberate the alizarin by enzymatic fermentation or hydrolysis, and the same applies to other anthraquinone glucosides. Galiosin is in fact less stable than ruberythric acid and easily hydrolysed. Fermentation takes place if, for example, dry ground madder is stored for some time. Enzymatic activity and hydrolysis of the glucosides can also take place in the dye-bath, depending on the temperature at which it is maintained and what other materials, such as bran, may be present. One English 14th- or 15th-century madder dyeing recipe recommends that the bath should be maintained at the temperature of milk (about 36–38 °C); in other cases the bath is heated to a higher temperature or even boiled, which would denature any enzymes present.[33] The composition of the bath

Table 3 Examples of madder lakes.

Artist and school	Title and date	Area examined	Alizarin content
English School	*The Virgin*, *c.*1380 (Colchester Castle Museum)	Virgin's robe	present[a]
Garofalo (Benvenuto Tisi)	*The Virgin and Child with Saints Dominic and Catherine of Siena*, *c.*1500 (London, National Gallery, NG 3102)	red canopy of throne[b]	present
	Two Couples with Cupid, *c.*1535–45 (London, National Gallery, NG 1362)	red drapery around figures on right-hand side[c]	none detectable
Raphael	*The Crucified Christ with the Virgin Mary, Saints and Angels* (*The Mond Crucifixion*), *c.*1503 (London, National Gallery, NG 3943)	St John's red cloak[b]	(present)[d]
	Saint Catherine of Alexandria, *c.*1507–08 (London, National Gallery, NG 168)	red drapery[b]	none detectable
Master of the Saint Barbara Legend	*A Man and his Wife* (Upton House, via the Hamilton Kerr Institute, University of Cambridge)	red dot over gold leaf	small amount
Workshop of Marinus van Reymerswaele	*Two Tax Collectors*, *c.*1540 (London, National Gallery, NG 944)	(i) red sleeve of man, right-hand side[b]	none detectable
		(ii) purple sleeve of man, left-hand side[b]	none detectable
Hans Eworth	*Portrait of Sir John Luttrell*, 1550 (London, Courtauld Institute Gallery)	Sir John Luttrell's colours	present

Notes
a: Traces of ruberythric acid and other glucosides present.
b: Madder lake in underpaint.
c: Kermes lake mixed with madder lake.
d: Alizarin crimson from retouching may be present as a contaminant.

is thus influenced by many factors and may have been quite variable, even before any dyeing took place. The composition may be altered initially as a result of the dyeing process if alizarin or, alternatively, pseudopurpurin/purpurin is taken up preferentially during the first and subsequent dyeings.[34] This might influence the composition of any sediment subsequently formed: as a hypothesis, if hydrolysis of the glucosides was complete and alizarin was then removed, the residual liquor in the bath and any precipitated material would effectively be richer in pseudopurpurin/purpurin.

There is in fact some scanty documentary evidence for the use of sediment or scum from the dye-bath as a pigment. The statutes of the painters' guild of Tournai (Doornik), 1480, include *fleurs de garance* (flowers of madder) in a list of pigments that panel painters were permitted to use.[35] This is not the 19th-century *fleurs de garance* (ground madder treated with dilute sulfuric acid until it began to ferment, then packed for use); the term may be analogous to the 15th- and 16th-century term 'flowers of woad' or *florey* (*florie* in French), referring to the indigo floating on top of the woad vat, which was skimmed off and used as a pigment. A very few 17th-century sources also refer to the practice.[36] The dyestuff content of such a material could be variable in the case of a madder or, indeed, kermes dye-bath and it is not clear what inorganic elements (relating to the alum mordant used on the cloth, or other substances added during dyeing) it might contain. The substrate of red lake pigments is usually hydrated alumina, sometimes with the addition of calcium salts or other materials, but it is interesting that the elemental analysis of the kermes lake used in the painting from Marinus van Reymerswaele's workshop was slightly unusual from this point of view, containing unexpectedly high amounts of potassium and sulfur. The sulfur and the presence of flavokermesic acid could relate to the use of wool clippings rather than silk as the source of dyestuff, the use of strong alkali, or, perhaps, a later dyeing from the bath. However it is possible to imagine circumstances – poorly rinsed mordanted textile in the dye-bath or a second treatment of textile dyed once – where unfixed alum in solution and dyestuff components would be able to combine and precipitate, giving rise to sediment or scum. This might be particularly so for the rather insoluble madder anthraquinones: pseudopurpurin, for example, although soluble in hot alum solution, reacts with alum to form an insoluble red precipitate, which can be observed coming out of solution if the temperature falls.

Chromatographic analysis of the dyestuff from the first experimental madder dyeing (described above) revealed the presence of pseudopurpurin and alizarin (about 30%), with small amounts of other anthraquinones. The pattern of constituents observed in the pigment prepared by extraction of the dyed wool with 0.1M alkali was the same. Allowing for the fact that the acid hydrolysis method of extraction used for the Bruges textiles causes hydrolysis of any pseudopurpurin present to purpurin, the average dyestuff profiles are similar. The fact that 15th- and 16th-century madder pigment samples may be so different in composition leads to the suspicion that madder-dyed wool was not the only source of dyestuff for their manufacture.

Madder lakes containing alizarin as well as pseudopurpurin are sometimes identified. An example is in the underpaint of the red canopy in Garofalo's *Virgin and Child with Saints Dominic and Catherine of Siena* (1500–1505, London, The National Gallery; see Table 3). However, like the kermes lake in the *Two Tax Collectors*, mentioned above, it has an unusual pattern of constituents in the substrate, including much sulfur. This may relate to the extraction of the dyestuff from dyed wool, or, as discussed above, there may be some other explanation.

The dyestuff used to make madder lakes need not have been a product of the dye-bath at all: it could have been obtained from the root itself. The presence of certain non-dyeing constituents, such as residual traces of anthraquinone glucosides, might be evidence for this, but this would depend on the conditions of extraction of the dyestuff. Their absence would not necessarily rule out the root as the immediate source of dye, particularly as the extract from a small paint sample might contain them in amounts too small to be detectable. However, a lake of this type may have been used for the Virgin's robe in the English painting of 1380 (Table 3), which contains traces of ruberythric acid and other glucosides.

Conclusion

Analysis of the dyestuff extracted from lake pigments used during the 13th–16th centuries has revealed clear evidence for the method of manufacture of lac lakes, by alkaline extraction from sticklac, and kermes lakes, from shearings of dyed silk. The presence of traces of ellagic acid and mixtures of kermes and cochineal (Old and New World) dyestuffs in lake pigments provides further evidence for the strong links between pigment and dye-bath. Madder lakes of this period are very variable in composition, sometimes containing no detectable alizarin and thus unlike madder-dyed wool in their dyestuff content. The intriguing possibility of the use of dye-bath sediment – if this is the material referred to in the Tournai guild statutes – requires investigation. More detailed analytical results are needed to clarify the methods of manufacture of madder lakes.[37]

Acknowledgements

Thanks are due to Melanie Brown (then a second year chemistry student at Edinburgh University), for assistance with the experimental work, and to Marika Spring, Catherine Higgitt and Rachel Grout (then Radcliffe Research Fellow), of the National Gallery Scientific Department, for examinations of pigment substrates; and not least to Spike Bucklow (Hamilton Kerr

Institute, University of Cambridge), Unn Plahter (Universitetets Oldsaksamlingen, Oslo) and Aviva Burnstock (Department of Technology and Conservation, Courtauld Institute of Art). The HPLC equipment (including the microflow processor), computer and software have been most generously lent to the National Gallery Scientific Department by Hewlett-Packard Ltd.

Appendix

Distilled water was used for all washing, dyeing and pigment preparations to avoid contamination by iron, calcium or other metal ions. All lakes were filtered, washed and dried in a desiccator after making.

Lac lakes

Sticklac was obtained from the Royal Botanic Gardens, Kew, and Ashill Colour Studio, Shefford, Bedfordshire (1991, now incorporated into P&M Woolcraft, Milton Keynes). For the Strassburg Manuscript lake, 3 g ground sticklac was left overnight in 0.2M sodium carbonate (Na_2CO_3) solution. The following day the solution was boiled, 1 g alum ($AlK(SO_4)_2{\cdot}12H_2O$) was added; the precipitate formed was filtered off and sufficient alum was added to complete the precipitation of the lake. For the Bolognese Manuscript lake, 50 g ground sticklac was treated with 200 mL approximately 0.3M hot potassium carbonate (K_2CO_3) solution, stirred and filtered rapidly. The residue was washed and the washings added to the original solution. The lake was precipitated by the addition of alum. For the lake made by aqueous extraction, 40 g ground sticklac was warmed gently in 800 mL water before the addition of 25 g K_2CO_3 in water, giving an approximately 0.2M solution. Sufficient of a solution of 60 g alum in water was added to precipitate the lake.

Kermes dyeing and lakes

The kermes used for dyeing (from Algeria) was supplied by Ashill Colour Studio in 1991. Silk yarn was obtained from the Handweavers' Studio, Haroldstone Road, London. For the dye-bath, 20 g of insects was simmered in distilled water for about an hour. 20 g silk was washed in a solution of soap flakes at 30–40 °C, rinsed well in several changes of water at the same temperature and mordanted in a solution of 5 g alum (so 25% by weight) in water at about 60 °C for about 45 minutes. It was left to cool in the mordant solution before rinsing well and was kept damp until it was dyed. The mordanted silk was placed in the dye-bath overnight. The bath was brought to a temperature of about 60 °C over one hour and allowed to dye for a similar time. It was left to cool, rinsed well and allowed to dry naturally. To make the lake, 10 g dyed silk was cut finely and warmed gently in a solution of 0.1M K_2CO_3 aq for about one hour, giving a purple solution. It was filtered to remove the silk residue and alum solution was added to precipitate the lake.

Madder dyeing and lakes

The wool used was supplied by J. Hyslop Bathgate & Co., Galashiels, in 1990. It was washed thoroughly in a hot soap flakes solution and rinsed well. Ground '*krap*' grade madder, grown and prepared by Dr D.J. Hill, Department of Environmental Biology, University of Bristol in 1997, was used for dyeing. 20 g wool was mordanted in a solution of 5 g alum in water at just below 100 °C for one hour, allowed to cool in the mordant solution and rinsed well. The madder was soaked overnight in distilled water with the mordanted wool. The following day, the bath containing the wool was heated to about 60 °C over a period of about one hour, maintained at this temperature for 45 minutes then allowed to cool. The wool was rinsed well and allowed to dry. To make the lakes, 10 g wool was divided into two 5 g portions which were clipped as before. The procedure described above was then followed, 0.1M K_2CO_3 aq being used to extract the dyestuff from one portion and 1m KOH aq for the other.

High-performance liquid chromatography

Analysis was carried out using Hewlett-Packard (now Agilent) HP1100 series binary pumps and vacuum degasser, modified for use with a capillary microbore column by the incorporation of an LC Packings Acurate™ microflow processor. This splits off a calibrated proportion (20%) of the solvent stream, reducing a flow rate of 100 μl min^{-1} at the pump to 20 μl min^{-1} at the injector. A 25 cm 800 μm i.d. Zorbax ODS (C18) column, with 5 μm packing, supplied by Presearch, was used. Detection was performed using the HP1100 diode array detector, fitted with a 0.5 μL flow cell, path length 10 mm, 4 nm slit width, monitoring signals at 254, 275, 330, 491 and 540 nm. The eluents used were (A) 99.9% water/0.1% trifluoroacetic acid; (B) 94.9% acetonitrile/5% methanol/0.1% trifluoroacetic acid; initial flow rate 20 μl min^{-1}. The gradient was as follows: 95% A: 5% B for 5 min; 5–30% B in 100 min; 30–45% B in 45 min; 45–6% B, reduce flow rate to 16 μl min^{-1} in 4 min, then 46–75% B in 136 min; maintain at 75% B for 30 min, 75–95% B and increase flow rate to 20 μl min^{-1} in 10min, 95% B 20min, 95–5% B 10 min; post run equilibration time 15 minutes. HP Chemstation software (version 8.01) was used to process the data.

Editor's note

Address for correspondence: Scientific Department, The National Gallery, Trafalgar Square, London, WC2N 5DN, United Kingdom (jo.kirby@ng-london.org.uk). Paper received 23 July 2003; revised version received 8 June 2004.

Notes and references

1. Merrifield, M.P. (1849) *Original Treatises dating from the XIIth to XVIIIth Centuries on the Arts of Painting*, 2 vols, London (reprinted New York, Dover 1967), vol. I, Manuscripts of Jehan Le Begue, *Experimenta de coloribus*, no. 12, pp. 50–51; vol. II, *Segreti per colori* (the Bolognese Manuscript), no. B.137, pp. 454–5 (kermes). For early occurrences see Smith, C.S. and Hawthorne, J.G. (1974) 'Mappae clavicula: a little key to the world of medieval techniques', *Transactions of the American Philosophical Society* N.S., 64, 4, pp. 3–128; see, for example, nos. 174A–6, pp. 52–3 (kermes), nos. 88–9, p. 54 (madder).
2. Rosetti, G. (1969) *The Plictho of Gioanventura Rosetti* (Venice 1548) tr. S.M. Edelstein and H.C. Borghetty, Cambridge (Mass.) and London, pp. 17 and 107; Wallert, A. (1991) '"Cimatura de grana": identification of natural organic colorants and binding media in mediaeval manuscript illumination', *Zeitschrift für Kunsttechnologie und Konservierung* 5, 1, pp. 74–83, esp. p. 74.
3. See, for example, Ploss, E.E. (1962) *Ein Buch von Alten Farben: Technologie der Textilfarben im Mittelalter mit einem Ausblick auf die festen Farben*, Heidelberg and Berlin 1962; the Nuremburg Kunstbuch, no. li, pp. 113–14.
4. Smith and Hawthorne 1974 (see note 1 above), no. 111, p. 43, no. 223, pp. 61–2, no. 253, p. 66.
5. Vest, M. and Wouters, J. (1999) 'Dyestuffs and pigments in 12th to 18th century alum-tawed bookbinding leather in European collections', in J. Bridgland (ed.) *12th Triennial Meeting of ICOM Committee for Conservation, 29 August–3 September 1999* (preprints), London, vol. II, pp. 714–20; Wouters, J. (1996) 'Dye analysis of Florentine borders of the 14th to 16th centuries', *Dyes in History and Archaeology* 14, pp. 48–59.
6. Thomson, R.H. (1971) *Naturally Occurring Quinones*, 2nd edn, London, pp. 418, 443, 455–6 and 462–71.
7. Merrifield 1849 (see note 1 above), vol. II, the Bolognese Manuscript, no. 129, pp. 446–7.
8. Berger, E. (1897) *Quellen und Technik der Fresko-, Oel und Tempera-Malerei des Mittelalters*, Munich, pp. 143–76, esp. pp. 160–61; *The Strasburg Manuscript: A Medieval Painter's Handbook* (1966), tr. V. and R. Borradaile, London, pp. 34–5.
9. Kirby, J. and White, R. (1996) 'The identification of red lake pigment dyestuffs and a discussion of their use', *National Gallery Technical Bulletin* 17, pp. 56–80, esp. p. 60; Campbell, L., Dunkerton, J., Kirby, J. and Monnas, L. (2001) 'Two panels by Ercole de' Roberti and the identification of "*veluto morello*"', *National Gallery Technical Bulletin* 22, pp. 29–41, esp. pp. 32,

38 and 39, note 13. 4% BF_3/MeOH, used for its efficiency at breaking up the paint film by methylation, dissolves out the dyestuff. It also extracts mordant dyestuffs from textiles evenly – that is, preferential methylation of one acid rather than the other is not observed – but slowly. It has been found that carboxylic acid groups present in anthraquinones such as the laccaic acids are methylated, at least partly. *Under the conditions used* it does not appear to methylate other anthraquinones (alizarin, for example, is unaffected). The chromatographic conditions used were not designed for quantitative work and the methylation procedure renders quantitative estimates of components inconvenient.

10. Thomson 1971 (see note 6 above), p. 470. Since this paper was written it has been possible to compare the principal alkali-soluble constituent with a sample of erythrolaccin supplied by Professor K. Venkataraman, National Chemical Laboratory, Poona, India, in 1976, but unavailable when the present work was carried out. This confirmed the identity of the constituent as erythrolaccin, using HPLC–DAD; mass-spectrometric analysis has yet to be carried out.
11. Examination of the pigments prepared in the laboratory and the sample from the Thornham Parva altarpiece by Fourier transform infrared microspectrometry (FTIR microscopy) showed bands attributable to shellac constituents: ester carbonyl bands appear at 1735 (doublet) and 1240 cm^{-1} ; carboxylic acid carbonyl bands at 1715 and 1163 cm^{-1}; also small C–O band at 1040 cm^{-1}. This analysis was carried out by Catherine Higgitt: C.L. Higgitt and J. Kirby, in preparation for publication.
12. *Instruction générale pour la teinture des laines* (1671), Paris, pp. 24–5, 31 and 73.
13. Hofenk de Graaff, J.H. (1996) 'Recycling in the 17th-century textile industry', *Dyes in History and Archaeology* 14, pp. 60–69, reprinted in Hofenk de Graaff, J.H. (2004) *The Colourful Past*, London (Archetype), pp. 344–52.
14. Merrifield 1849 (see note 1 above), vol. II, the Bolognese Manuscript, no. 109, pp. 432–5.
15. Ploss 1962 (see note 3 above).
16. Rosetti 1969 (see note 2 above), pp. 44–5 and 134–7, 52–3 and 144–5.
17. Dalby, G. (1985) *Natural Dyes, Fast and Fugitive*, Alcombe (Minehead, Ashill Publications; repr. 1989), p. 25; Goodwin, J. (1982) *A Dyer's Manual*, London (repr. 1990), pp. 52–4.
18. The amide I C–O absorption at *c*.1654 cm^{-1} and amide II amino deformation at *c*.1630 cm^{-1} appear as one broad band at *c*.1650 cm^{-1}. The amide II band occurs as a doublet at 1548 and 1518 cm^{-1} and is seen at *c*.1540 cm^{-1}. The N–H stretch at *c*.3290 cm^{-1} is masked by –OH and other bands attributable to the pigment substrate. These absorptions are very much more difficult to see in the pigments prepared by the more gentle alkaline extraction, or in examples from paintings: C.L. Higgitt and J. Kirby, in preparation for publication.
19. Birkmaier, U., Wallert, A. and Rothe, A. (1995) 'Technical examination of Titian's *Venus and Adonis*: a note on early Italian oil painting technique', in *Historical Painting Techniques, Materials and Studio Practice: Preprints of a Symposium, University of Leiden, The Netherlands, 26–29 June 1995*, ed. A. Wallert, E. Hermens and M. Peek, Malibu, pp. 117–26, esp. p. 123; Murrell, J. (1993/94) 'John Guillim's book: a heraldic painter's *vade mecum*', *The Walpole Society* LVII, pp. 1–51, esp. p. 25: 'To Make a Red Rose'. This describes the use of fabric to 'blot' an application of glaze paint.
20. Kirby and White 1996 (see note 9 above), pp. 70–73. The earliest example of kermes lake identified at the National Gallery to date was used in the late 14th-century Westminster Abbey Chapter House wall paintings, London: see Howard, H. (2003) *Pigments of English Medieval Wall Painting*, London (Archetype), pp. 116–21.
21. Wouters, J. and Verhecken, A. (1989a) 'The scale insect dyes (Homoptera: Coccoidea). Species recognition by HPLC and diode-array analysis of the dyestuffs', *Annales de la Société Entomologique de France* (N.S.) 25, 4, pp. 393–410, esp. pp. 398–400; Wouters, J. and Verhecken, A. (1989b) 'The coccid insect dyes: HPLC and computerized diode-array analysis of dyed yarns', *Studies in Conservation* 34, 4, pp. 189–200, esp. pp. 192–3.
22. Insects collected for Dr Harald Böhmer in Anatolia in 1989, samples of which were given to the author by Dr John S. Mills and Cheryl Porter; insects from the D.V. Thompson Jr. Collection (acquired 1937 from Professor O. Dimroth), Department of Technology and Conservation, Courtauld Institute of Art.
23. Wouters and Verhecken 1989b (see note 21 above), pp. 196–8; Wouters 1996 (see note 5 above), pp. 50–52.

24. This suggestion was made by Maarten van Bommel, Instituut Collectie Nederland, Amsterdam.
25. Wouters and Verhecken 1989b (see note 21 above), p. 198.
26. Kirby, J., Saunders, D. and Cupitt, J. (1997) 'Colorants and colour change', *Early Italian Paintings, Techniques and Analysis; Symposium, Maastricht, 9–10 October 1996*, eds T. Bakkenist, R. Hoppenbrouwers and H. Dubois, Maastricht, pp. 65–71 and 88–90, esp. pp. 68 and 89, plates VII and VIII.
27. Ackroyd, P., Billinge, R., Campbell, L. and Kirby, J. (2003) 'The "Two Tax-Gatherers" by Marinus van Reymerswaele: original and replica', *National Gallery Technical Bulletin* 24, 50–63, esp. pp. 59 and 63 (notes 44 and 46).
28. Rosetti 1969 (see note 2 above), pp. 45 and 136, in relation to kermes dyeing.
29. This observation has also been made by Judith Hofenk de Graaff: see Ackroyd *et al.* 2003 (see note 27 above), notes 44 and 46, p. 63.
30. Wouters, J. (1987) 'Analyse des colorants des tapisseries brugeoises des XVIe et XVIIe siècles', in G. Delmarcel and E. Duverger, *Bruges et la tapisserie*, Bruges, pp. 515–26, esp. pp. 523–6; Wouters, J. (2001) 'The dye of *Rubia peregrina*. I. Preliminary investigations', *Dyes in History and Archaeology* 16/17, pp. 145–57, esp. pp. 153–5.
31. *Bruges et la tapisserie*, 31. Wouters 2001 (see note 30 above).
32. *Bruges et la tapisserie*, 32. Schweppe, H. and Winter, J. (1997) 'Madder and alizarin', in *Artists' Pigments: A Handbook of their History and Characteristics*, vol. 3, ed. E. West FitzHugh, Washington DC and Oxford, pp. 109–42, esp. p. 119.
33. Compare *Early English Miscellanies in Prose and Verse, selected from an inedited manuscript of the fifteenth century* (1855) ed. J.O. Halliwell, London (Wharton Club), no. X, pp. 72–91, esp. p. 88; 'Die Conste van alderhande substancien te verwen' (*T'Bouck van wondre*, Brussels 1513), in *Middelnederlandse verfrecepten voor miniaturen en 'alderhande substancien'* (1986), ed. W.L. Braekman, Scripta 18, Brussels, text 2, no. 4, p. 64; Rosetti 1969 (see note 2 above), pp. 27–8, 117–18 and other places. See also Sanyova, J. (2000/1) *Contribution à l'étude de la structure et des propriétés des laques de garance,* PhD thesis, Université libre de Bruxelles, p. 156.
34. Taylor, G.W. (1992) 'The chemistry of madder dyeings: the application of semi-quantitative TLC', *Dyes in History and Archaeology* 10, pp. 35–7; Wouters 2001 (see note 30 above), pp. 153–4.
35. Goovaerts, A. (1896) 'Les ordonnances données en 1480, à Tournai, aux métiers des peintres et des verriers', *Compte rendu des séances de la Commission royale d'histoire, ou Recueil de ses Bulletins* 5e série, VI, pp. 97–182; clause 13, pp. 156–8.
36. See, for example, Caniparius, P.M (1718) *De atramentis cujuscunque generis*, Rotterdam (reprint of 1st edn, Venice 1619), pp. 292 and 294.
37. This is discussed further in Kirby, J., Spring, M. and Higgitt, C. (2005) 'The technology of red lake pigment manufacture: study of the dyestuff substrate', *National Gallery Technical Bulletin* 26, pp. 71–85; Kirby, J., Higgitt, C. and Spring, M. (2005) 'Madder lakes of the 15th–17th centuries: variability of the dyestuff content', paper presented at the 24th Meeting of *Dyes in History and Archaeology*, Liverpool, 3–4 November 2005, accepted for publication.

Monitoring of Damage to Historic Tapestries (MODHT): A Newly Initiated EU Project*

Anita Quye

Introduction

Tapestries woven in renowned European centres between the 15th and the 18th century are among the most valuable testimonies of European cultural heritage. Their survival, however, is jeopardised by the degradation processes operating in coloured fibres and metal threads. Tapestry preservation relies on expert damage assessment to improve conditions of display, storage and proposed conservation treatment, and to assist with decisions concerning conditions of loan. In practical terms, conservators and restorers handling tapestries find that these often large and heavy textiles are susceptible to damage if the fibre and inter-fibre strengths are weak, but the potential for damage can prove difficult to assess by visual examination alone. Better objective methods for identifying existing and potential damage can be gained through enhanced knowledge of the physical and chemical properties of the methods and materials used for tapestry construction, and the effect of ageing on them. Some work has been performed on the history of tapestries and their conservation,[1] and on the characterisation of pigments and dyes,[2] but no known studies have addressed the relationship between general deterioration and changes in material properties.

The project 'Monitoring of Damage to Historic Tapestries' (MODHT) seeks to improve the care and protection of tapestries by identifying physical and chemical indicators of damage through the application of established and novel scientific analytical techniques to historical and replica tapestries. The indicators being developed will relate molecular level damage to a tapestry to its construction materials and techniques, as well as to ageing processes. This relationship involves documenting the 'history' of a representative number of historic tapestries in order to correlate display environments, length of display and past treatments, in addition to chemical and physical properties, with present condition. This paper outlines the work programme for the MODHT project and describes the materials for replicating tapestries to model the effects of construction and ageing.

* The MODHT project, which took place under the European Commission Fifth Framework Programme, Contract No. EV4K-CT-2001-00048, was completed in 2005. The work carried out over the three years of its duration has been of considerable value, greatly increasing our understanding of the materials used in the construction of historic tapestries and their degradation. This paper gives a useful and interesting introduction to the project; see the addendum for an update to the text following completion of the project.

Outline of work

MODHT is an interdisciplinary project involving collaboration between experts in the fields of tapestry conservation, curation, conservation science and analytical and textile chemistry. The three-year project started in April 2002 and is funded by the European Commission under the Fifth Framework Programme. The partners are Historic Royal Palaces (HRP), UK (David Howell, project coordinator); National Museums of Scotland, UK (Anita Quye); Patrimonio Nacional, Spain (Concha Herrero Carretero and Angel González); Royal Institute for Cultural Heritage (KIK), Belgium (Jan Wouters); University of Edinburgh, UK (Alison Hulme and Hamish McNab); University of London (Birkbeck College), UK (Marianne Odlyha); and University of Manchester Institute of Science and Technology (UMIST), UK (Chris Carr). The work programme is composed of six work packages as described in Table 1. The ultimate aim of the project is for its outcomes to benefit practitioners in tapestry preservation, although the results from MODHT are also expected to be relevant to the preservation of other groups of historical textiles.

Historic tapestries

The MODHT project focuses on northern and southern European collections of European tapestries made between the 15th and the 18th century. High status tapestries from Belgian, Spanish and British collections with exceptionally good historical records for their display and conservation/restoration have been selected for sampling, and the production centres of Brussels, Arras, Bruges and Antwerp are represented. The tapestries being studied are in varying degrees of preservation, so information held in records on their place and date of manufacture or commission, past conservation treatment, conditions of storage and periods of display is important for comparing and interpreting their level of damage. To date, 17 tapestries have been sampled from their reverse side. This has yielded 402 threads, of which 325 are dyed wool and silk yarns and 77 are metal threads. Sampling

Table 1 The six work packages of the MODHT programme.

1 Preparation of model tapestries	2 Field campaigns	3 Analysis of physicochemical state of metal threads in tapestries	4 Characterisation of the physicochemical state of wool and silk	5 Study of mordants and dyes	6 Compilation of parameters for damage assessment
a. Production of model samples based on traditional European materials and processes b. Accelerated ageing of woven, dyed model samples c. Documentation of materials and techniques	a. Complete set of historic samples for analysis from the three European sampling campaigns b. Full documentation of the historic tapestries under study in the project, including materials, techniques and art-historical information	a. Development of sample preparation methodology for micro-scale analysis b. Assessment of metal composition c. Characterisation of metallic corrosion products and micro-mechanical data	a. Calibrated amino acid analysis of wool and silk b. Molecular weight distribution c. Surface chemical composition of fibres d. Thermal stability indices for wool and silk e. Viscoelastic properties and tensile strength f. Damage assessment of fibres taken from metal threads	a. Mordant identification procedures b. Dye degradation profiles c. Typical dye degradation products	a. Final report based on collation, tabulation of data from the combined analytical approach, demonstrating similarities and trends and giving recommendations b. Workshop and dissemination to end users and the public

from the front of a select number of tapestries is anticipated during the project term.

Model tapestries

The production of model samples of woven tapestry and of yarns, including metal threads, serves two important purposes for the study. Firstly, the replicas, which are based on traditional materials and techniques, enable an understanding of how the originals were constructed. Secondly, ageing processes can be modelled by exposing the replicas to accelerated ageing conditions in controlled environments.[3] The ageing studies are being used to model light-induced damage for dyes and yarn fibres and to investigate the effect of past conservation treatments on the changes observed. The dye degradation studies will examine neoflavonoids, which include light-sensitive brazilwoods; the anthraquinone components of madder; and will extend existing knowledge of flavonoid dye photodegradation.[4] The effect of alum and iron mordants on dye photodegradation is also being addressed. Ultimately, molecular changes to the dyes will be correlated with changes in fibre and tensile strength as a consequence of ageing, influenced by environments and conservation treatments.

Two accelerated light ageing regimes are being used. One is a commercial xenon arc climate chamber, based at Hampton Court Palace, intended to induce fibre damage under conditions which cause severe photodegradation of the dyes. The other, set up at the University of Edinburgh, uses a fluorescent light chamber of a design used in some other museums and galleries[5] in order to model dye photodegradation under conditions already

Table 2 Dyes and dyeing agents selected for the project.

Dyes	Mordants	Dyeing assistants and dye-bath additives
brazilwood (*Caesalpinia sappan* L.)	alder bark (*Alnus glutinosa* (L.) Gaertn.)	copper turnings
cochineal (*Dactylopius coccus* Costa)	potassium aluminium sulfate (alum)	gum arabic
copper(II) sulfate	oak galls (*Quercus infectoria* Oliv.)	iron filings
dyer's greenweed (*Genista tinctoria* L.)		potassium carbonate
iron(II) sulfate		sanderswood (*Pterocarpus santolinus* L. f.)
madder (*Rubia tinctorum* L.)		sodium chloride
weld (*Reseda luteola* L.)		starch paste
woad (*Isatis tinctoria* L.)		tartaric acid
young fustic (*Cotinus coggygria* Scop.)		turmeric (*Curcuma domestica* Valeton)

Table 3 HPLC–PDA conditions used for the analysis of flavonoids and turmeric.

Analytes	Column	Eluent	Eluent programme	Eluent flow rate	PDA parameters
flavonoids	5μm Sphericlone ODS2 (Phenomenex), 150 × 4.6 mm with guard column	20% (v/v) MeOH(aq) (A), 100% MeOH (B) and 5% w/v ortho-phosphoric acid (aq) (C)	isocratic 67A:23B:10C from 0 min to 3 min, linear gradient to 0A:90B:10C from 3 min to 29 min, linear gradient to 67A:23B:10C from 29 min to 30 min, and 5 min equilibration at starting conditions	1.2 mL min^{-1}	2.4 nm resolution; 1 point s^{-1} sampling rate; spectral range 250–750 nm
turmeric	as for flavonoids	50:50 acetonitrile:1%(aq) citric acid adjusted to pH3.4 with sodium hydroxide	isocratic	1 mL min^{-1}	as for flavonoids

demonstrated to provide meaningful models for museum-displayed historical dyed textiles.[6]

Following experiments carried out to ascertain the best way to produce technically accurate tapestries, woven models of each dyed yarn have been made at UMIST for accelerated light ageing. The models have been made with traditional European materials and methods to create woven tapestries with weave structure,[7] yarns,[8] and dyeing methods[9] typical for the historical period of interest. Their construction is thus comparable to the historical tapestries, within the limitations of available weaving equipment.[10] A pragmatic approach was adopted for the conversion of mass, volume and time in the historical dye recipes to modern scientific units of measurement: for example each 'pound' in the texts was assumed to be equivalent to 454 g, and the ingredients were normalised for the dyeing of 100 g of yarn to allow direct comparison between the recipes. The dyes chosen for study are weld, dyer's greenweed, young fustic (for the silk cores of metal threads), madder, brazilwood, cochineal, woad/indigo and metallic sulfates (Table 2). The yarns are Italian silk (2-ply, 66 Tex) and English wool (3-ply, 158 Tex).

All dyestuffs purchased or donated for the project were analysed by reversed-phase high-performance liquid chromatography with photodiode array detection (HPLC–PDA) to identify their botanical or entomological source. The gradient HPLC method used at KIK was modified slightly by the NMS to maximise flavonoid information,[11] and a reversed-phase method was developed for turmeric (Table 3). Supplies of weld (*Reseda luteola* L.), dyer's greenweed (*Genista tinctoria* L.), young fustic (*Cotinus coggygria* Scop.), turmeric (*Curcuma domestica* Valeton), madder (*Rubia tinctorum* L.), and Mexican cochineal (*Dactylopius coccus* Costa) were verified as suitable for dyeing the model tapestries. Figures 1–4 show examples of chromatograms and UV-visible spectra obtained for the first four of these dyes; the principal constituents of each are listed in Table 4.

Comparison of commercially acquired brazilwood and sanderswood with botanical refer-

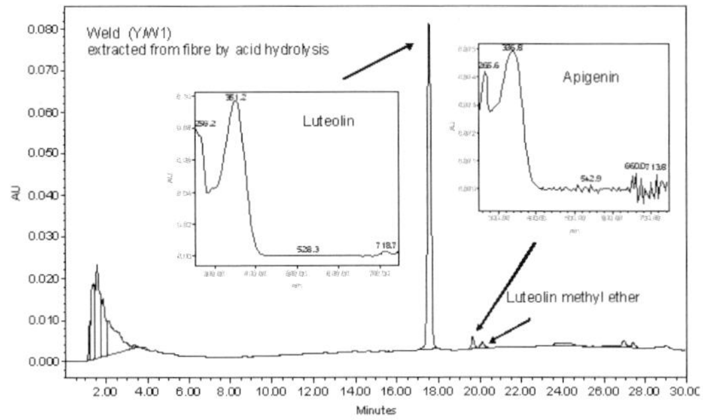

Figure 1 Chromatograms and UV-visible spectra for weld (*Reseda luteola* L.) used for replica tapestries. Acid-hydrolysed extracts from dyed wool analysed by the NMS HPLC method.

(a)

(b)

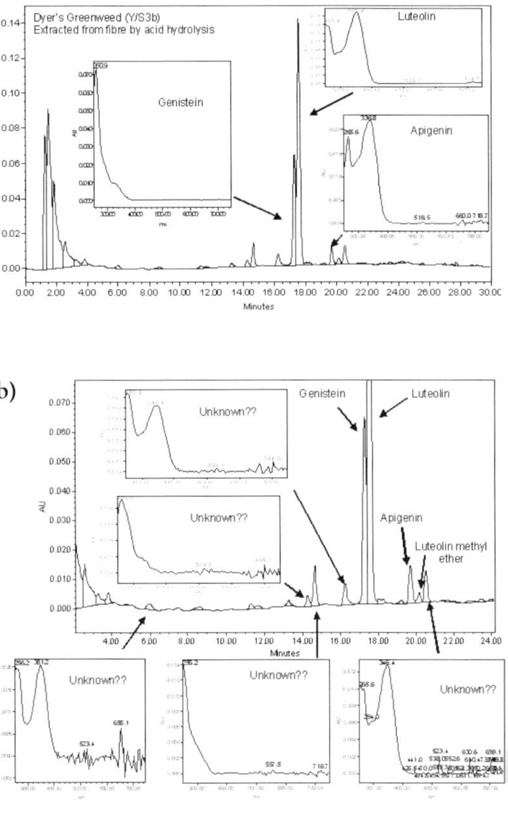

Figure 2 Chromatograms and UV-visible spectra for (a) dyer's greenweed (*Genista tinctoria* L.) used for replica tapestries and (b) detail showing minor constituents. Acid-hydrolysed extracts from dyed wool analysed by the NMS HPLC method.

ences of *Caesalpinia sappan* L. and *Pterocarpus santilinus* L. f. respectively show similar, but not identical, common chromatographic and spectrometric features (Figs 5–7). Because redwood dyes are chemically complex,[12] and the MODHT project includes an analytical investigation of characteristic compounds to distinguish between species in extracts from dyed yarns (data will

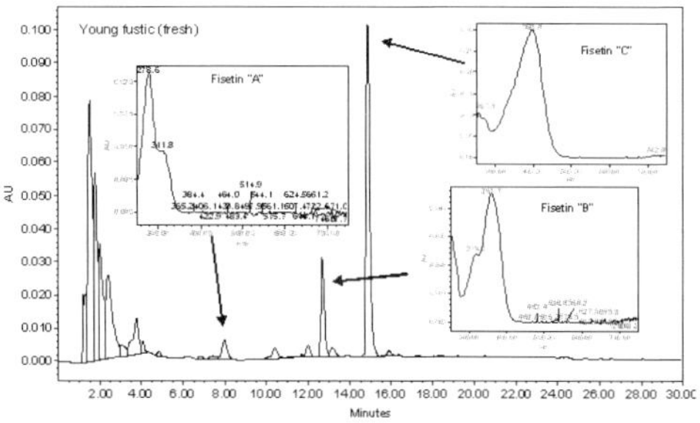

Figure 3 Chromatograms and UV-visible spectra for young fustic (*Cotinus coggygria* Scop.) used for replica tapestries. Fresh material analysed.

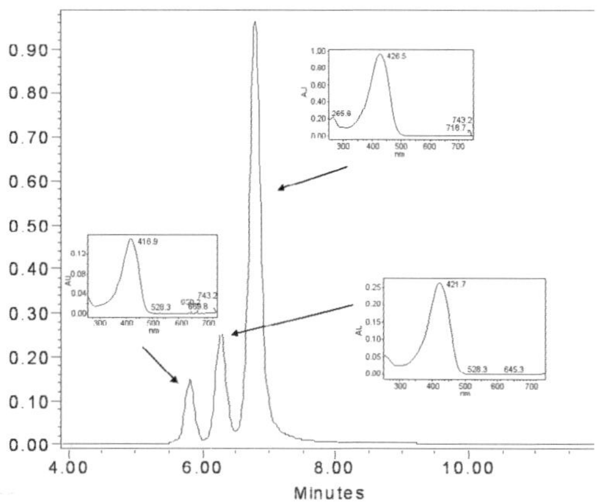

Figure 4 Chromatograms and UV-visible spectra for curcumin (*Curcuma domestica* Valeton) used for replica tapestries. Acid-hydrolysed extracts from dyed wool analysed by the NMS HPLC method.

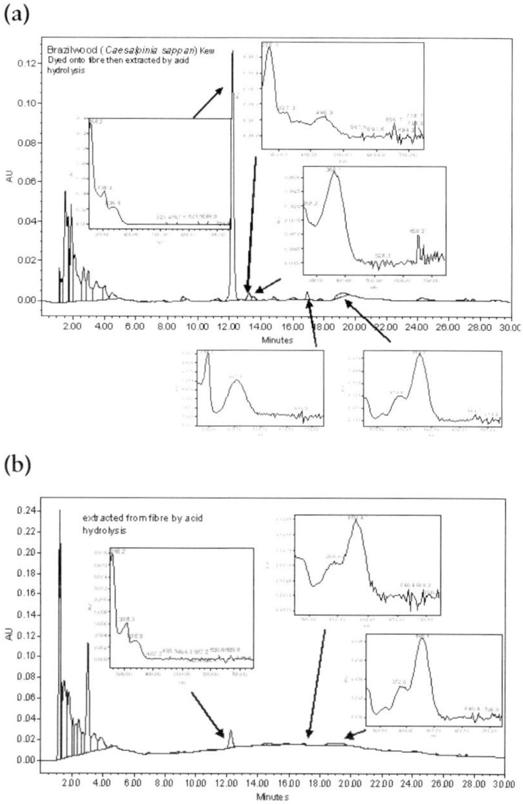

Figure 5 Chromatograms and UV-visible spectra for commercial brazilwood and *Caesalpinia sappan* L. dyed onto fibres and then extracted following the standard NMS acid hydrolysis procedure: (a) reference sample of *Caesalpinia sappan* (Royal Botanic Gardens, Kew) and (b) commercial sample.

Table 4 Details of constituents found in yellow dyes used for MODHT replica tapestries (see Figs 1–4).

Dye	Principal characteristic constituents observed[13]	Chemical class
weld	luteolin	flavonoid
	apigenin	
	luteolin methyl ether (characterised by LC–MS)[14]	
dyer's greenwood	genistein	flavonoid
	luteolin	
young fustic	fisetin	flavonoid
	myricetin (none observed in botanical reference sample supplied)	
turmeric[15]	curcumin	curcuminoid

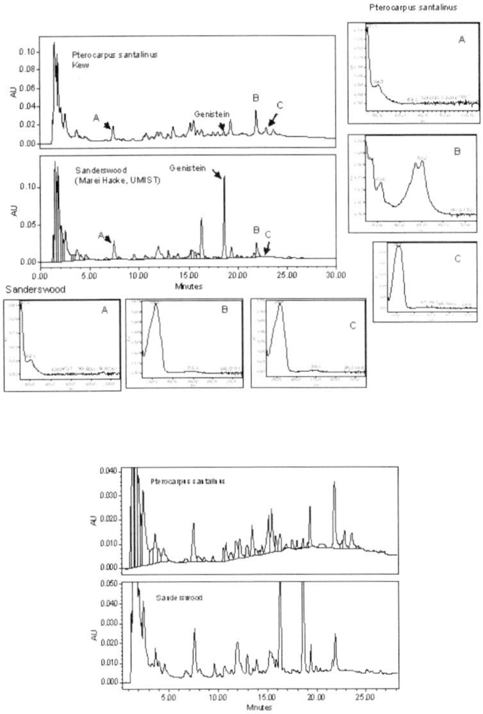

Figure 6 Chromatograms and UV-visible spectra for commercial red sanderswood obtained from Marei Hacke (UMIST) and *Pterocarpus santalinus* L. f. from the Royal Botanic Gardens, Kew, dyed onto fibres and then extracted following the standard NMS acid hydrolysis procedure.

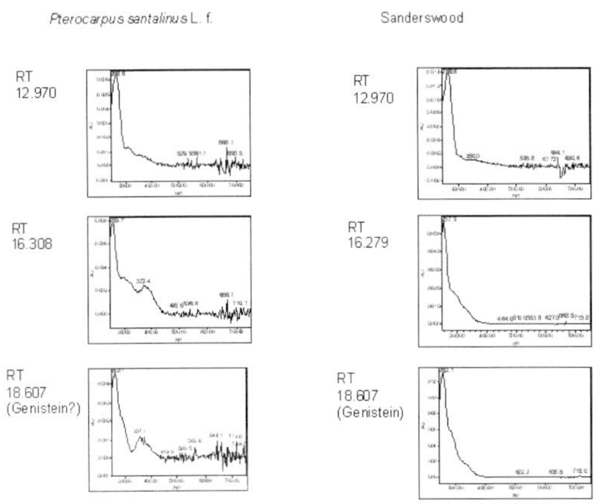

Figure 7 Important comparisons between chromatograms and UV-visible spectra for *Pterocarpus santalinus* L. f. and commercial red sanderswood, dyed onto fibres and then extracted following the standard NMS acid hydrolysis procedure.

be published in due course), the commercial sources of brazilwood and sanderswood have been used for the time being, with the provision that new tapestries will be prepared and aged if the sources turn out to be botanically unsuitable. Analytical data for oak gall dye is similarly under current analytical investigation and will be discussed in subsequent publications.

Work in progress

The key aims of the project are outlined in the work packages. Work in progress includes HPLC–PDA analysis of dyed yarns from historic tapestries; the characterisation of neoflavonoid (redwood) dyes by HPLC–PDA and LC–ESI–MS^n; and light-ageing experiments for flavonoid dyes. Analytical methods for elemental characterisation of metal threads are under development and techniques for the identification of inorganic mordants are being explored.

Fibre studies include improvements to a gel permeation chromatography (GPC) method for determining the molecular weight distribution for silk with microsamples, and development of an amino acid analysis protocol and methodology to detect tryptophan and optimise cystine/cysteine analysis. Solid state ^{13}C NMR is being used to characterise the carbon states of peptide amino acids and carboxylic acids for wool and silk fibres, with molecular changes to the peptide chain being measured by Fourier transform infrared (FTIR) spectrometry. Two thermoanalytical (TA) techniques, differential scanning calorimetry (DSC) and thermogravimetric analysis (TGA), show thermal transitions under inert and oxidising conditions that may relate to fibre structure changes: for example, loss of the helical structure for wool. TA techniques are also providing thermal stability indices to compare the extent of damage in accelerated light-aged samples with that in historical samples, while micro-TA measurements on sample cross-sections are providing spatially resolved thermal information at the micron level. Differences between the viscoelastic properties of light-aged silk and wool, including glass transition temperatures (T_g), are being measured with dynamic mechanical thermal analysis (DMA), while measurements on the degree of crystallinity are being undertaken by X-ray diffraction (XRD) with synchrotron radiation. The application of DMA in reverse configuration is proving useful for *in-situ* measurements of the viscoelastic and mechanical properties of wool and silk in response to environmental changes, for instance the effect of variable relative humidity in the range 20% to 80% RH.[16]

Results will continue to be reported and published during the course of the project, with the MODHT website as the central point for information. Dissemination workshops, public lectures and exhibitions will be arranged when the study is completed in 2005.

Acknowledgements

MODHT researchers: Thomas Campbell (for HRP), Johan Denis (KIK), Marei Hacke (UMIST), Kathryn Hallett (HRP), Marie-Christine Maquoi (KIK), David Peggie (UoE/NMS) and Ina Vanden Berghe (KIK). Also John Edmonds for assistance with the woad vat; Ester Ferreira (AMOLF, Amsterdam) for helpful discussion of flavonoid analysis and light-ageing experiments; Michel Garcia (Association de Garance, Lauris, France), Dr Peter Gasson (Jodrell Laboratory, Royal Botanical Gardens, Kew) and Hew Prendergast (Centre for Economic Botany, Royal Botanical Gardens, Kew) for supplying botanical specimens; and Jo Kirby and David Saunders (National Gallery, London) for helpful discussion of light-ageing method and instrumentation.

MODHT was supported by the European Commission under the Fifth Framework Programme, contributing to the implementation of the Energy, Environment and Sustainable Development Programme Key Action 4, 'The City of Tomorrow and Cultural Heritage', subsection 4.2.1 'Improved damage assessment of cultural heritage,'; Contract No. EV4K-CT-2001-00048, 2002–2005.

Dyeing materials suppliers

- Potassium aluminium sulfate dodecahydrate (alum; $AlK(SO_4)_2 \cdot 12H_2O$), copper turnings, iron(II) sulfate ($FeSO_4 \cdot 7H_2O$), gum arabic, iron filings, potassium carbonate (K_2CO_3), (+)-tartaric acid ($(CHOH \cdot COOH)_2$): Fisher Scientific UK Ltd, Loughborough, UK.
- Brazilwood, cochineal, madder, oak galls, weld: Verfmolen De Kat, Zaandam, The Netherlands.
- Dyer's greenweed, sanderswood, turmeric: George Weil Fibrecrafts Ltd, Guildford, UK.
- English wool yarn manufactured by W. & J. Whitehead, Ltd, Bradford, UK and A. Orme & Co, Elland, West Yorkshire, UK.
- Italian spun silk: H.T. Gaddum & Co. Ltd, Macclesfield, UK.
- Woad (couched): John Edmonds, Reading University, UK in cooperation with SPINDIGO.
- Young fustic (*Cotinus coggygria* Scop.): Association de Garance, Lauris, France.
- Black alder (*Alnus glutinosa* (L.) Gaertn.): Royal Botanic Gardens, Kew, UK. Sappanwood (*Caesalpinia sappan* L.) and red sanderswood or sandalwood (*Pterocarpus santilinus* L. f.) reference samples: Royal Botanic Gardens, Kew, UK.

Editor's note

Address for correspondence: Conservation and Analytical Research, National Museums Scotland, Chambers Street, Edinburgh, EH1 1JF, Scotland (A.Quye@nms.ac.uk). Paper received 22 November 2003.

Notes and references

1. Campbell, T.P. (2002) 'Report on the display history of the tapestries at Hampton Court

Palace', MODHT report; Herrero Carretero, C. (2002) 'La colección de tapices de la Corona de España. Selección estudio. Propuesta e análisis de la herencia textil española', MODHT report; Thomson, W.G. (1906) *A History of Tapestry from the Earliest Times until the Present Day*, London; Howell, D. (1996) 'Some mechanical effects of inappropriate humidity on textiles', *ICOM Committee for Conservation, 11th Triennial Meeting, Edinburgh, Scotland, 1–6 September 1996: Preprints*, ed. J. Bridgland, 2 vols, London, vol. II, pp. 692–8; Cooke, W.D. and Howell, D. (1988) 'Diagnosis of deterioration in a tapestry using scanning electron microscopy', *The Conservator* 12, pp. 47–51; Bilson, T., Cooke, B. Thomas and Howell, D. (1997) 'Mechanical aspects of lining "loose hung" textiles', *Fabric of an Exhibition: An Interdisciplinary Approach: Preprints / L'Étoffe d'une exposition: une approche pluridisclipliaire: Prétirages* (1st North American Textile Conservation Conference, Canadian Conservation Institute, Ottawa, September 1997), Ottawa, pp. 63–9.

2. Wouters, J. (1987) 'Kleurstofanalyse van Brugse wandtapijten uit de 16de en 17de eeuw (Analyse des colorants des tapisseries brugeoises des XVIe et XVIIe siècles)', in G. Delmarcel and E. Duverger (eds), *Brugge en de tapijtkunst/ Bruges et la tapisserie*, Bruges, pp. 515–26.
3. Ferreira, E.S.B. (2002) *New Approaches towards the Identification of Yellow Dyes in Ancient Textiles*, PhD thesis, University of Edinburgh.
4. Ferreira, E.S.B., Quye, A., MacNab, H. and Hulme, A.N. (2002) 'Photo-oxidation products of quercetin and morin as markers for the characterisation of natural flavonoid yellow dyes in ancient textiles', *Dyes in History and Archaeology* 18, pp. 63–72.
5. Saunders, D. and Kirby, J. (2001) 'A comparison of light-accelerated ageing regimes in some galleries and museums', *The Conservator* 25, pp. 95–104.
6. Ferreira 2002 (see note 3 above).
7. Bilson *et al.* 1997 (see note 1 above); Campbell, T.P. (2002) *Tapestry in the Renaissance: Art and Magnificence*, exhibition catalogue, The Metropolitan Museum of Art, New York; Rapp Buri, A. and Stucky-Schürer, M. (2001), *Burgundische Tapisserien*, Munich, p. 420; Rapp Buri, A. and Stucky-Schürer, M (1990) *Zahm und Wild: Basler und Strassburger Bildteppiche des 15. Jahrhunderts*, Mainz, p. 29.
8. Bilson *et al.* 1997 (see note 1 above); Rapp Buri and Stucky-Schürer 2001 (see note 7 above).
9. Rosettti, G. (1969) *The Plictho of Gioanventura Rosetti. Instructions in the Art of the Dyers which teaches the dyeing of woolen cloths, linens, cottons, and silk by the great art as well as by the common*, translation of the 1548 (first) edition, by S.M. Edelstein and H.C. Borghetty, Cambridge, MA, pp. 131–52, 186–99; De Nie, W.L.J. (1937) *De ontwikkeling der Noord-Nederlandsche textielververij van de veertiende tot de achttiende eeuw*, Leiden; *T'bouck va[n] wondre* (Brussels, 1513) and Hs 1317, University Library Ghent, Texts II and III in W.L. Braekman (ed.) (1986), *Middelnederlandse verfrecepten voor miniaturen en "alderhande substancien"*, Brussels (OMIREL), pp. 58–105 (Scripta – Mediaeval and Renaissance Texts and Studies – 18).
10. Wouters 1987 (see note 2 above).
11. Wouters, J. and Verhecken, A. (1989) 'The coccid insect dyes. HPLC and computerized diode-array analysis of dyed yarns' *Studies in Conservation* 34, pp. 189–200.
12. Nowik, W. (2001) 'The possibility of differentiation and identification of red and blue "soluble" dyewoods: determination of species used in dyeing and chemistry of their dyestuffs', *Dyes in History and Archaeology* 16/17, pp. 129–44; Seshadri, T.R. (1972) 'Polyphenols of *Pterocarpus* and *Dalbergia* woods', *Phytochemistry* 11, pp. 881–98.
13. Cardon, D. and de Chateneuf, G. (1990) *Guide des teintures naturelles*, Paris-Lausanne, pp. 61–2, 65–6, 67–8 and 129–30.
14. Ferreira 2002 (see note 3 above).
15. Hiserodt, R., Hatman, T.G., Chi-Tang, H. and Rosen, R.T. (1996) 'Characterization of powdered turmeric by liquid chromatography-mass spectrometry and gas chromatography-mass spectrometry', *Journal of Chromatography A* 740, pp. 51–63.
16. Odlyha, M., Cohen, N.S., Foster, G.M. and Campana, R. (2002) 'Characterisation of historic and unaged parchments using thermomechanical and thermogravimetric techniques', in R. Larsen (ed.), *Microanalysis of Parchment*, London, pp. 73–89.

Addendum

Update following the completion of the MODHT project

Kathryn Hallett

The model tapestries proved to be an invaluable resource for the advancement of the chosen scientific analysis techniques, some of which were applied to historic tapestry materials for the first time during this project. A range of surface and bulk, thermal and chemical analytical techniques has been applied and evaluated in terms of their effectiveness at assessing damage in both model tapestry materials subjected to accelerated ageing and historic tapestries. The structural and chemical modifications to wool and silk fibres have been related to the macro properties of tapestry strength and integrity. The information obtained from amino acid compositions, surface chemical compositions, molecular weight distributions, infrared spectra and mechanical properties of fibres has been correlated with the historical tapestry conditions as recorded by conservators. Sensitive damage micro-markers have been successfully identified and these provide a basis for the analysis of wider tapestry collections. Analysis of dye degradation products and mordants, also at the micro level, has led to further information on the dyes and mordants originally used in these textiles. Dyeing methods were shown to influence the strength properties of unaged fibres. Since the dyeing method is dependent on the natural source, accurate identification of the original dye source and/or mordant is thus an important factor in damage prediction. The development of new methods for characterising dye sources in historic samples has provided vital support in the attainment of this goal.

This research project has produced two PhD theses and the findings have been published in a wide variety of forums (see below). In addition to presentations at international conferences, a final MODHT workshop was held at Hampton Court Palace in June 2005 to disseminate the objectives, results and conclusions to potential end-users. Over 70 delegates attended, with representation from heritage organisations around the world. The results and a more expansive discussion of the project are currently being prepared for publication in a book for a wider audience (conservators, curators and collectors): *Preserving the Art of Historical Tapestries* will be published shortly by National Museums Scotland, Edinburgh.

Editor's note

Address for correspondence: Conservation and Collection Care, Historic Royal Palaces, Hampton Court Palace, Surrey, KT8 9AU, UK (Kathryn.Hallett@hrp.org.uk). Received 29 February 2008.

Selected publications

Hacke, A-M. (2006) *Investigation into the Nature and Ageing of Tapestry Materials*, PhD thesis, University of Manchester.

Hallett, K. and Howell, D. (2005) 'Size exclusion chromatography as a tool for monitoring silk degradation in historic tapestries', in P. Wyeth and R. Janaway (eds), *Scientific Analysis of Ancient and Historic Textiles: Informing Preservation, Display and Interpretation; AHRB Research Centre for Textile Conservation and Textile Studies, First Annual Conference, 13–15 July 2004*, London, pp. 143–50.

Hallett, K. and Howell, D. (2005) 'Size exclusion chromatography of silk – inferring the tensile strength and assessing the condition of historic tapestries', in I. Verger (ed.), *ICOM Committee for Conservation, 14th Triennial Meeting, The Hague, 12–16 September 2005: Preprints*, 2 vols, London, vol. II, pp. 911–19.

Hulme, A.N., McNab, H., Peggie, D.A. and Quye, A. (2005) 'Negative ion electrospray mass spectrometry of neoflavonoids', *Phytochemistry* 66, 23, pp. 2766–70.

Hulme, A.N., McNab, H., Peggie, D.A. and Quye, A. (2005) 'The application of liquid chromatography–mass spectrometry and accelerated light ageing for the analytical identification of yellow flavonoid dyes in historical tapestries', in P. Wyeth and R. Janaway (eds), *Scientific Analysis of Ancient and Historic Textiles: Informing Preservation, Display and Interpretation; AHRB Research Centre for Textile Conservation and Textile Studies, First Annual Conference, 13–15 July 2004*, London, pp. 208–15.

Hulme, A.N., McNab, H., Peggie, D.A. and Quye, A. (in press) 'The chemical characterisation by PDA–HPLC and HPLC–ESI–MS of unaged and aged fibre samples dyed with sawwort (*Serratula tinctoria* L.)', *Dyes in History and Archaeology* 22, in course of publication.

Hulme, A.N., McNab, H., Peggie, D.A., Quye, A., Vanden Berghe, I. and Wouters, J. (2005) 'The analytical characterisation of the main component found in logwood dyed textile samples after hydrochloric acid extraction', in I. Verger (ed.), *ICOM Committee for Conservation, 14th Triennial Meeting, The Hague, 12–16 September 2005: Preprints*, 2 vols, London, vol. II, pp. 783–8.

Odlyha, M., Wang, Q., Foster, G.M., de Groot, J., Horton, M. and Bozec, L. (2005) 'Monitoring of damage to historic tapestries: the application of dynamic mechanical thermal analysis to model and historic tapestries', in P. Wyeth and R. Janaway (eds), *Scientific Analysis of Ancient and Historic Textiles: Informing Preservation, Display and Interpretation; AHRB Research Centre for Textile Conservation and Textile Studies, First Annual Conference, 13–15 July 2004*, London, pp. 126–36.

Odlyha, M., Wang, Q., Foster, G.M., de Groot, J., Horton, M. and Bozec, L. (2005) 'Thermal analysis of model and historic tapestries', *Journal of Thermal Analysis and Calorimetry* 82, 3, pp. 627–36.

Peggie, D.A. (2006) *The Development and Application of Analytical Methods for the Identification of Dyes on Historical Textiles*, PhD thesis, University of Edinburgh.

Peggie, D.A., Hulme, A.N., McNab, H. and Quye, A. 2007 'Towards the identification of characteristic minor components from textiles dyed with weld (*Reseda luteola* L.) and those dyed with Mexican cochineal (*Dactylopius coccus* Costa)', *Microchimica Acta*, published online 10.12.2007, DOI 10.1007/s00604-007-0866-0.

Vanden Berghe, I. and Wouters, J. (2005) 'Identification and condition evaluation of deteriorated protein fibres at the sub-microgram level by calibrated amino acid analysis', in P. Wyeth and R. Janaway (eds), *Scientific Analysis of Ancient and Historic Textiles: Informing Preservation, Display and Interpretation; AHRB Research Centre for Textile Conservation and Textile Studies, First Annual Conference, 13–15 July 2004*, London, pp. 151–8.

Dye Analysis of Braids Used for Japanese Armour Dating from the 14th to the 16th Century

Yoshiko Sasaki,* Masanori Sato, Mari Omura and Ken Sasaki

Introduction

Historical textiles have been dyed various colours, and there are some in which the colour tone can still be seen. In Japan, the ancient textiles of the Shōsōin (Imperial repository) and Hōryū-ji Temple, both in Nara, are famous for their advanced weaving techniques with various colours.[1] The textiles dating from before the 6th century have notably degraded and it is seldom that colour remains clearly in textiles of this age in Japanese collections.

The dye analysis for such historical textiles has in the past been carried out by dyeing specialists, by observation or using simple chemical tests.[2] Due to the development of chemical analytical instruments in recent years, many results have been obtained for the identification of colouring matters in very tiny samples.[3] We have been trying to identify natural dyestuffs used for ancient textiles using various analytical techniques.[4] As these textiles are a cultural property, analyses by non-destructive methods are required. If the state of preservation is good, the reflectance spectrum measured on the surface of the textile is a useful way to identify the dyestuff. However, when the apparent colour tone has been changed by time, circumstances of preservation and the use of a mordant, it is difficult to identify the dyestuff molecules remaining at the textile surface by the reflectance spectrum alone. There are some possibilities for sampling if the sample taken is very small and does not spoil the value of a historical textile: if, perhaps, it is taken from a detached fragment or an inside part of the garment or fabric.

Japanese armours (*Yoroi*, *Haramaki*, and *Doumaru*) are made of many iron plates connected by braid and/or leather straps. A selection of single or multicoloured dyed silk braid fragments, which had fallen from *Yoroi*, *Doumaru* and *Haramaki* armours from Homuri, Nara and Osaka, dating from the 14th to the 16th century (see lists of samples below), were available for spectroscopic dye analysis. The major parts of the braids were becoming degraded and fragile due to ageing in the hot and humid climate, but some parts could still be used for direct visible reflectance and fluorescence spectrum analyses without untying the braids.

In this report, some braids that had fallen off the armour, or had been removed during repair but were stored together with the armour to which they belonged, were offered for dye analysis. The part of the braid positioned inside the armour is not heavily degraded, so the remaining state of colour is good and suitable for dye examination. It was possible to perform the reflectance spectroscopic analysis non-destructively, without loosening the braid and dividing the treads for each colour string-braided using two or more differently coloured threads. A good spectrum, indicative of the kind of dyestuff, could be obtained. In those braids in which the construction and the organisation of the colour threads had collapsed, sampling was performed in these areas, for which permission was obtained; more detailed analysis could thus be tried.

Experimental

Dye analysis

Visible reflectance spectra were measured using a Hitachi Spectrophotometer U-3500 equipped with an integrating sphere attachment. The reflectance spectrum obtained using this equipment was automatically transformed into an absorption spectrum by means of the Kubelka-Munk equation, making it easy to compare with other results obtained directly as absorption spectra. These spectra were compared with those given by standard materials. The absorption maxima for the standards were 556 nm for munjeet, 607 nm for gromwell, 463 nm for amur cork tree and 667 nm for indigo.

Three-dimensional fluorescence spectra were measured using a Hitachi Fluorescence Spectrophotometer F-4500. Examples of the results obtained at the different conditions used are shown in Figures 3b and 4b, discussed in more detail below. In Figure 3b, the excitation wavelength range was 200–650 nm and the emission was continuously scanned from 200 nm to 800 nm. In Figure 4b, the excitation wavelength was 455nm and the emission was scanned from 400 nm to 700 nm.

A gas chromatograph–mass spectrometer (GC–MS; Shimadzu QP-5000) was used, modifying the operational mode to meet our special requirements by removing the separation column from the apparatus. The injected sample solution containing the mixed dyestuff components was vaporised and sent directly to the mass spectrometer without prior separation. Since the absolute amount of sample is very small and hence the sensitivity of the apparatus is insufficient to detect the components of the mixture separately, the mass spectrogram of the sample mixture was recorded directly and the peak of the molecular ion of the required molecule detected. We propose here the tentative designation of this usage of modified GC–MS as gas–mass spectrometry (gas–MS). For the gas–MS measurement, dyestuffs were extracted from the braids with N,N′-dimethylformamide (DMF). The equipment was used in electron impact mode at 70 eV, with a source temperature of 300 °C.

Samples

Samples were taken from examples of *Yoroi*, *Doumaru* and *Haramaki* armours kept in the Kushibiki-Hachimangu Shrine, Aomori Prefecture, the Kasuga-Taisha Shrine, Nara Prefecture, the Hase-dera Temple, also in Nara Prefecture and the Kongo-ji Temple, Osaka. These are listed in Table 1. A 14th-century Yoroi armour (Ku-4 in Table 1) from the Kushibiki-Hachimangu shrine is illustrated in Plate 23; the armours examined from the other three sites, as described in Table 1, are illustrated in Plates 24–26.

Results and discussion

The results obtained, given in Table 2, indicate the usage of munjeet (*Rubia cordifolia* L.) or madder group (including *Rubia tinctorum* L.), gromwell (*Lithospermum officinale* L.), indigo (probably from *Polygonum tinctorium* Ait.) and amur cork tree (*Phellodendron amurense* Rupr.). Analyses of the dye extracts from the braids by gas–MS also suggested the presence of munjeet and indigo.

Analysis of single coloured braids

Red

Distinction needs to be made between the dye plants safflower (*Carthamus tinctorius* L.), the

Table 1 Samples.

Location of armour	Armour	Braids sampled
Kushibiki-Hachimangu Shrine, Aomori	Ku-1: *Yoroi* (*Tsumadori* pattern) with white braid lacing (14th century)	red lace braid
		purple, blue and green edge braid
	Ku-2: *Doumaru* (pattern with red shoulder) with white braid lacing (14th–16th century)	red lace braid
		purple, dark blue and green edge braid
	Ku-3: *Yoroi* (pattern with white and green shoulder) with purple braid lacing (14th century)	purple and green lace braid
	Ku-4: *Yoroi* with red lacing (14th century), Plate 23	red lace braid
Kasuga-Taisha Shrine, Nara	Ka-1: *Yoroi* (pattern with bamboo and tiger) with red braid lacing (14th century), Plate 24	red lace braid
Hase-dera Temple, Nara	Ha-1: *Yoroi* with red braid lacing (15th century), Plate 25	red lace braid
		yellow, purple and green edge braid
Kongo-ji Temple, Osaka	Ko-1: *Haramaki* No. 6 with blue leather lacing (14th–16th century), Plate 26	red *Hishinui* braid (lower part, decorative stitching)
	Ko-2: *Haramaki* No.12 (pattern with white shoulder) with black leather lacing (14th–16th century)	red *Hishinui* braid (lower part, decorative stitching)
	Ko-3: *Uname* braid	white, purple and green braid

Table 2 Results of dye analysis.

Sample number	Type of braid	Thread colour	Dyestuff identified
Ku-1	lace braid (from area of design)	red	munjeet (or madder)
	edge braid	purple	gromwell
		blue and green	indigo and amur cork tree
Ku-2	lace braid	red	munjeet
	edge braid	green	indigo and amur cork tree
		purple	gromwell
		dark blue	indigo
Ku-3	Lace braid	purple	gromwell
		green	indigo and amur cork tree
Ku-4	lace braid	red	munjeet (and indigo)
Ka-1	lace braid	red	munjeet (or madder)
Ha-1	lace braid	red	munjeet (or madder)
	edge braid	yellow	amur cork tree
		yellow and purple	amur cork tree and gromwell
		purple	gromwell
		green	amur cork tree and indigo
Ko-1	*Hishinui* braid (lower part, decorative stitching)	red	munjeet (or madder)
Ko-2	*Hishinui* braid (lower part, decorative stitching)	red	madder group?
Ko-3	*Uname* braid (tied)	multicoloured	gromwell, amur cork tree, indigo?

madder group as a whole and sappanwood (*Caesalpinia sappan* L.); furthermore within the madder group, between madder of a Japanese type (munjeet, *Rubia cordifolia* L.) and those of a Western type (here designated madder, *Rubia tinctorum* L.). The distinction between safflower and the madder group can be made from the differences in their visible reflectance spectra, if the sample of dyed material has been kept in sufficiently good conditions.

The visible reflectance spectrum of the sample threads dyed with sappanwood changes significantly depending on the kind of mordant. Moreover, it is difficult to use the fluorescence spectrum for the identification of samples dyed with sappanwood because some types of mordant may disturb the fluorescence emission due to brazilin (and its oxidised form brazilein), the main colouring matter of sappanwood.

In general, the identification of safflower is difficult because it discolours easily and is not soluble in organic solvents. Furthermore, gas–MS is unsuitable for detecting the safflower dye molecule because of its lack of heat stability and low solubility in volatile solvents.

In the case of the madder group dyes, after visible reflectance spectra have been recorded, the dye is extracted and the mass spectra are measured. These show the presence of compounds of the substituted anthraquinone type characteristic of the madder group dyes. However, these vary in their solubility in organic solvents and their heat stability and not all of the compounds on the cloth can be extracted perfectly. Some compounds, such as sugars or ions that do not evaporate or have low heat stability, cannot be examined by this gas-MS method. Therefore comparison of the compounds of the madder group dyes is limited by the experimental conditions. This means that the compositional ratio of a madder group dye as recorded in the gas–MS spectra is not the same as that of a sample dyed with a madder group dyestuff: in other words, the content of the madder group dyestuff itself. Furthermore, bearing in mind the possibilities of thermal degradation or the existence of isomers, the dye molecules having the molecular weight presumed from the mass spectrum recorded might not exist, because they are not separated and identified.

The visible reflectance spectra of the red braids from Ku-2, Ka-1, Ha-1 and Ko-1 were measured without sampling (Fig.1a). The spectra of Ka-1 and Ko-1 showed that they were dyed with a madder group dye, but the spectra of Ku-2 and Ha-1 were broader and shifted to a longer wavelength compared with that of standard munjeet due to the passage of time and circumstances, and thus no distinction could be made between safflower dye and the madder group (munjeet) dye.

The madder group dyes were easily extracted from the threads and could be examined by the gas–MS technique (with the proviso mentioned above). Safflower dye, on the other hand, could hardly be extracted by organic solvents and the gas–MS method of examination could not be used. Figure 1b shows the gas–MS spectrum of

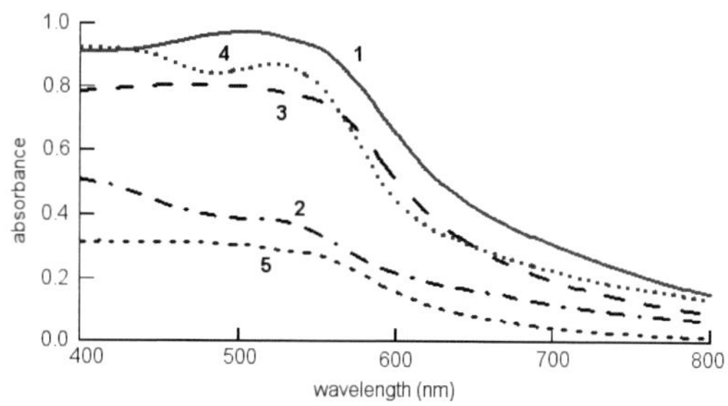

Figure 1a The visible reflectance spectra of red braids from Ku-2, Ka-1, Ha-1 and Ko-1, compared with that of silk dyed with munjeet used as a standard. 1: munjeet; 2: Ku-2; 3: Ka-1; 4: Ha-1 5: Ko-1.

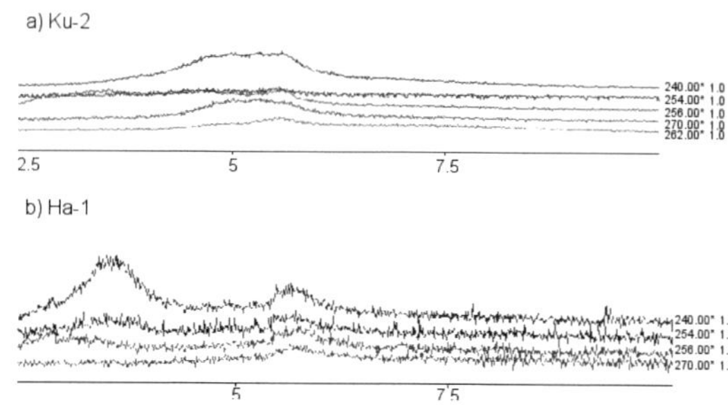

Figure 1b Gas–MS spectra of the DMF extract from red braids from Ku-2 and Ha-1.

an extract from the madder-dyed red threads. The peak at *m/z* 240 attributable to alizarin, the dyestuff constituent of the madder group dye, could be observed although its intensity was very low in the extracts of Ku-2. The identification of madder rather than safflower from this measurement procedure suggests that a madder dye may have been used. In the gas–MS spectrum of the extract from the lace braid red sample from Ha-1, the peak at *m/z* 240 showed the presence of alizarin and thus of a madder group dye; the peak pattern of the anthraquinone groups suggests that the dyestuff used may have been munjeet.

Purple

Although gromwell was assumed to have been used for the purple colour in these armours, the possibility of the use of a double dyeing of a blue and a red must be considered. The visible reflectance spectrum of Ha-1 suggested the usage of gromwell (Fig. 2). Since the mass spectrum of the DMF extract from the purple threads of Ha-1 showed no constituent with an *m/z* of 262, which would indicate the presence of indigo, the dye used was probably gromwell.

Blue

Due to the effects of the passage of time and preservation circumstances, the visible reflectance spectra of indigo often varies, but indigo was easily identified by its gas–MS spectrum.

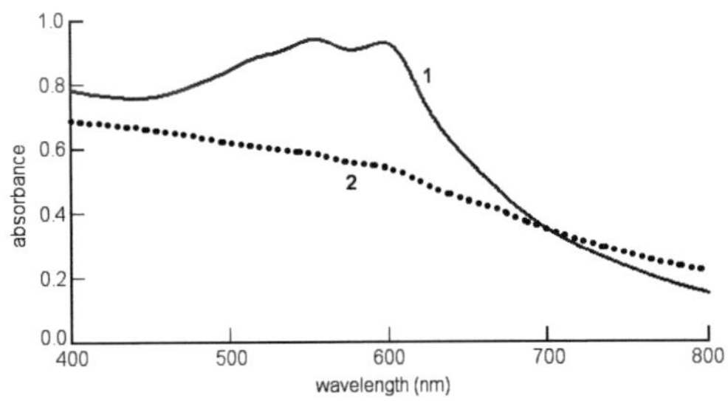

Figure 2 Visible reflectance spectra of purple edge braid from Ha-1 and a gromwell standard. 1: gromwell standard; 2: purple edge braid.

Yellow

The use of the dyestuffs from amur cork tree, *Gardenia* sp. and *Kariyasu* (*Muscanthus tinctorius* Hack.) were all possible and the measurements of their visible reflectance spectra and fluorescence reflectance spectra are effective for the distinction of these dyestuffs. The yellow threads of Ha-1 were identified as having been dyed with amur cork tree by the use of both methods (Fig. 3a, 3b).[5]

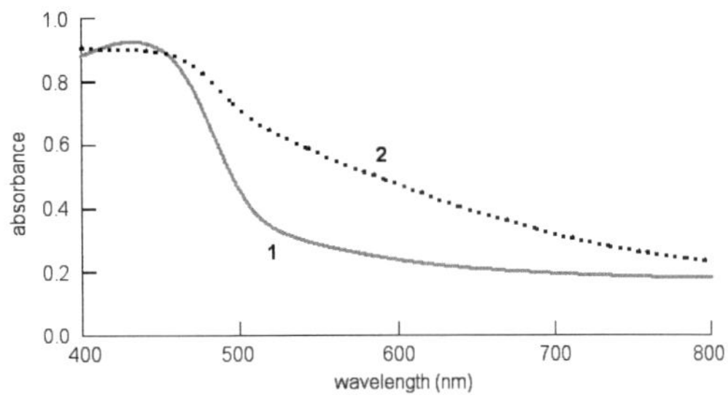

Figure 3a Visible reflectance spectra of yellow edge braid from Ha-1 and an amur cork tree standard. 1: amur cork tree standard; 2: yellow edge braid.

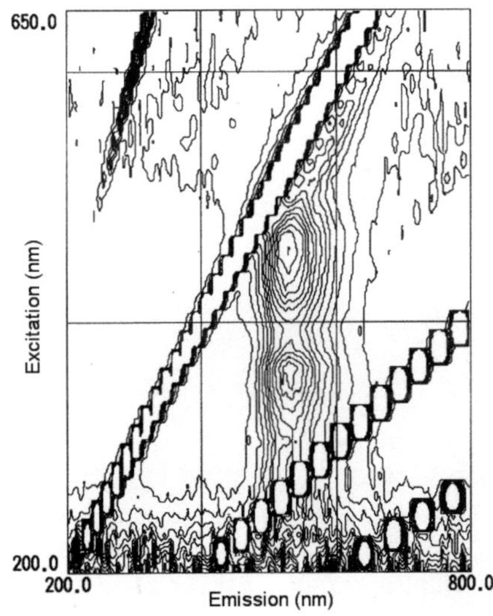

Figure 3b The fluorescence spectrum of the Ha-1 yellow edge braid; excitation wavelength range 200–650 nm, emission scanned from 200 to 800 nm.

Green

Because green was achieved by double dyeing of blue (indigo) and a yellow dye, the sum peak of indigo and of a yellow dyestuff is observed on measuring the visible reflectance spectrum (Fig. 4a). The yellow dyestuff used in this experiment appeared to be that from amur cork tree from its fluorescence spectrum, although spectrum broadening was observed. The fluorescence spectrum was used for the identification of amur cork tree because the dyestuff has a strong fluorescence (Fig. 4b). When a sample dyed with indigo and amur cork tree was examined, it was found that the fluorescence intensity in the vicinity of 510 nm does not change by using an excitation wavelength of 455 nm, although the fluorescence of amur cork tree was weaker due to the double dyeing with indigo. As a result of the measurement of the visible reflectance spectrum and fluorescence spectrum, the yellow dyestuff was identified as amur cork tree and the presence of the blue dye suggested indigo. The gas–MS spectrum of the DMF extract of the sample showed the presence of the *m*/*z* 262 peak due to indigo (Fig. 4c).

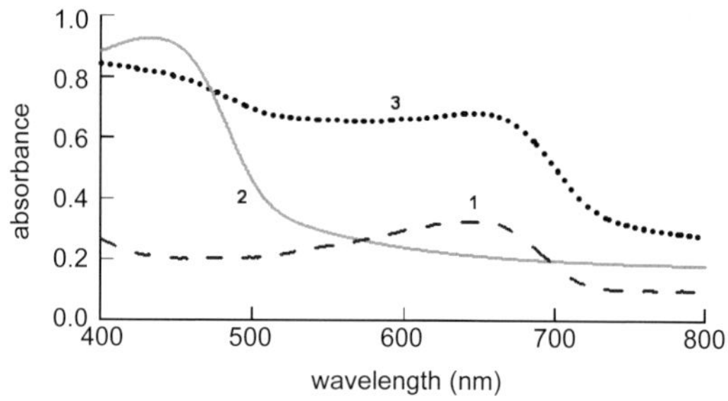

Figure 4a Visible reflectance spectra of green edge braid from Ha-1 and indigo and amur cork tree standards. 1: indigo standard; 2: amur cork tree standard; 3: green edge braid.

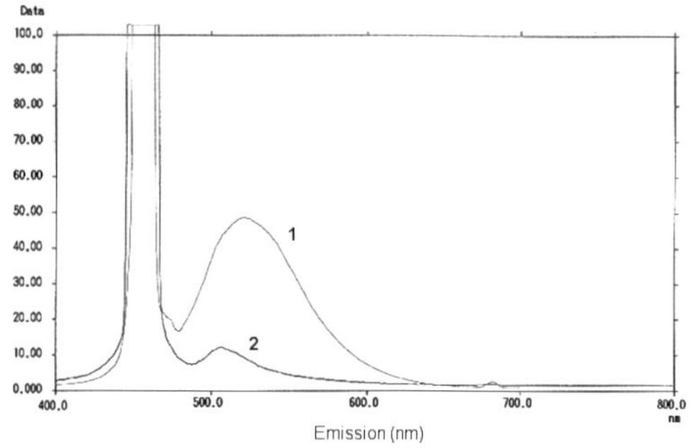

Figure 4b The fluorescence spectrum of the Ha-1 green edge braid; excitation wavelength 455 nm, emission scanned from 400 to 700 nm.

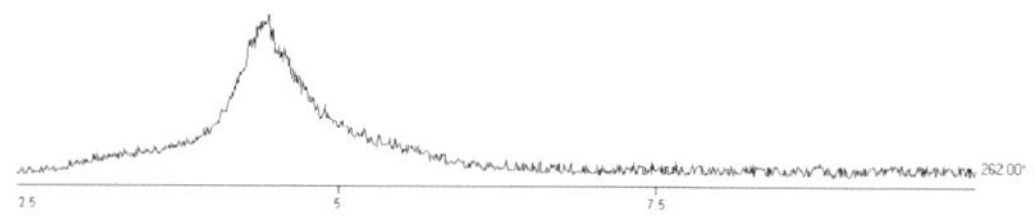

Figure 4c Gas–MS spectrum of the DMF extract of the green edge braid from Ha-1.

Analysis of multicoloured braids

Since it is generally difficult to obtain permission to untie the braid, a non-destructive procedure is required for dye analysis. The reflectance spectra of the tied multicoloured braids are then the sum of each component colour in the braids. The visible reflectance and fluorescence spectra of the still-tied *Uname* braid from the armour from Konogo-ji (Ko-3, white, purple and green) were measured (Fig. 5).

The *Takuboku* design for the *Uname* braid consists of four colour strings. The visible reflectance spectrum and the fluorescence spectrum, composed of the mixed spectra of purple and green (blue and yellow) showed that gromwell was used for purple and amur cork tree was used for yellow, one component of the green. Although it was thought that indigo was usually used as one component of the green colour, the

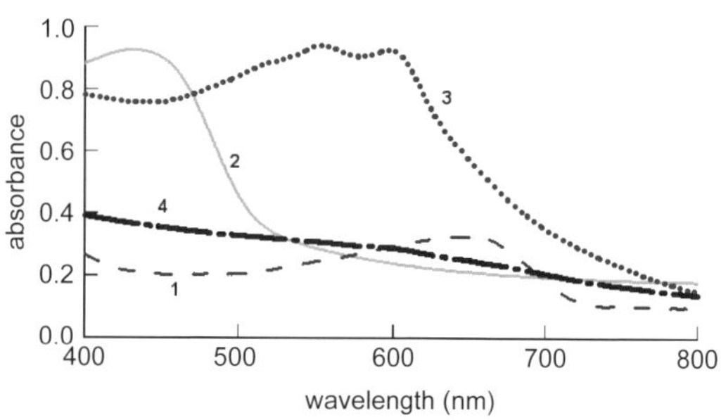

Figure 5 The visible reflectance spectra of *Uname* braid (tied) from Ko-3 and three standards. 1: indigo standard; 2: amur cork tree standard; 3: gromwell standard; 4: multicoloured *Uname* braid (mixed spectrum).

reflectance visible peak of indigo overlaps with that of gromwell; thus the identification of each component dyestuff without separation is not possible using visible reflectance spectrometry.

Storage conditions

In the case of the red lace braid of sample Ku-4, the colour varied in some areas. The dye technician suggested that the traditional double-dyed colour (munjeet and gromwell) might have been used for this sample. The gas–MS spectrum of the DMF extract of the sample recorded immediately after extraction showed a weak peak at m/z 262 (Fig. 6). This peak gradually disappeared after one week. Indigo is very easily dissolved in DMF as the extraction solvent and the peak at m/z 262 could be attributable to indigo. However, the chemical analysis does not show whether the double-dyed braid was intentionally dyed with indigo or whether the colour change was due to some unknown reason. The use of a madder group dye (perhaps a Japanese madder, munjeet) was suggested by the gas–MS spectrum of the sample measured one week later.

When the sample of variably coloured red lace braid was checked again, it was found that there was a colour difference on one braid. However, it appears that there is no necessity for changing the colour only in one portion from the point of view of the pattern design. The armour was originally stored in a wooden box called *Yoroibitu*. This armour (Ku-4) had once been kept in very poor storage conditions. Therefore, the reason for the presence of indigo is considered to be due to contamination rather than to double dyeing.

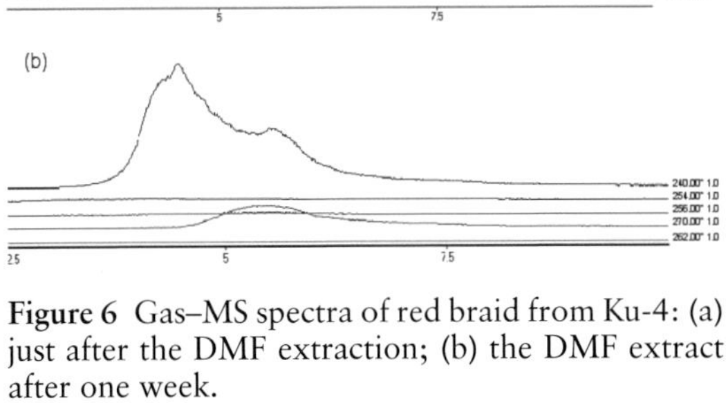

Figure 6 Gas–MS spectra of red braid from Ku-4: (a) just after the DMF extraction; (b) the DMF extract after one week.

Naming of Japanese armour and the relationship to the dyestuff

In Japan, the armours are usually named after their colours and the design of the lace braid. For example, *Doumaru* (pattern with red shoulder) with white braid lacing with a helmet and a large sleeve (K-2) is explained as follows: 'it is made of many plates connected by white lace braid and at the shoulder red lace braid is used for the requirement of the design. The helmet and the large sleeve for protecting the shoulder are also a set.' The armour name shows the colour name and does not indicate the dyestuff used for the braids. Thus for red lace braid, one cannot distinguish from the name whether, for example, it was dyed by safflower or by a madder group dye.

The Kongo-ji Temple published the repair report for the important cultural properties held there in 1969, including some red *Hishinui* braids for *Haramaki* with leather lacing. The report stated that some were dyed using a madder group dye and some were dyed with safflower. In this study, the presence of munjeet was identified for *Haramaki* No.6 (Ko-1; see Table 2) and 7 by visible reflectance spectrometry and gas–MS spectrometry, the results thus correspond with their report. The red braids for *Haramaki* Nos.12 (Ko-2; Table 2) and 13 were reported to have been dyed with safflower, but they were discoloured and this could not be confirmed by visible reflectance spectrometry. The possibility that a madder group dye had been used was suggested by the results of gas–MS measurement (Fig. 7). Since chemical analysis was never performed in the past, the suggestion in the report that the red braids were dyed with safflower is uncertain. They have lost all colour and it was impossible to extract any trace of dyestuff for identification. In general, once the name of the dyestuff was documented in the ancient literature, the conclusion that it

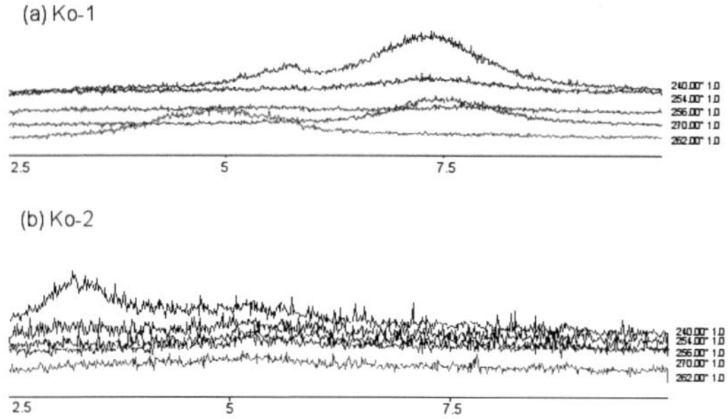

Figure 7 Gas–MS spectra of red braids from Ko-1 and Ko-2.

had therefore been used was accepted, even without chemical analysis. It is difficult to identify these dyes completely by chemical analysis and observation by the dyeing specialist offers important information. It is important to apply chemical analysis and then to consider reports about what is thought to have been identified, or suggested from the data in the light of the dyeing specialist's observations.

Acknowledgements

We would like to thank the Kushibiki-Hachimangu Shrine, the Kasuga-Taisha Shrine, the Hase-dera Temple and the Kongo-ji Temple for the donation of the samples and for permission to publish the results of this research, and we are also grateful to Akiko Miyazaki, the dyeing specialist, for her donation of the silk dyed with munjeet as the standard sample.

This research is partly sponsored by Grants-in-Aid for Scientific Research from the Japan Society for the Promotion for Science (Y. SASAKI No. 11610366).

Editor's note

*Author for correspondence: Museum and Archives, Kyoto Institute of Technology, Matsugasaki, Sakyo-ku, Kyoto 606-8585, Japan (ysasaki@kit.ac.jp). When this research was carried out the author was employed at the Conservation and Restoration Science Laboratory, Nara National Research Institute for Cultural Properties, Nara 630-8577, Japan.
Paper received 25 March 2004; revised version received 5 July 2004.

Notes and references

1. Matsumoto, K. (1993) *Shōsōin Textiles*, Kyoto.
2. Yamazaki, S. (2001) *A Secret of Ancient Dye-Techniques for 2000 Years* [in Japanese], Tokyo; Yoshioka, S. (2002) *Dye for Japanese Colour* [in Japanese], Tokyo; Uemura, R. and Takagi, Y. (1952, 1959, 1967, 1970) 'Bulletin, Study on the Japanese Culture in Relation to the Imperial Family and Court' [in Japanese], Nara (Japan), vols 11, 14, 19, 22.
3. Noda Y., Shimoyama, S. and Kasamatsu, M. (2003) 'The reproduction of a traditional votive figure based on the non-destructive analysis of colorants', *Dyes in History and Archaeology* 19, pp. 67–73.
4. Sasaki, Y. *et al.* (2002) 'Mass spectrometric identification of natural dyestuffs for ancient textiles', paper presented at the International Congress on the Conservation and Restoration for Archaeological Objects, Nara, 14–16 February 2002; Sato, M. and Sasaki, Y. (2003) 'Blue and purple dyestuffs used for ancient textiles', *Dyes in History and Archaeology* 19, pp. 100–105.
5. For a further example on the use of fluorescence spectroscopy in the identification of yellow dyestuffs, see Matsuda, Y. (1997) 'Non-destructive analysis of yellow natural dyestuffs by three-dimensional fluorescence spectroscopy', *Bunkazai Hozon-Syuhuku Gakkaishi* 41, pp. 54–63.

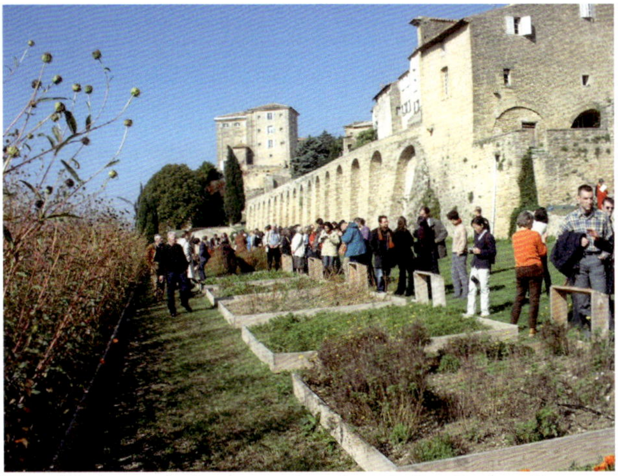

Plate 1 Delegates in the dye garden, Association Couleur Garance, Le Château, Lauris. (Photo: Chris Cooksey)

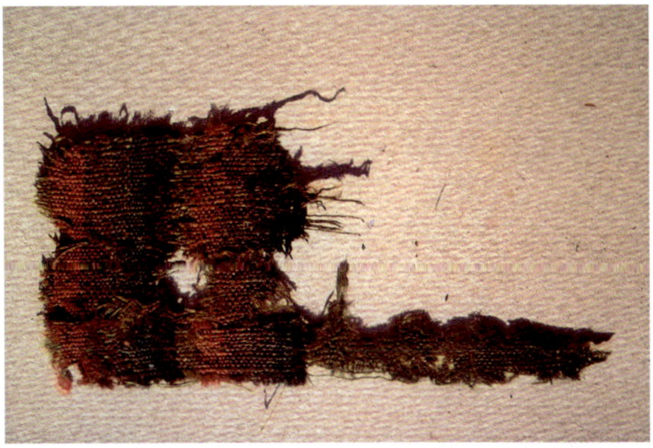

Plate 2 Fragment of tunic Z95.14011.2 from Maximianon. The green ground of the fabric and all the colours of the 'shaded band' are due to combinations, in different proportions, of an indigoid dye with madder of two different types. (Photo: D. Cardon)

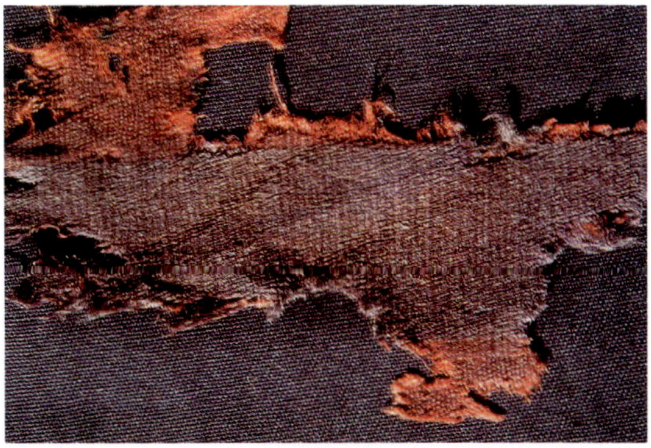

Plate 3 Fragment of tunic K3737.1 from Krokodilô. The mauve colour of the tapestry *clavus* is produced by a combination of royal purple, kermes and madder. (Photo: D. Cardon)

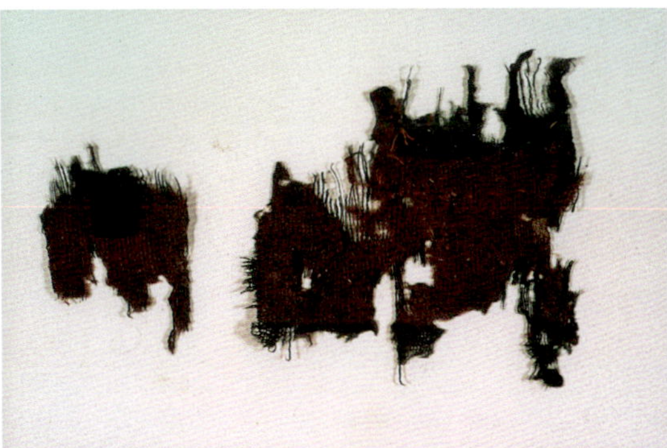

Plate 4 Two fragments of mantle Z95.32029.9 from the rubbish heap of Maximianon. The angle-shaped tapestry motifs are dyed with a combination of royal purple and kermes. (Photo: D. Cardon)

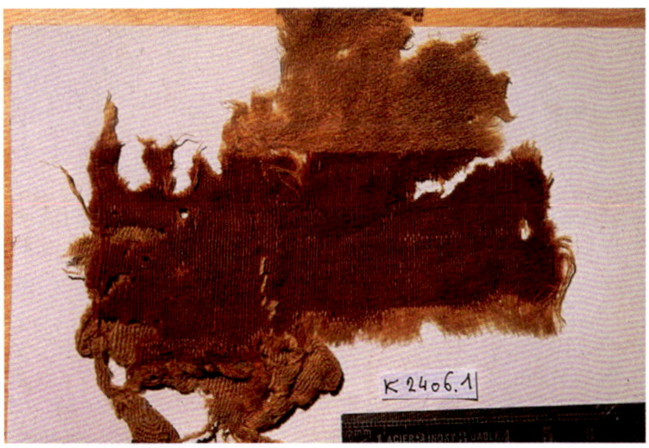

Plate 5 Fragment of tunic in damask weave from Krokodilô, K2406.1. The pink *clavus* is the result of a subtle combination of purpurin-rich madder, indigo and traces of tannin and weld. (Photo: D. Cardon).

Plate 6 (a) Right: fragment of an Egyptian garment with the vertical *clavi* and the transverse neck decoration, the neckbands A and B. The 19 fields of the *clavi* are numbered from 1-1 to 1-10 and 2-1 to 2-9; (b) below: detail of plate 1a, field 2-9: Adam and Eve. Museum für angewandte Kunst, T 8528/ 1928. (Photo: E. Hütter © MAK, Vienna)

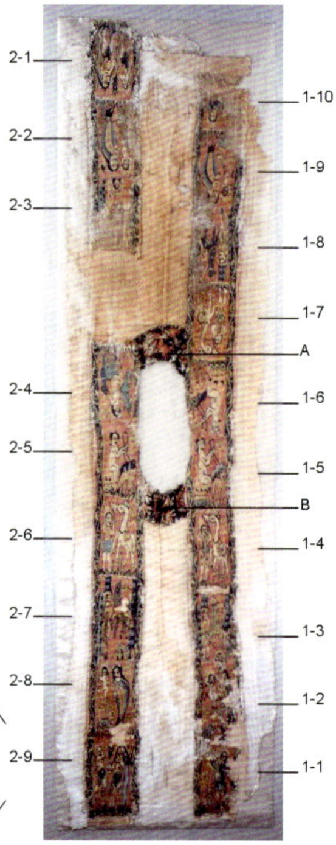

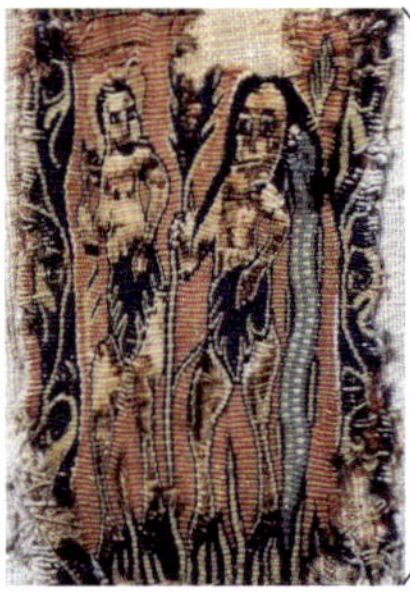

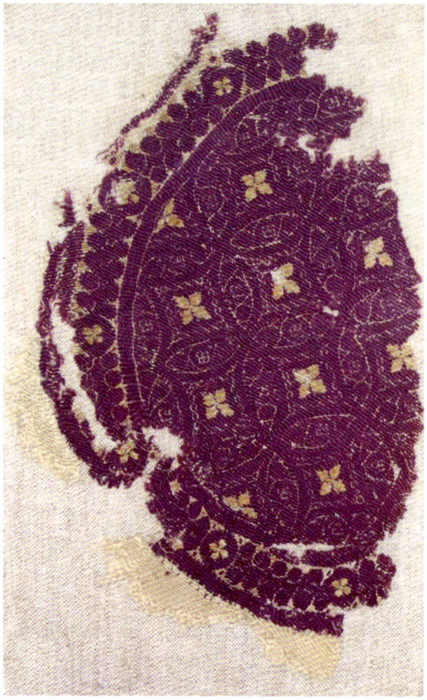

Plate 7 Fragment of a purple medallion, Museum für angewandte Kunst, T 6025/1905 = T 10069/1953 (© MAK, Vienna).

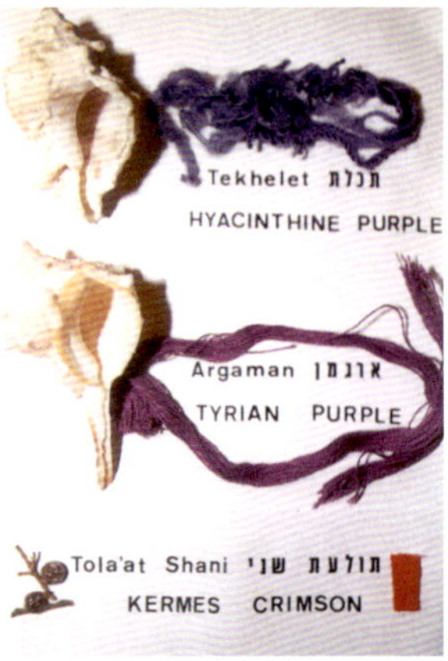

Plate 8 The three biblical dyes: *tekhelet* (hyacinthine purple, 'blue') from banded dye-murex; *argaman* (Tyrian purple) from spiny dye-murex; and *tola`at shani* (crimson) from kermes scale insect on oak branches.

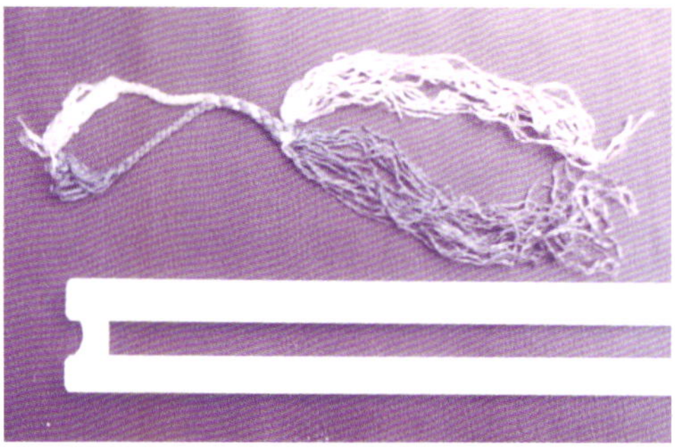

Plate 9 A single corner-tassel of a Karaite prayer shawl from Egypt; this tassel comprises 32 hanging cords, one half of which are blue. The loop at the left end of the tassel is torn open, where it had been tied to the garment corner. Donated to the Tekhelet Foundation by Curator Ezra Gorodetzky.

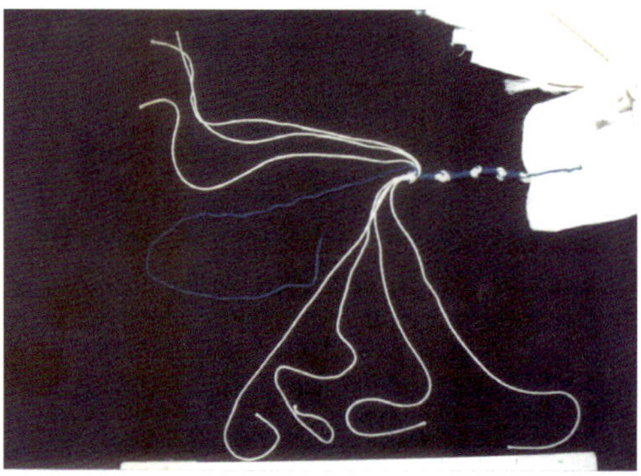

Plate 10 A single corner-tassel attached to a rabbinical Jewish prayer shawl. Here only one cord is *tekhelet*.

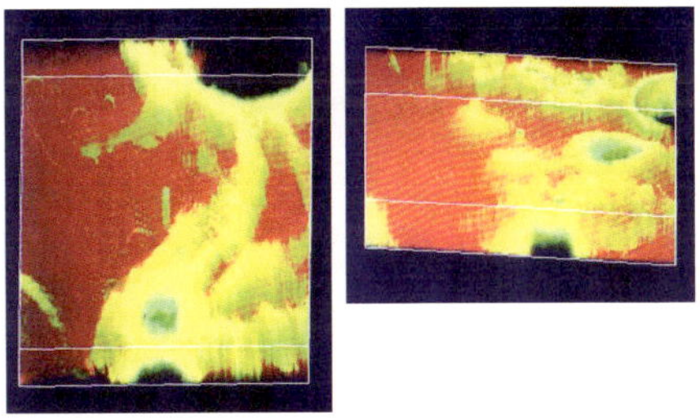

Plate 11 Two confocal scanning laser microscope views of a biofilm created by *Clostridium isatidis* growing on a slide to which indigo particles are attached. The bacterial colony fluoresces yellow, the indigo particles are blue-green and the slide surface fluoresces red.

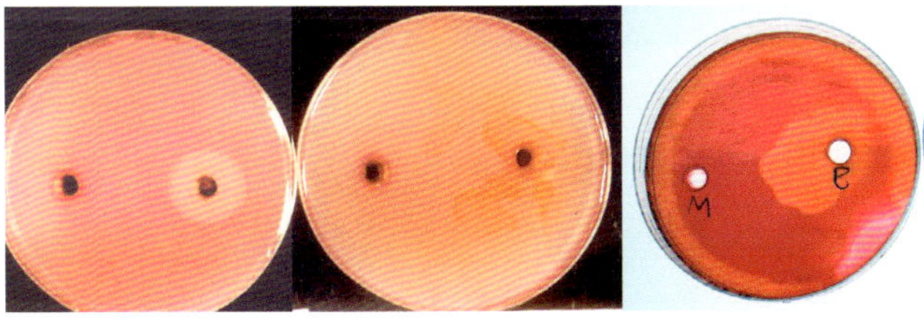

Plate 12 Detection of cellulase activity in bacterial cultures. Cellulase activity was determined by carboxymethylcellulose disappearance as detected by post-staining with Congo red. In each case, the left-hand wells contained a water control; the right-hand wells contained a sample of bacterial culture. From left to right, the cultures added to the plates were of (a) *Cellulomonas* sp., (b) *Clostridium isatidis* (grown with glucose), and (c) a sample of a woad vat.

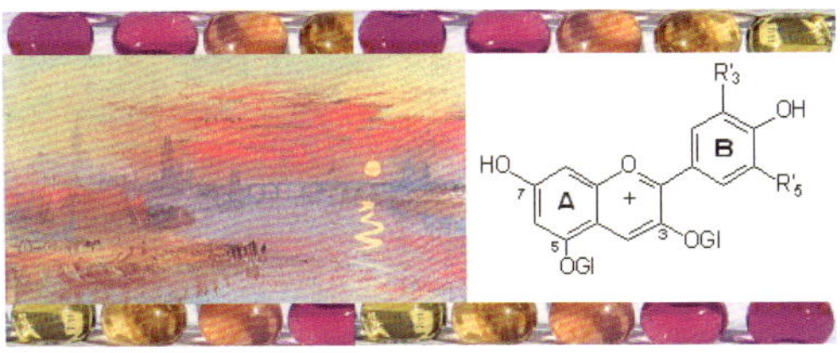

Plate 13 Left: J.M.W. Turner, *The Scarlet Sunset* (London, Tate Britain, D24666), *c*.1820–40. Watercolour and gouache on paper, 13.4 × 18.9 cm. (Reproduced by permission of Tate Britain.) Right: the basic chemical structure of anthocyanins, explained in greater detail in Figure 1. The two bands at top and bottom of the plate show the colour variation with pH of the anthocyanin, luteolinidin.

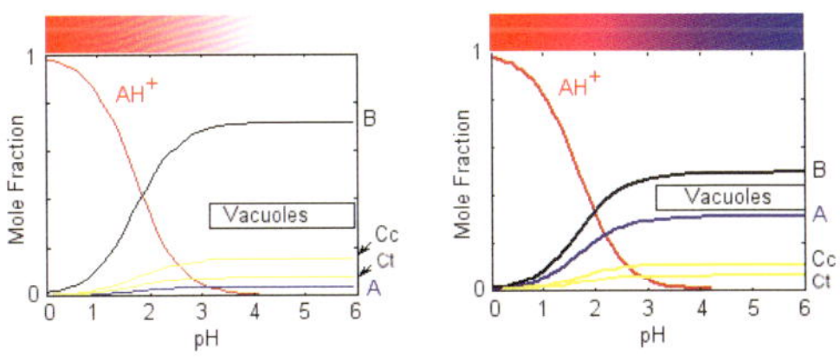

Plate 14 Mole fraction distribution of the several species of cyanin (cyanidin-3,5-diglucoside): a (left) free (pK_{ap}=1.7); b (right) simulation in the case of co-pigmentation with A for a co-pigment concentration of 0.1 M and an association constant, K'_{cp} =100 M^{-1}. The band at the top simulates the colour that is obtained at different pH values. For details about the physical meaning of pK_{ap}, K_{cp} and K'_{cp} (see Melo *et al.* 2000 (note 13)).

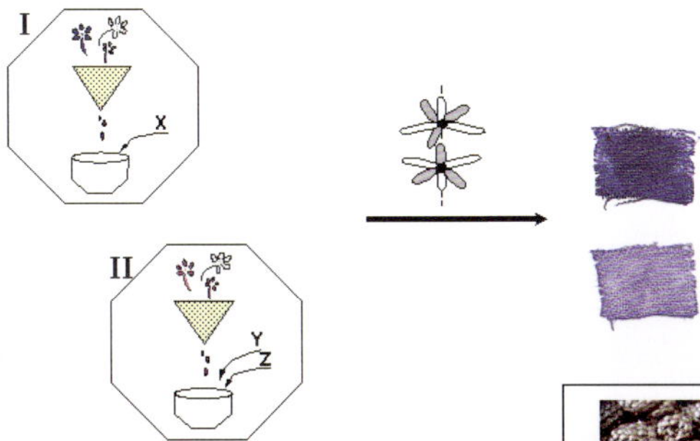

Plate 15 Making clothlets: X = ammonium chloride (sal ammoniac; NH_4Cl), Y = alum (aluminium potassium sulfate, $AlK(SO_4)_2 \cdot 12H_2O$), and Z = gum arabic. I. On the first day the flower petals were mashed to a pulp in a mortar; the pulp was pressed out through a clean cloth. A small amount of ammonium chloride was added to the juice from the petals, then clean linen cloths were steeped in the juice and after they had taken up enough of the colorant, they were hung to dry in the dark. II. The next day the whole procedure was repeated, but this time the juice was treated with gum arabic and alum, before the same cloths were steeped again in it. After that the cloths were thoroughly dried and finally carefully stored, away from humidity.

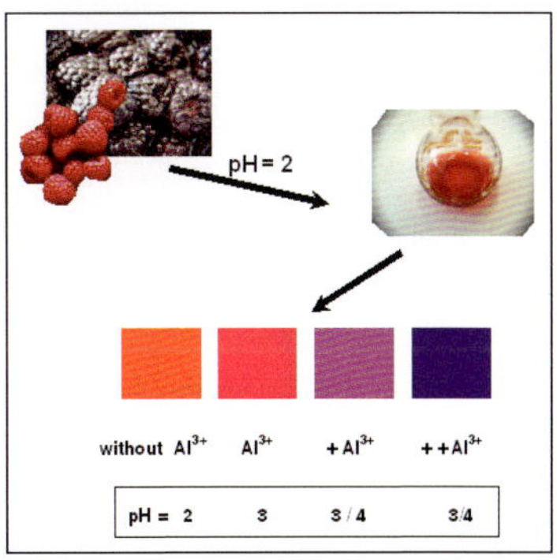

Plate 16 Cyanidin chromophore, with sugar substitution in position 3, can be extracted in moderately acidic medium from blackberries or raspberries. This solution can be used to dye parchment, where the final colour will depend on the pH of the solution and on the Al^{3+} concentration.

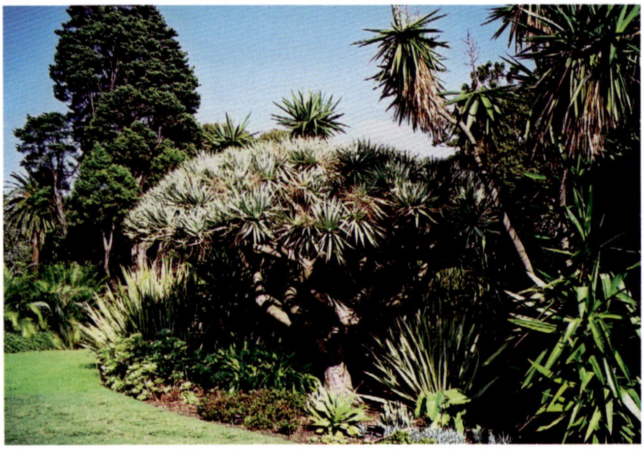

Plate 17 Dragon's blood tree, *Dracaena draco* L. (Liliaceae), Royal Botanic Gardens, Melbourne. (Photo: Jo Kirby)

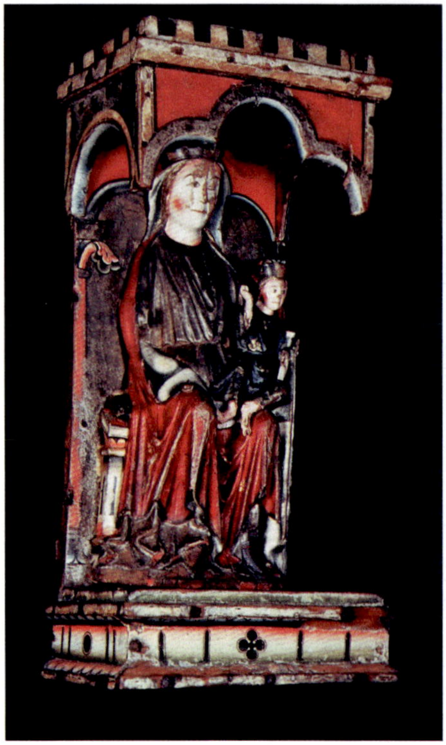

Plate 19 The *Virgin and Child* from Dal, Norway (Oslo, University Museum of Cultural Heritage, Inv. no. C 1410), *c*.1250–75. Polychromed wood, shrine 86.8 × 43.4 × 29.3 cm, sculpture 64 × 31 × 13 cm.

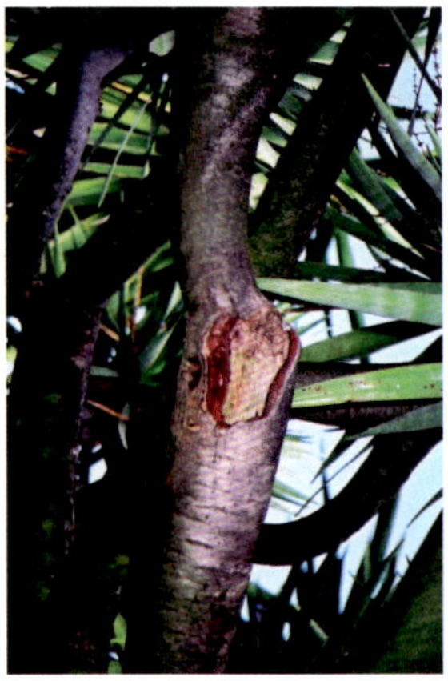

Plate 18 One of the branches of the dragon's blood tree illustrated in Plate 17, showing dried resin which had oozed from a damage in the bark. (Photo: Jo Kirby).

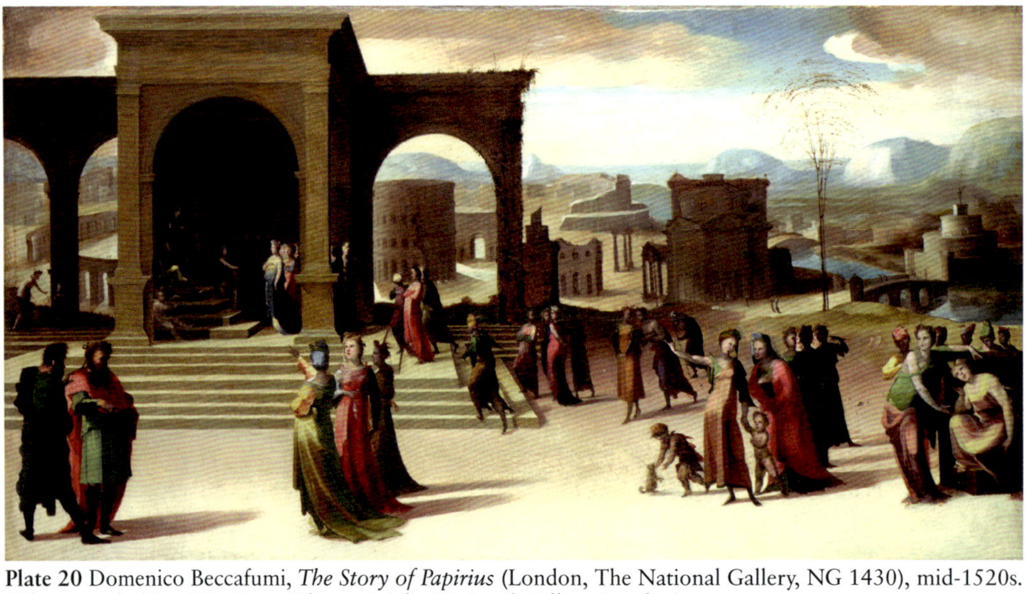

Plate 20 Domenico Beccafumi, *The Story of Papirius* (London, The National Gallery, NG 1430), mid-1520s. Oil on panel, 74 × 137.8 cm. (Photo: © The National Gallery, London)

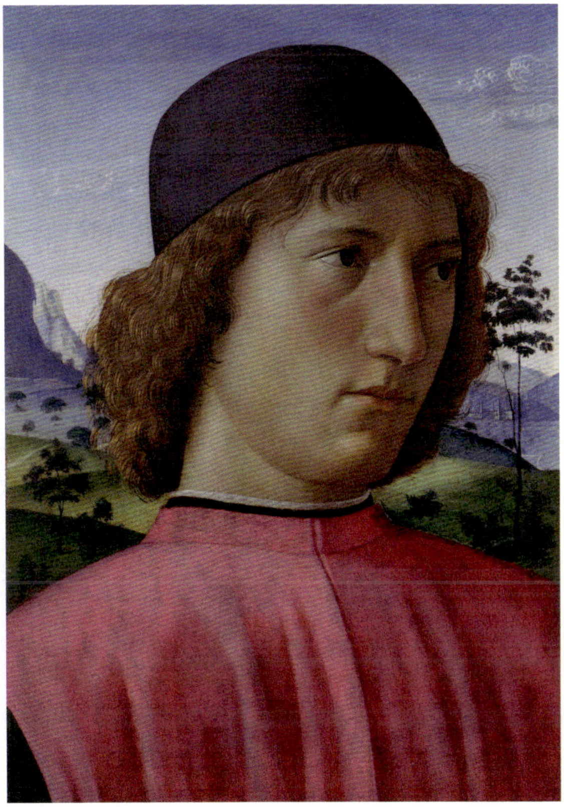

Plate 21 Domenico Ghirlandaio, *Portrait of a Young Man in Red* (London, The National Gallery, NG 2489), probably *c.*1480–90. Tempera on panel, painted surface 38.7 × 27.6 cm. (Photo: © The National Gallery, London)

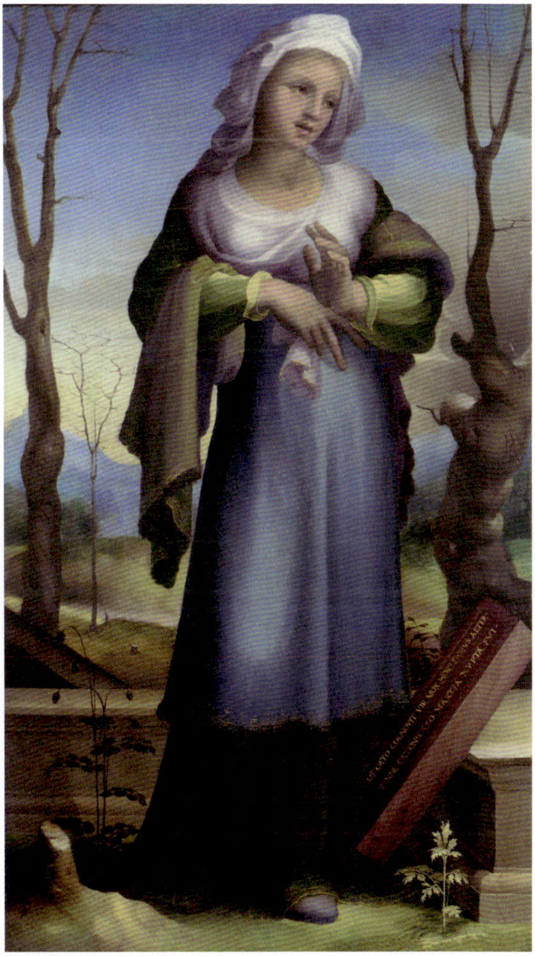

Plate 22 Domenico Beccafumi, *Marcia* (London, The National Gallery, NG 6369), *c.*1520–25. Oil on panel, 92.1 × 53.3 cm. (Photo: © The National Gallery, London)

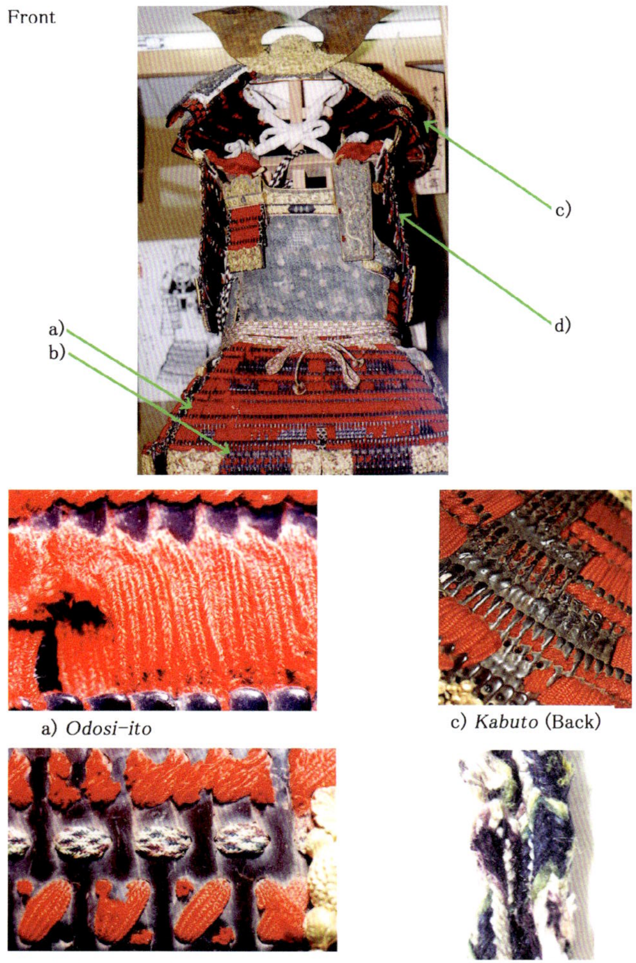

a) *Odosi-ito*

b) *Uneme* at *Kusazuri*

c) *Kabuto* (Back)

d) *Mimi-ito*

Plate 23 *Yoroi* with red lacing (Ku-4), 14th century, Kushibiki-Hachimangu Shrine, Aomori Prefecture, Japan.

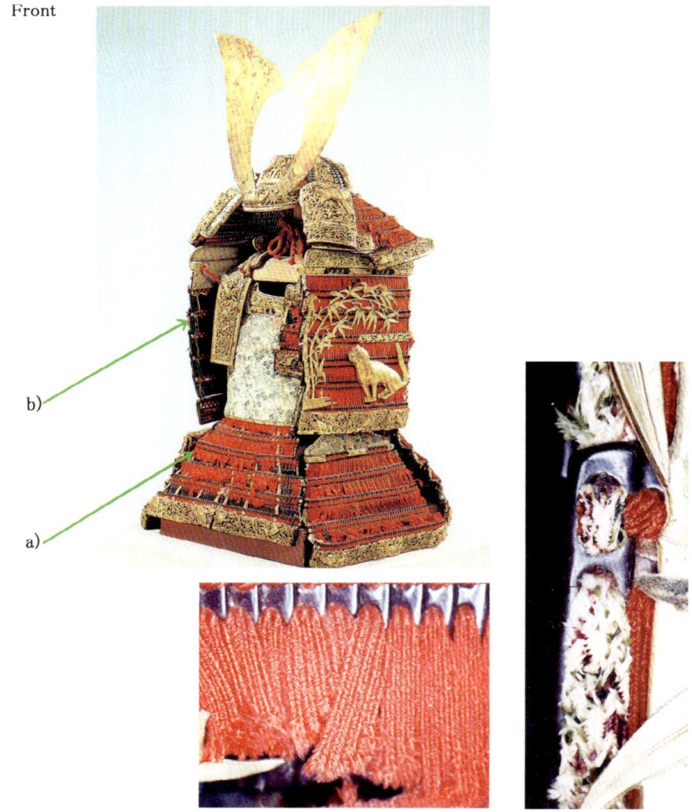

a) *Odoshi-ito*

b) *Mimi-ito*

Plate 24 *Yoroi* with red lacing (Ka-1), 14th century, Kasuga-Taisha Shrine, Nara Prefecture, Japan.

Front

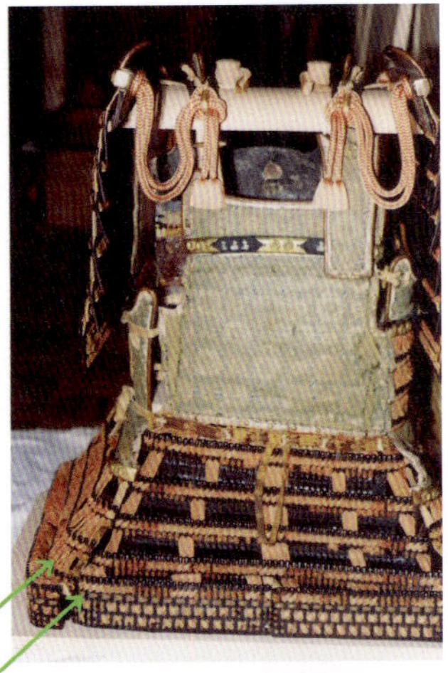

Back

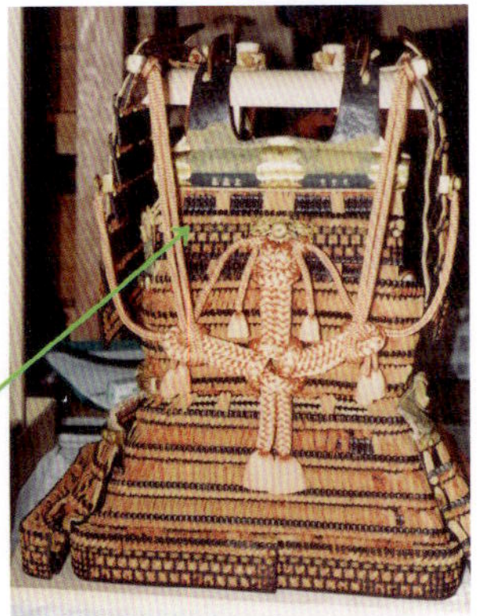

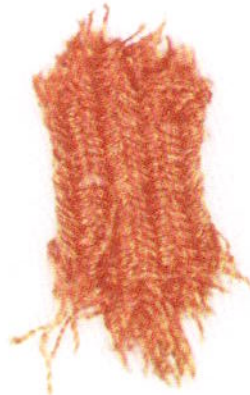

a) *Odosi-ito*

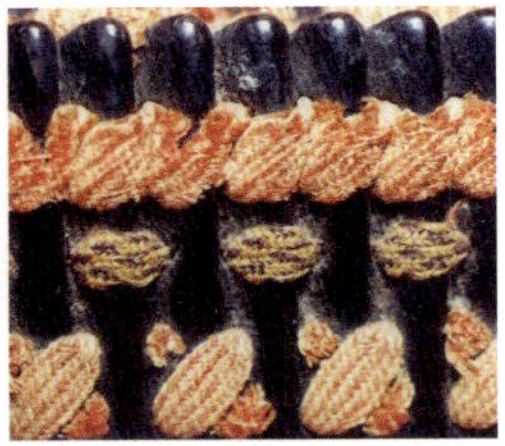

b) *Uname at Kusazuri*

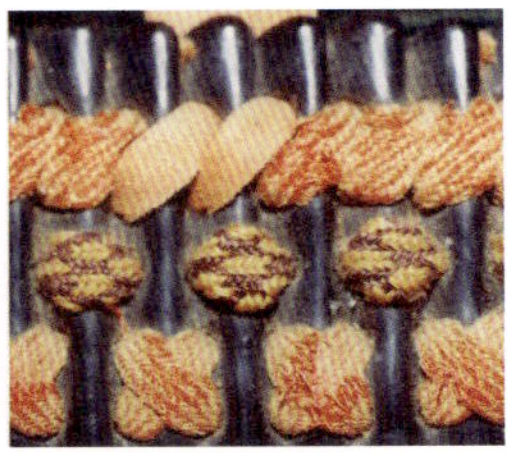

c) *Uname at Sakaita*

Plate 25 *Yoroi* with red lacing (Ha-1), 15th century, Hase-dera Temple, Nara Prefecture, Japan.

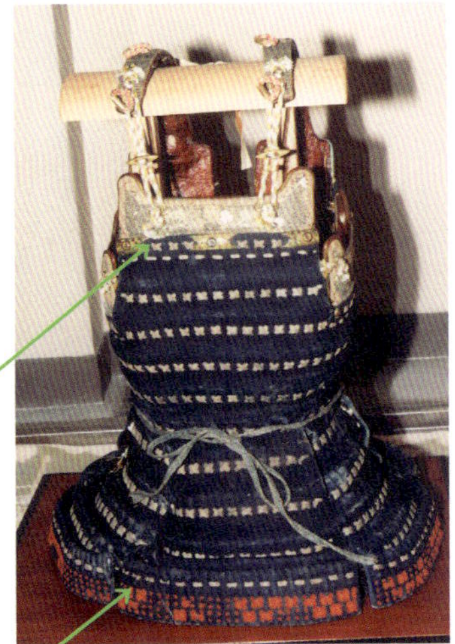

Front

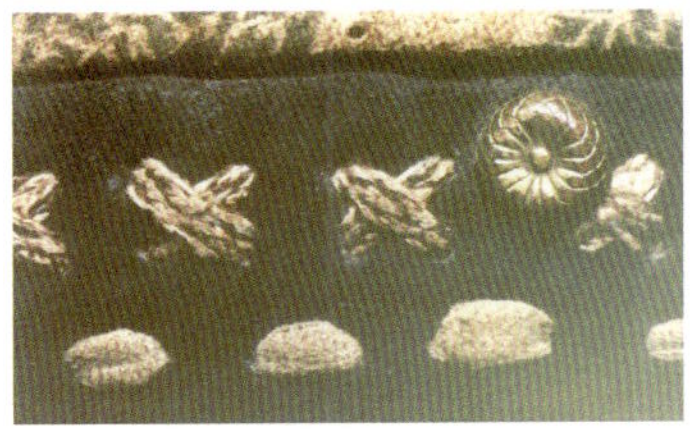

a) *Hanagarami*

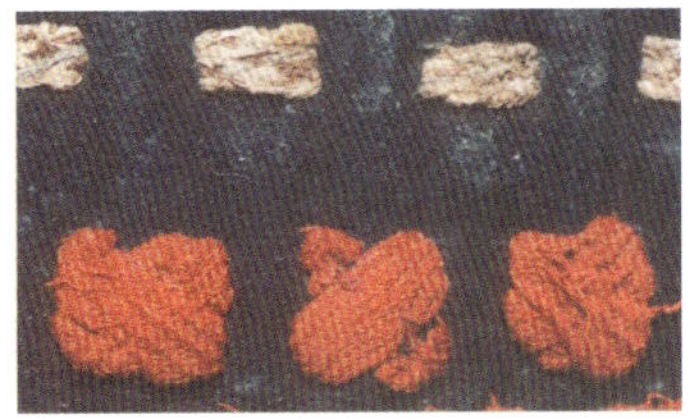

b) *Hisinui and Uname*

Plate 26 *Haramaki* with blue leather lacing (Ko-1), 14th–16th century, Kongo-ji Temple, Osaka, Japan.

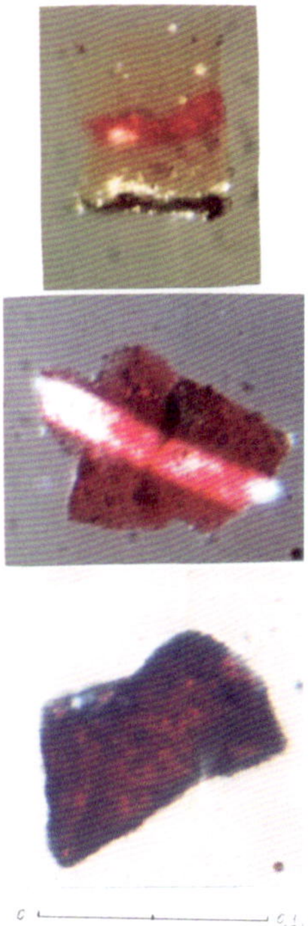

Plate 27 Photomicrographs of cross-sections of samples of 17th-century paint: (i) (top) protein-containing cochineal lake from 17th-century renovation of the 16th-century Greek icon, *The Mother of God*; (ii) (middle) sample of pure cochineal lake containing protein from the17th-century Russian icon *Saint Emmanuel*; (iii) (bottom) the same sample of lake paint stained with the histochemical dye Amido Black, which stains protein dark blue-black.

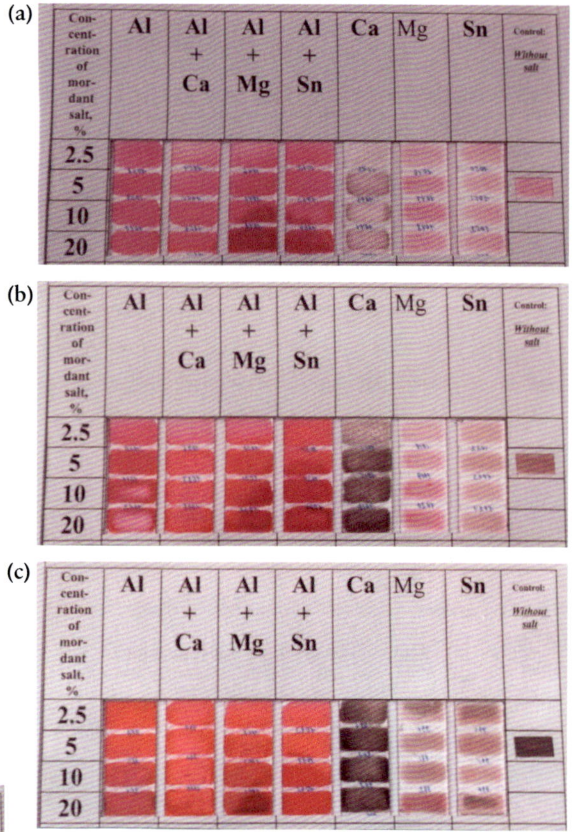

Plate 28 Dry films of three-component lakes, containing carminic acid, a mordant salt or salt mixture and the protein collagen, on glass. The films were prepared using four concentrations of each mordant salt solution and three concentrations of cochineal extract: (a) 0.06%; (b) 0.13%; (c) 0.26%. A control sample without mordant salt was also prepared.

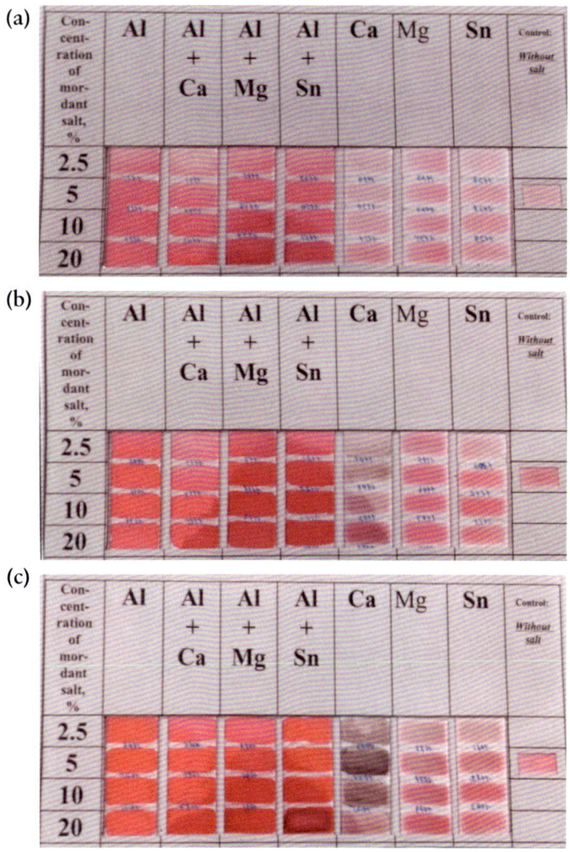

Plate 29 Dry films of four-component lakes, containing carminic acid, a mordant salt or salt mixture, collagen and cream of tartar, on glass. The films were prepared using four concentrations of each mordant salt solution and three concentrations of cochineal extract: (a) 0.06%; (b) 0.13%; (c) 0.26%. A control sample without mordant salt was also prepared.

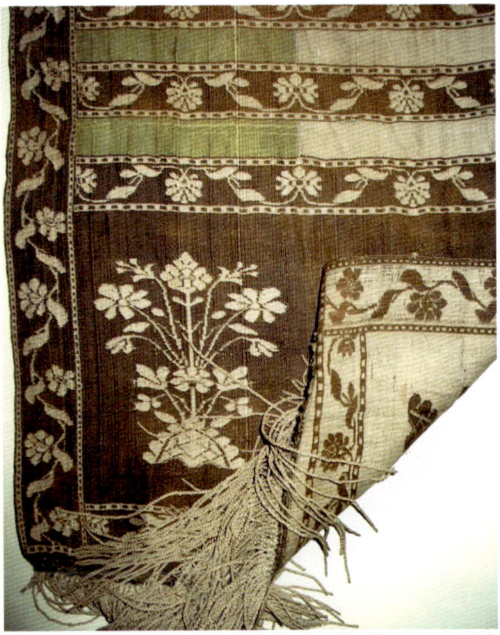

Plate 30 Waist sash of Lipków manfacture, 18th century: detail showing floral border, region of alternating narrow geometrical and floral-patterned stripes, a bouquet at the head of the sash and the two colours of the background between the stripes. The turned-back corner shows the reversed colours of the design on the other side of the sash. (Photo: Katarzyna Schmidt-Przewoźna)

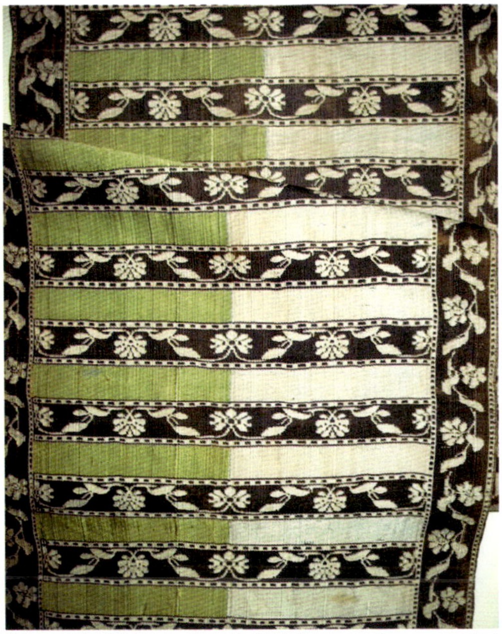

Plate 31 Waist sash of Lipków manfacture, 18th century: detail showing floral border and the floral-patterned stripes, bordered by a narrow geometrical-patterned stripe on each side. The division of the background into two colours, here green and creamy white, along the length of the sash is clearly seen. (Photo: Katarzyna Schmidt-Przewoźna)

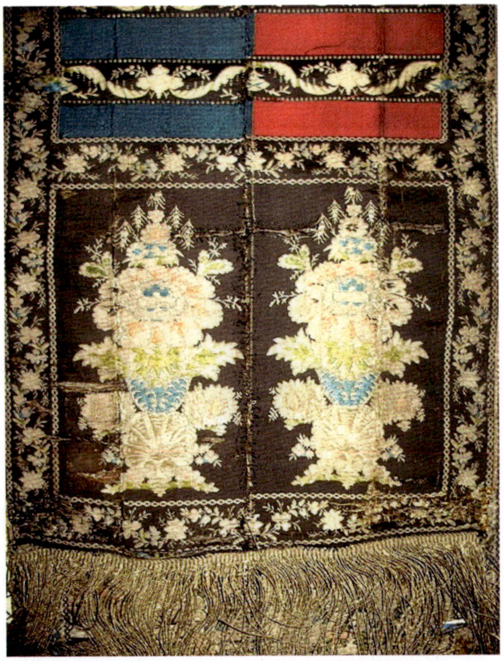

Plate 32 Detail of a waist sash from Kobyłki, showing paired elaborate designs (the 'head'), the floral border and stripes and the bright contrasting background colours. (Photo: Katarzyna Schmidt-Przewoźna)

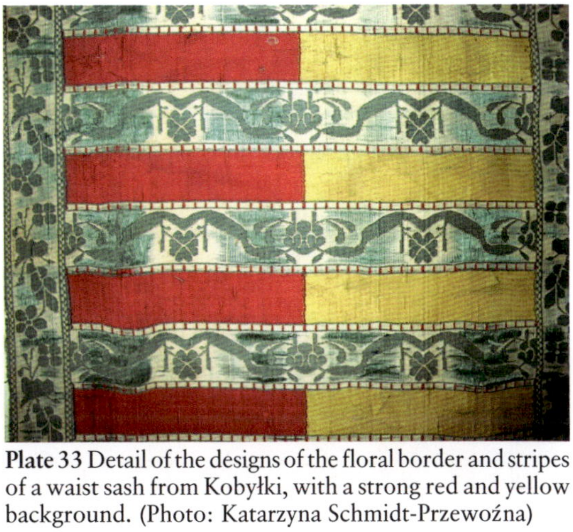

Plate 33 Detail of the designs of the floral border and stripes of a waist sash from Kobyłki, with a strong red and yellow background. (Photo: Katarzyna Schmidt-Przewoźna)

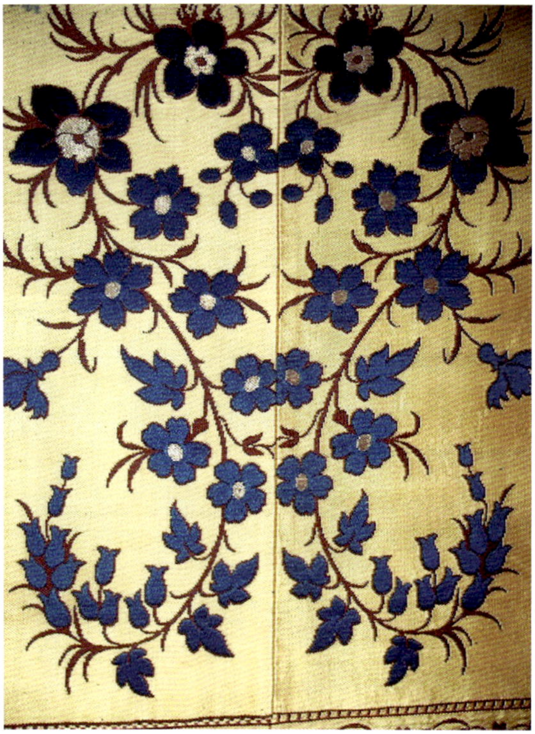

Plate 34 Detail of floral ornament of a Pashalis waist sash of the 18th century. (Photo: Katarzyna Schmidt-Przewoźna)

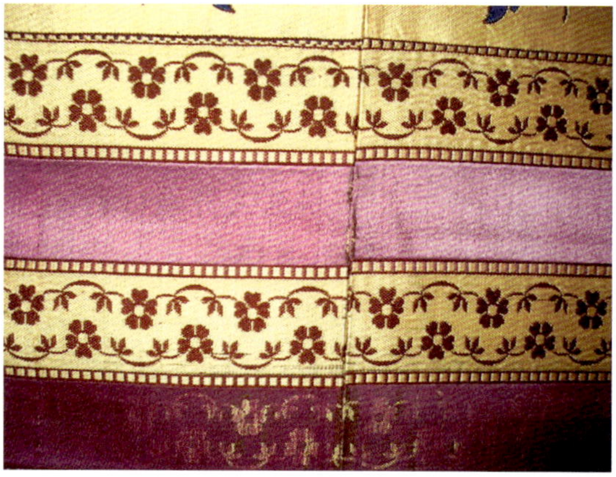

Plate 35 Detail of adjacent striped region from the same Pashalis waist sash. (Photo: Katarzyna Schmidt-Przewoźna)

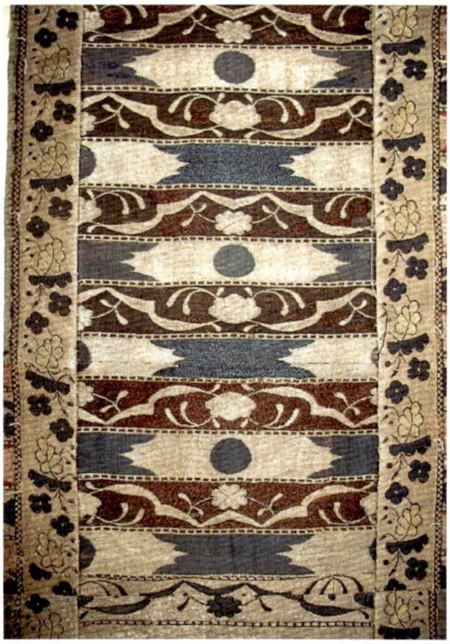

Plate 36 Detail of 18th-century waist sash of Słuck manufacture. (Photo: Katarzyna Schmidt-Przewoźna)

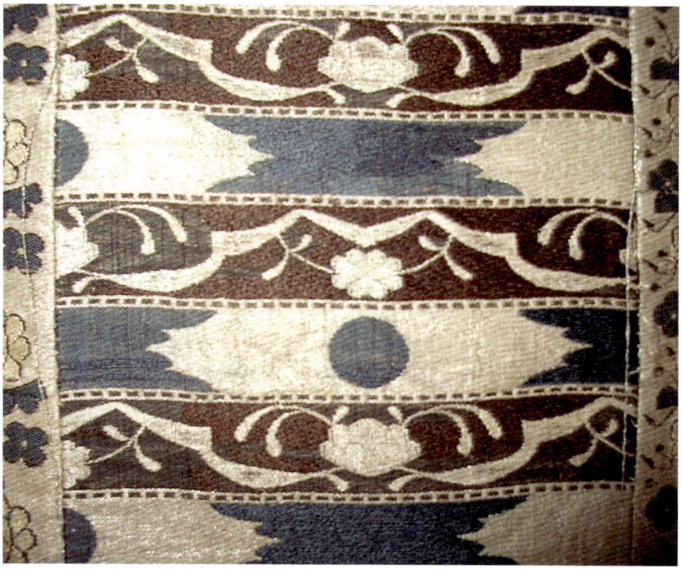

Plate 37 Enlarged detail of Słuck waist sash shown in Plate 36. (Photo: Katarzyna Schmidt-Przewoźna)

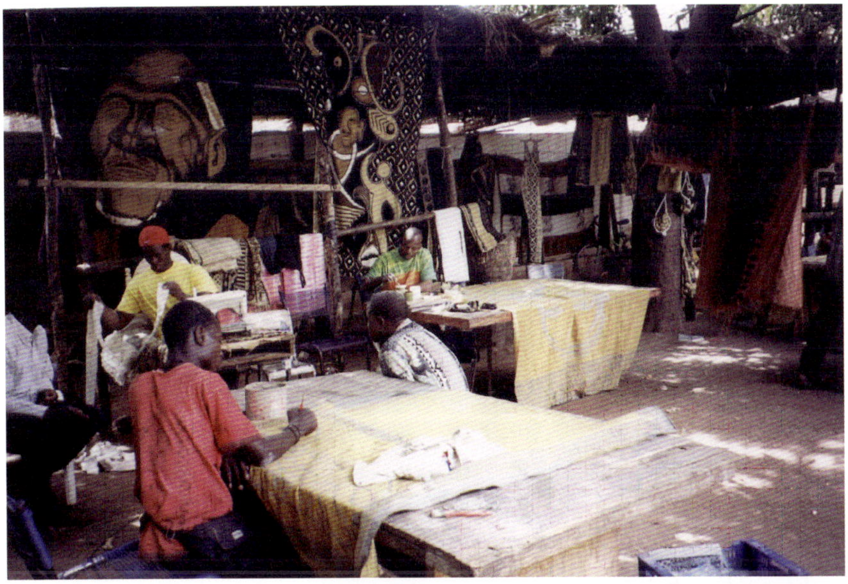

Plate 38 A Bogolan workshop in Bamako: young people drawing their own designs using traditional practices under the supervision of Augustin Sagara and Issa Niambélé, INA (National Institute of Arts). (Photo: Marie Marquet)

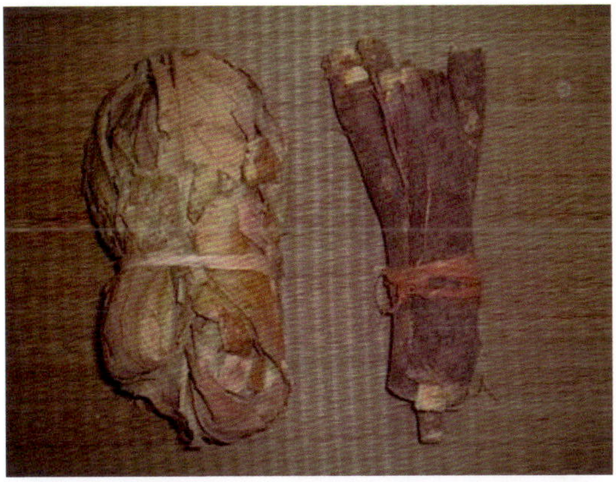

Plate 39 Leaves tied up in balls and bark of *Terminalia laxiflora*. (Photo: Marie Marquet)

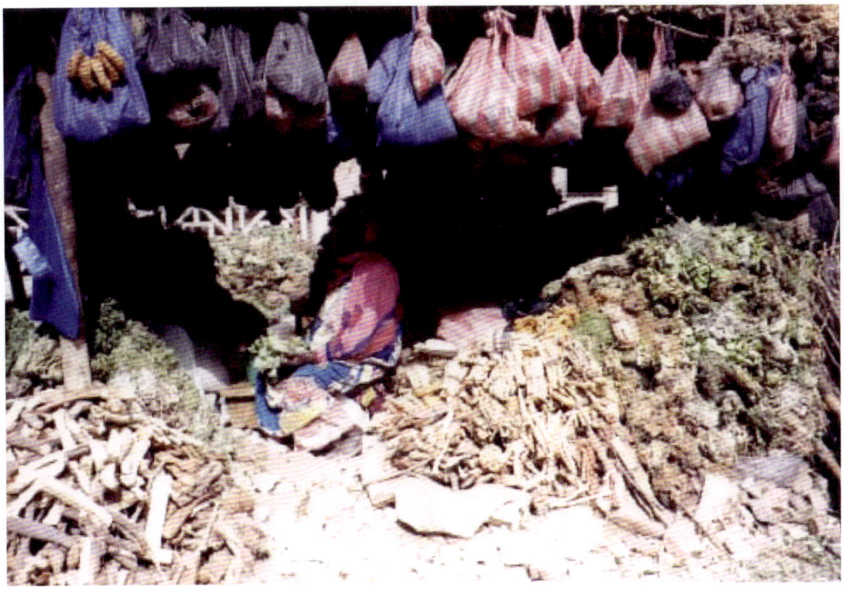

Plate 40 Plant shop at Dibida Market, Bamako. (Photo: Marie Marquet)

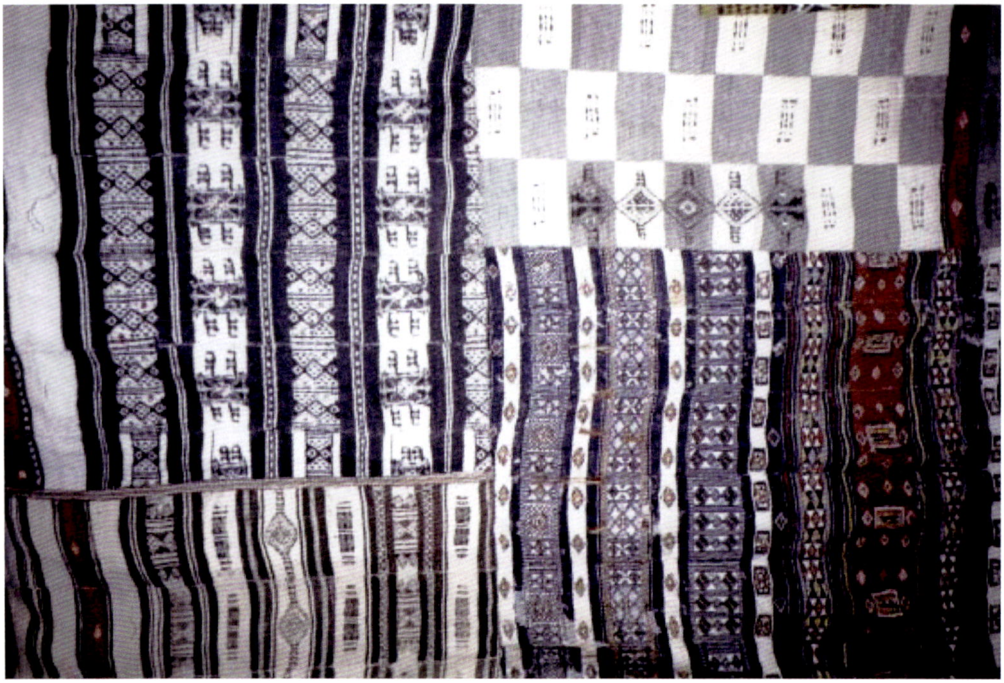

Plate 41 Fulani woollen blankets. (Photo: Marie Marquet)

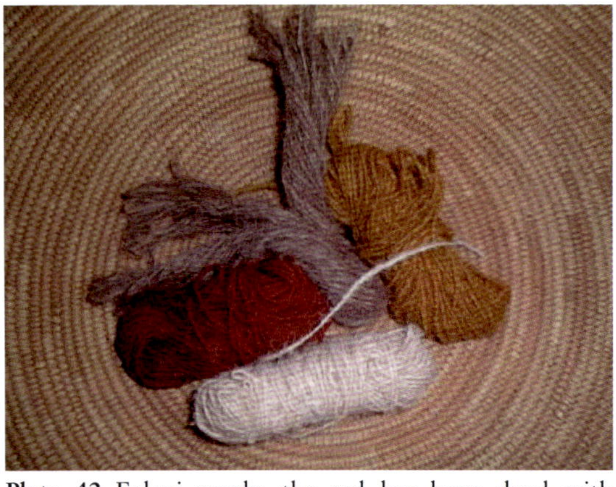

Plate 42 Fulani wools; the red has been dyed with *Hibiscus sabdariffa*, the yellow with *Podaxis pistilliaris*. (Photo: Marie Marquet)

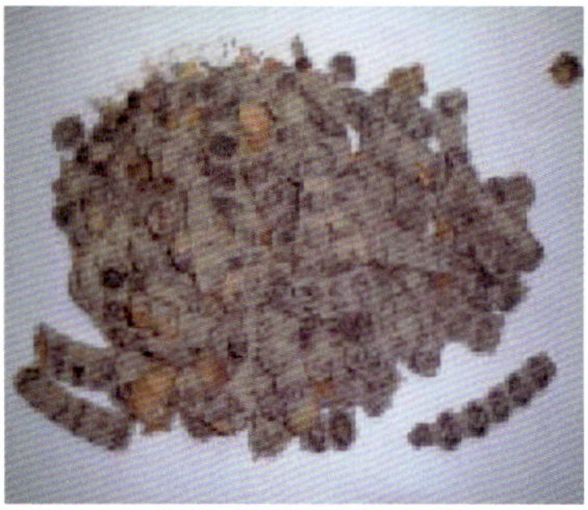

Plate 43 Pods of *Acacia nilotica* (L.) Willd. ex Del., used for dyeing black. (Photo: Marie Marquet)

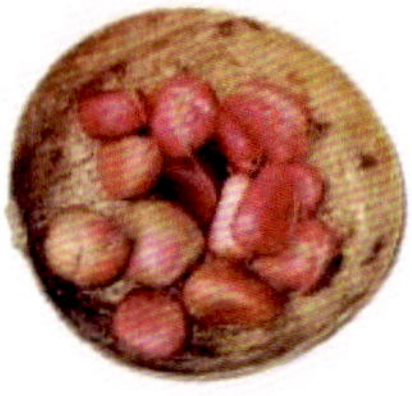

Plate 44 *Cola nitida* fruits are mashed to obtain an orange colour. (Photo: Marie Marquet)

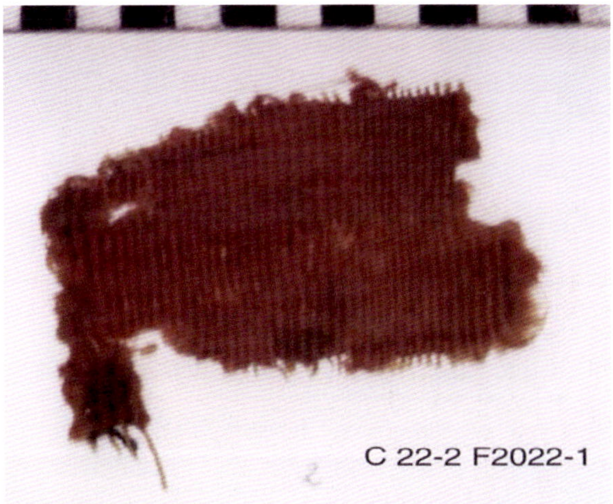

Plate 45 Woollen textile C22-2 with a surprising pink colour: the dye was extracted from lac insects. (Photo: Marie Marquet)

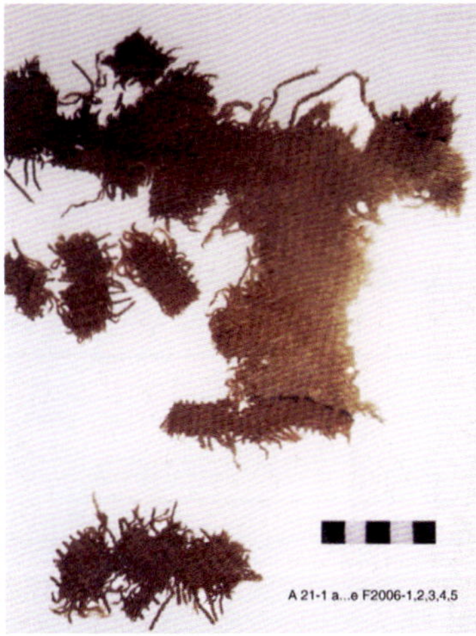

Plate 46 Woollen textile A21-1: black threads were dyed with ellagic acid associated with iron, and a yellow fragment which may have been been dyed with a *Sorghum* sp. They are associated with a third fragment, dyed with an unknown yellow colorant. (Photo: Marie Marquet)

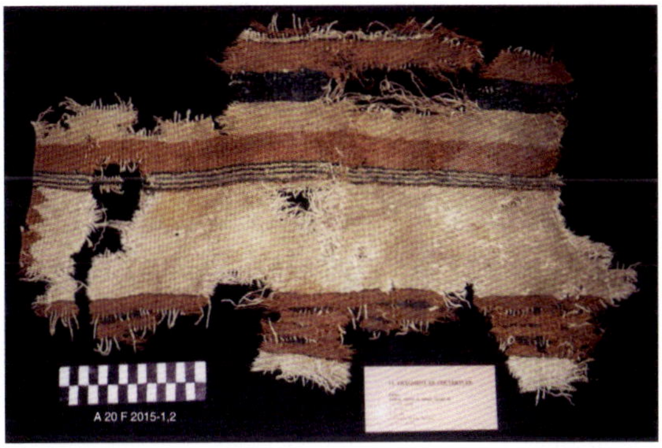

Plate 47 Textile A20: two samples, one red and one yellow, from this textile were found to contain apigenin, perhaps derived from a *Sorghum* sp. dye. (Photo: Marie Marquet)

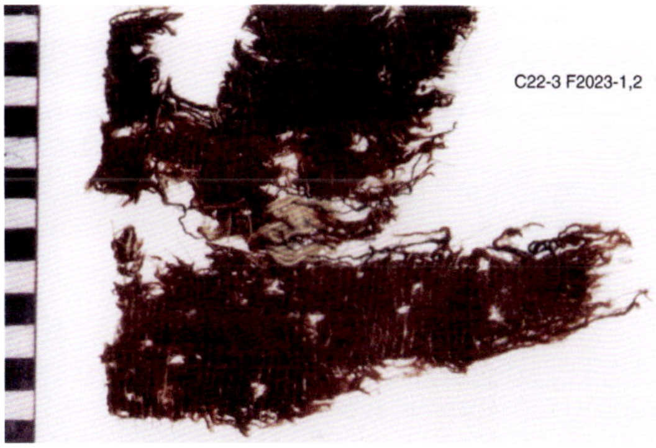

Plate 48 C22-3: an unidentified yellow colorant is present on this fine woollen textile from cave C. (Photo: Marie Marquet)

Plate 49 *Lo-kao*. From left to right: wool dyed with tin(II) chloride mordant (lilac); silk dyed with alum mordant (green); cotton dyed with weakly alkaline solution (blue-green); deep blue original dye deposited on cotton.

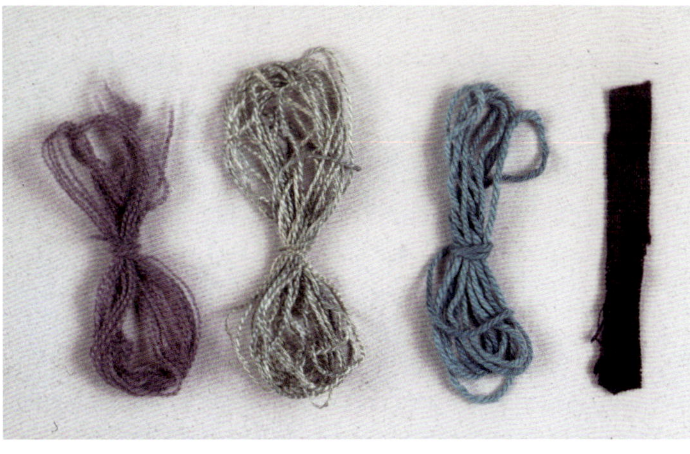

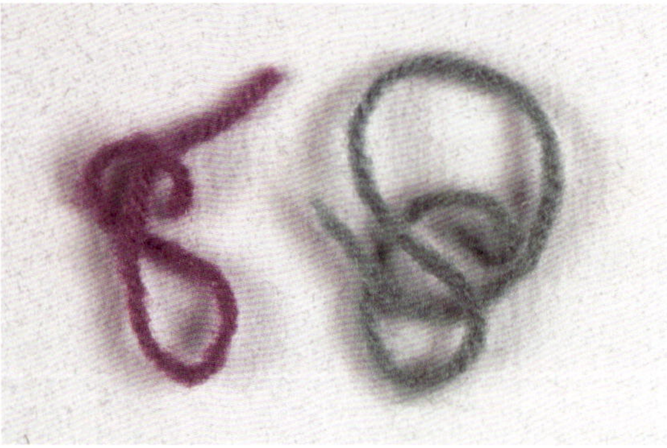

Plate 50 The dye from *Rhamnus* sp. (probably *R. cathartica* L.) bark on wool without mordant (purple) and with alum mordant (green).

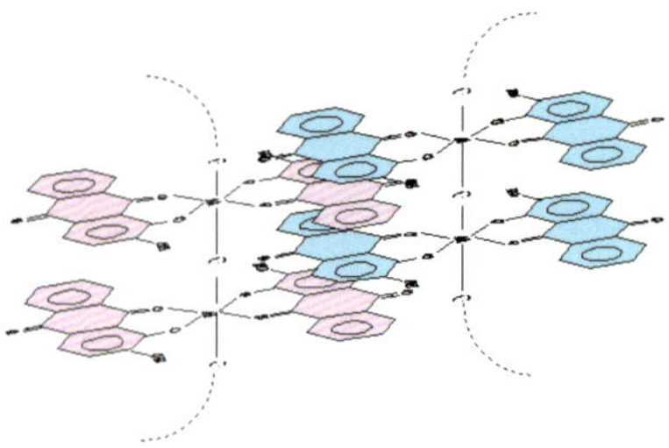

Plate 51 Model for alizarin: aluminium gels: 2:1 complexes, polymerised through Al–O–Al bonds, and the cross-linking of these polymers through interactions between their intercalated aromatic side rings.

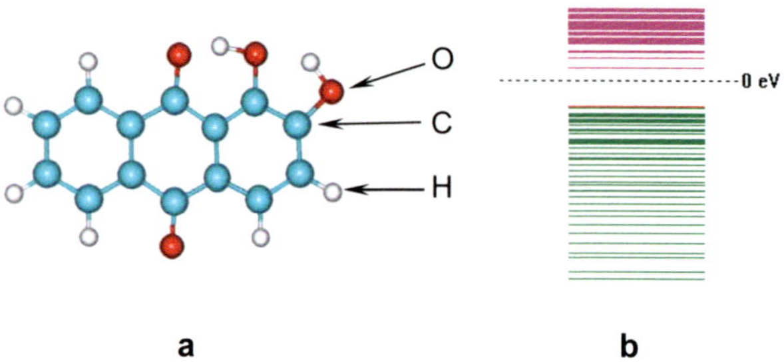

Plate 52 (a) Lowest energy conformation computed for an isolated alizarin molecule; (b) corresponding authorised computed energy levels at 0 °K.

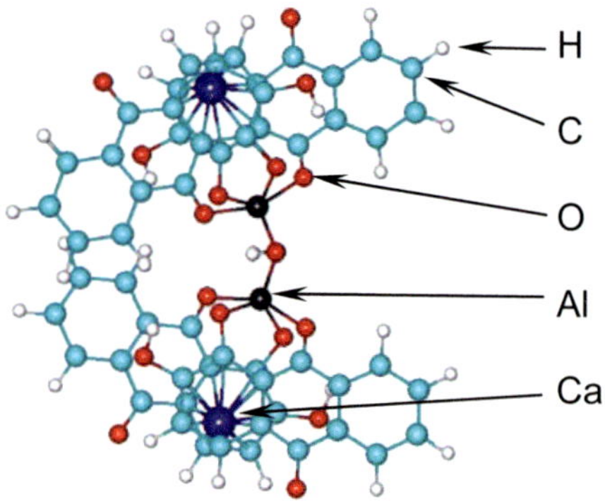

Plate 53 Computed top view of the structure of the $Ca_2Al_2(OH)_2Az_4^{4+}$ complex. Thick lines represent covalent bonds; thin lines, coordination bonds.

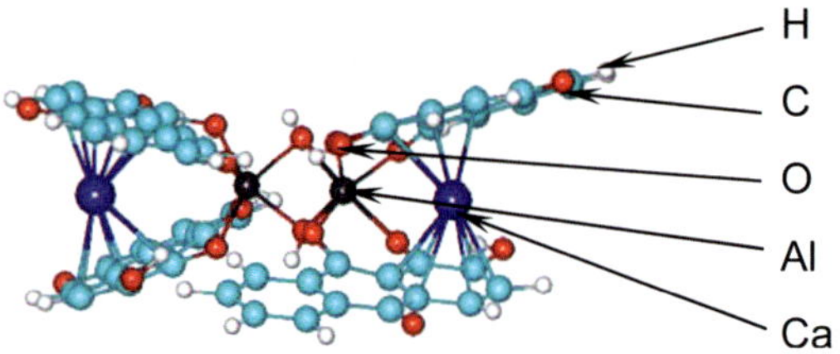

Plate 54 Computed side view of the structure of the $Ca_2Al_2(OH)_2Az_4^{4+}$ complex. Calcium cations are located in the median plane on the outside.

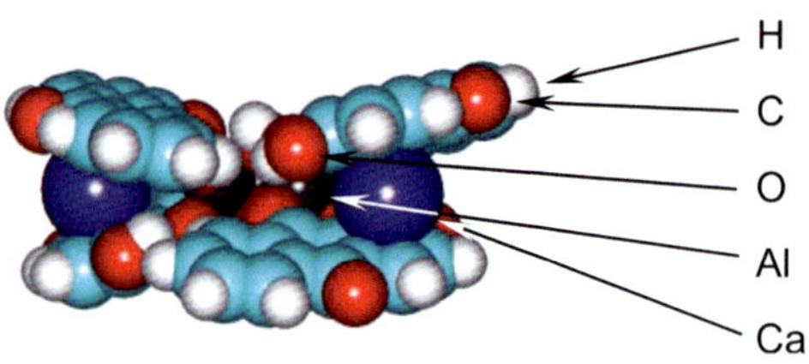

Plate 55 Same view as in Plate 54, but using a compact spheres model representation which respects the relative proportions of atomic diameters. It emphasises the role of the bulkiness of the calcium cation, causing the inclination of the alizarin planes. Note the aluminium ions hidden in the heart of the complex.

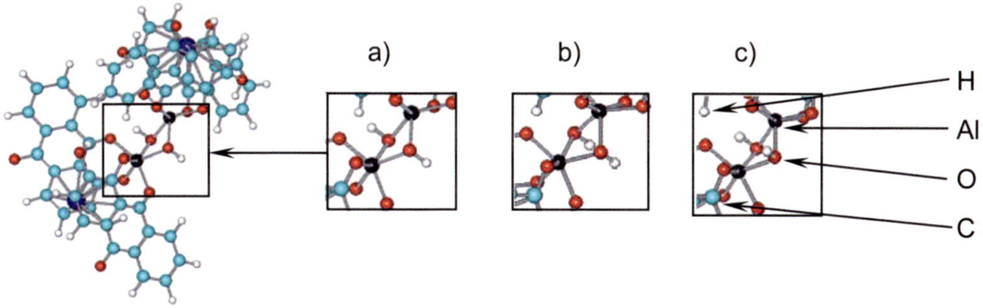

Plate 56 Structure of $Ca_2Al_2(OH)_2Az_4^{4+}$. Detailed views show the three positions possible for the central hydroxyls: a) symmetric and b), c) anti symmetric 1 and 2). (Figs 4–7, p. xx, correspond to the symmetric position.)

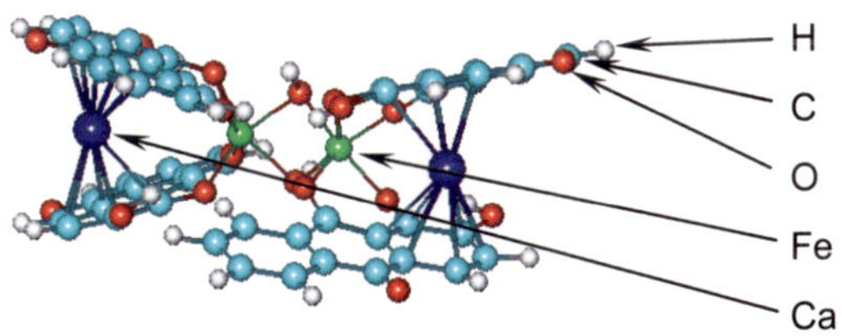

Plate 57 Computed structure of the $Ca_2Fe_2(OH)_2Az_4^{4+}$ complex, where Fe replaces Al: compare with Plate 54.

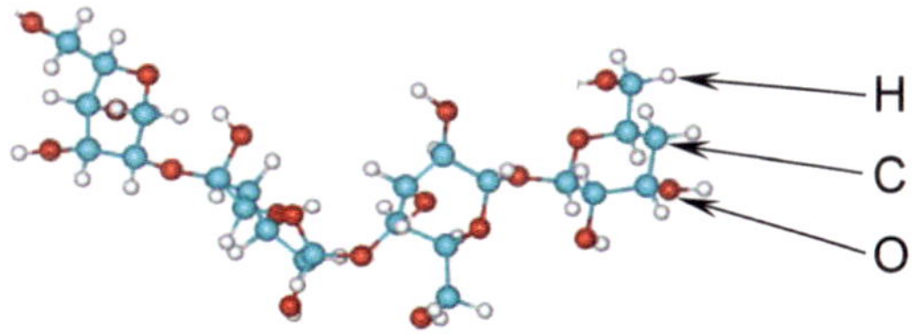

Plate 58 Non-planar structure of a cellulose fibril (here 81 atoms).

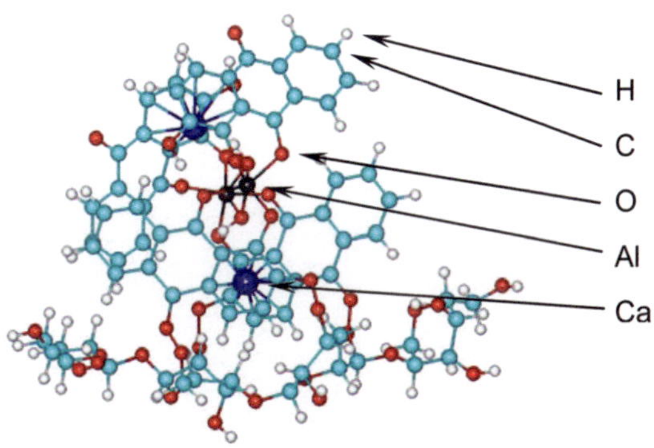

Plate 59 Chemisorption model of $Ca_2Al_2(OH)_2Az_4^{4+}$ on a cellulose fibril with creation of four peroxide bonds.

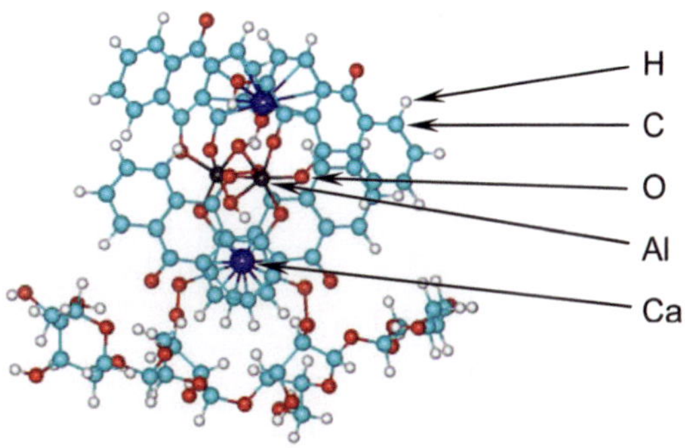

Plate 60 Chemisorption of $Ca_2Al_2(OH)_2Az_4^{4+}$ onto a cellulose fibril with creation of only two peroxide bonds.

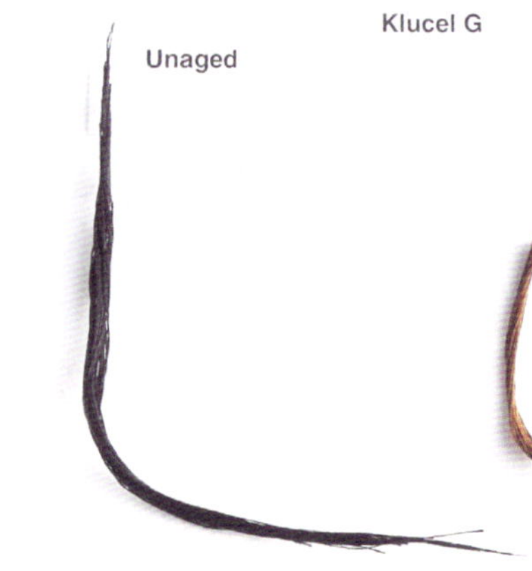

Plate 61 The effect of accelerated ageing on the colour of *P. tenax* fibres consolidated with 2% Klucel G.

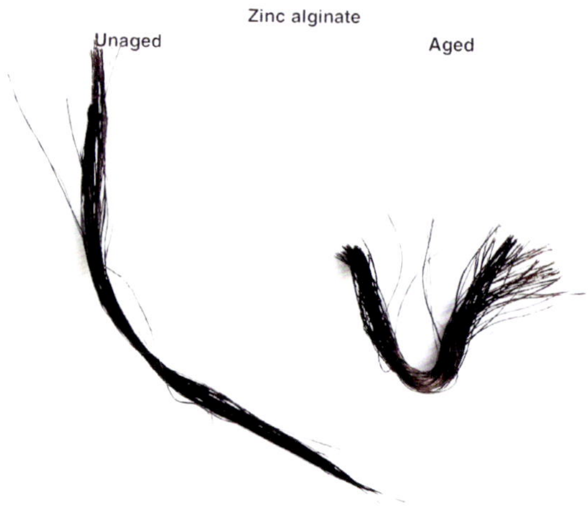

Plate 62 The effect of accelerated ageing on the colour of *P. tenax* fibres consolidated with 2% zinc alginate.

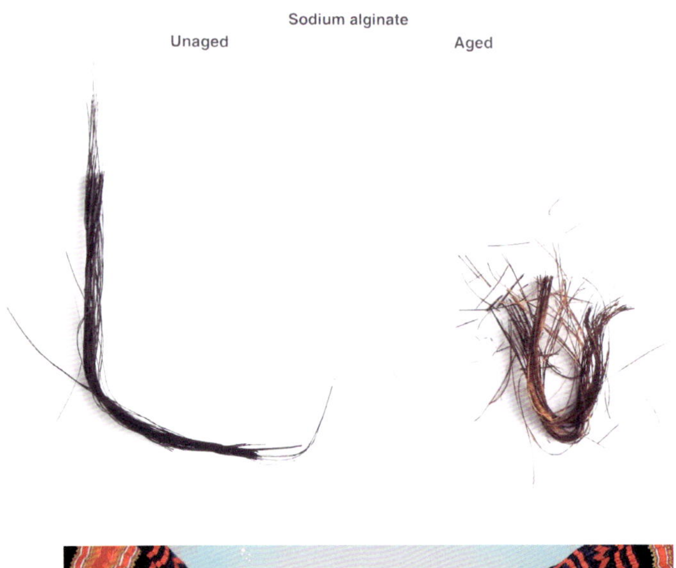

Plate 63 The effect of accelerated ageing on the colour of *P. tenax* fibres consolidated with 1% sodium alginate.

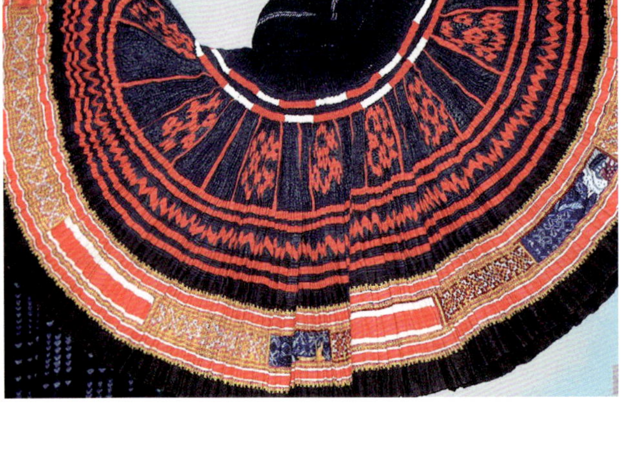

Plate 64 Traditional Lao skirt. (Photo: Witold Przewoźny)

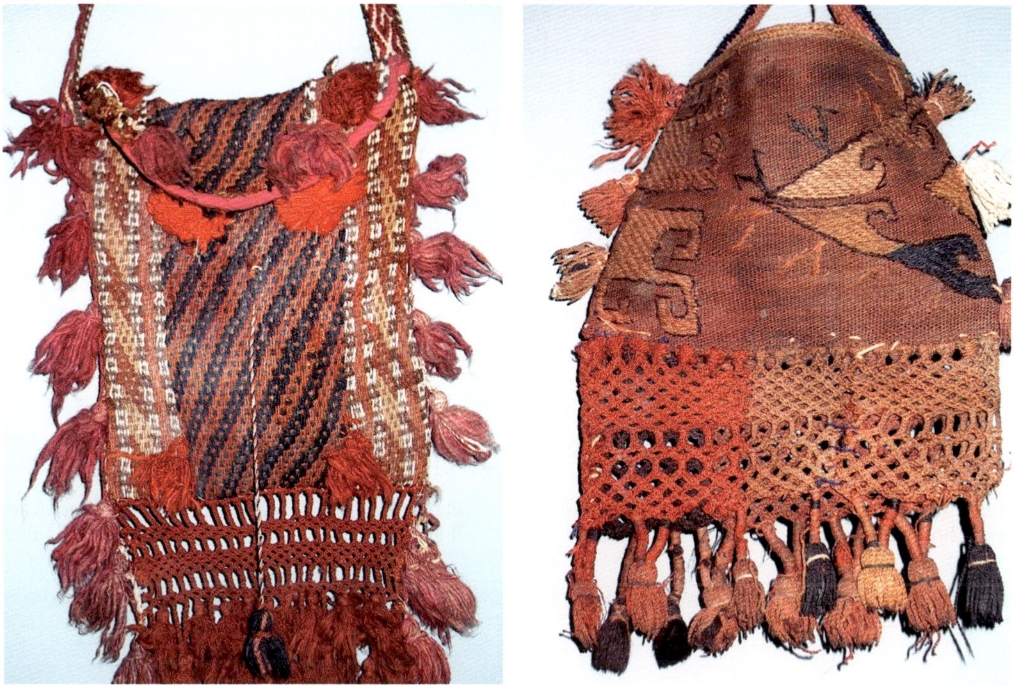

Plate 65 Pushtun bags from Afghanistan. (Photos: Katarzyna Schmidt-Przewoźna)

(a)

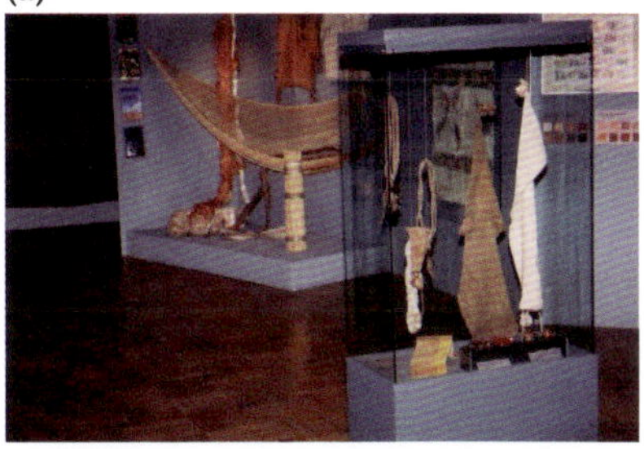

(b)

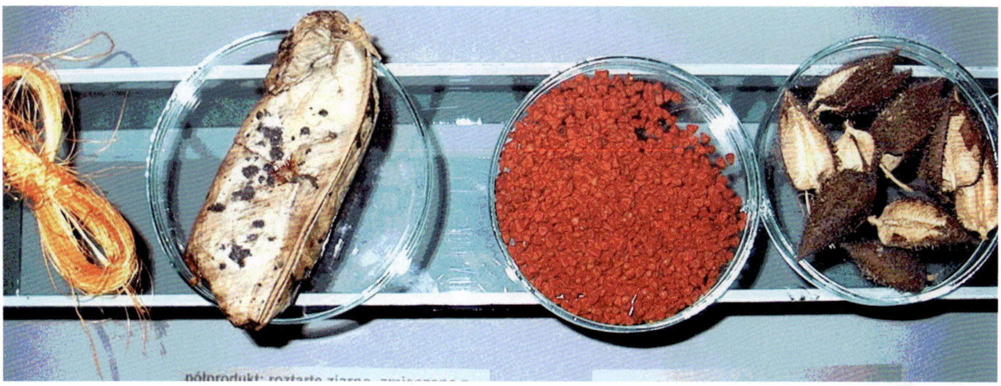

Plate 66 (a) View of the exhibition showing the textiles from South America. (b) Capsules and seeds of annatto (*Bixa orellana* L.). (Photos: Katarzyna Schmidt-Przewoźna)

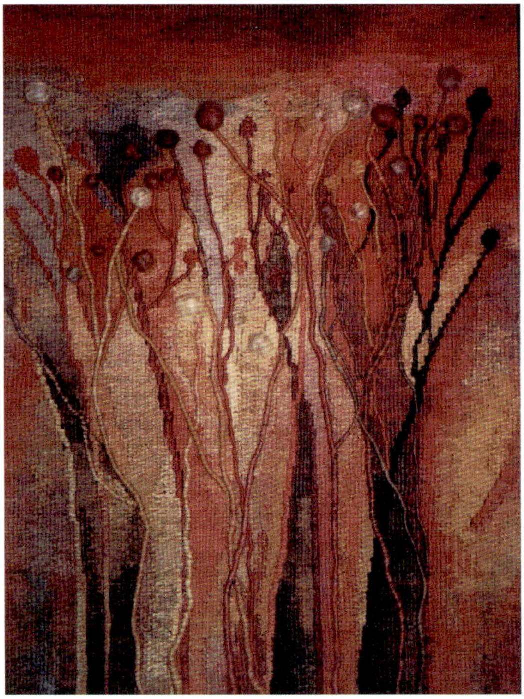

Plate 67 Katarzyna Schmidt-Przewoźna, *Flax*. Tapestry made with yarns dyed with natural dyes (see text), 80 × 120 cm. (Photo: Katarzyna Schmidt-Przewoźna)

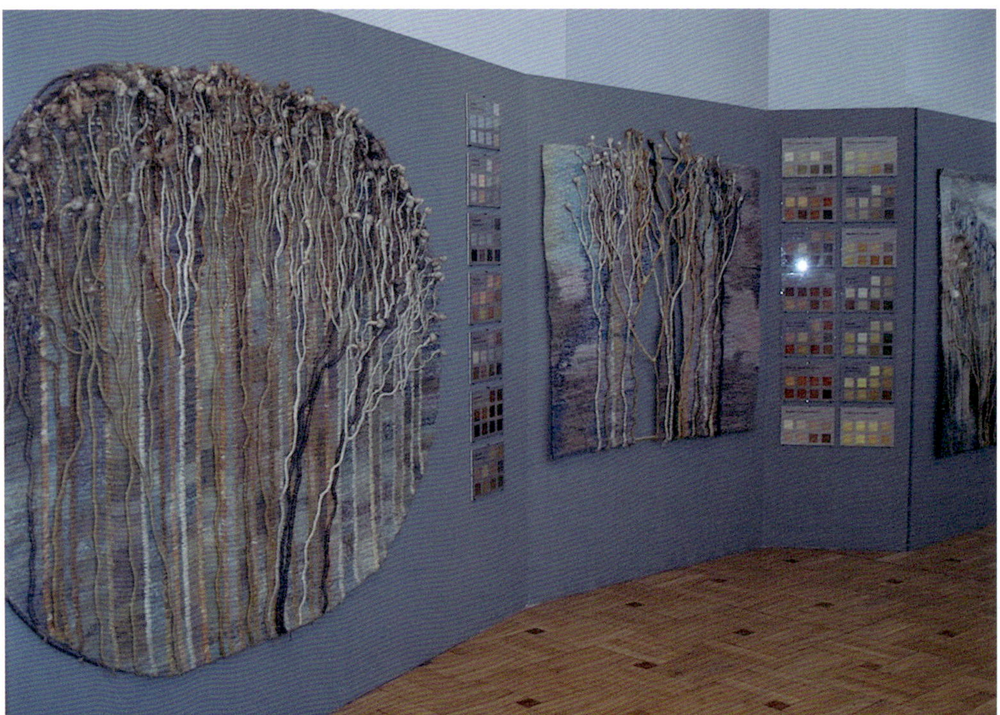

Plate 68 One of the exhibition displays showing colour trials and examples of tapestries produced using natural dyes. (Photo: Katarzyna Schmidt-Przewoźna)

Chromatographic and Spectroscopic Differentiation of Insect Dyes on East Asian Paintings

Jennifer Giaccai

Abstract Lac dye and cochineal have both been used as red pigments on East Asian paintings. Typically on objects that are difficult to sample, like the thin paint layers found on East Asian paintings, differentiating between the insect dyes can be difficult because the permissible sample size does not provide enough of the colorant for analysis with the typical methods used in the laboratory. High-performance liquid chromatography (HPLC) with diode array and fluorescence detection is being used to investigate the replacement of lac dye by cochineal in paints from China and Japan. In order to find the ideal chromatographic system two solvents, eight additives, two additive concentrations and four sample preparation methods were studied.

Introduction

The origin of the organic red colorant known as *enji* in Japan and *yan zhi* in China has been speculated upon for years. Examples of the colorant have been observed, but their exact chemical composition has only rarely been determined. A pigment census by Elisabeth West FitzHugh determined that 20% of the *ukiyo-e* school Japanese paintings at the Freer Gallery of Art contain an organic red colorant.[1] Organic red colorants have also been found by X-ray radiography on paintings dating from the 8th century and the first half of the 12th century in Japan.[2] In an Edo period (1603–1868) handscroll of the Tale of Genji, notations were made for painting areas of the picture using the symbol for *enji*; in these areas Taguchi and Taguchi (1977) found an insect dye by reflectance spectroscopy.[3] The authors explain that *enji* was one of many terms used during the Edo period and postulate that another option for the colorant would be a lake prepared from sappanwood or another red dyewood. Many authors have hypothesised a number of other options for *enji*, including safflower (*Carthamus tinctorius* L.), madder (*Rubia* spp.), *Perilla nankinensis* Decne., *Pterocarpus santalinus* L.f. and *Mirabilis jalapa* L. However, no scientific evidence for any of these colorants having been used on paintings has been identified to date.[4]

Recently, a study has begun in order to examine further the *enji* on paintings in the Freer Gallery collection. Non-invasive fibre optic reflectance spectroscopy (FORS) has determined that *enji* on the *ukiyo-e* school paintings is an insect dye.[5] Reflectance spectra for insect dyes show a double peak when not saturated; other organic red colorants tested, including alizarin lake, madder, brazilwood and safflower, show a single reflectance peak. A much smaller survey of 28 randomly selected Chinese paintings at the Freer and Sackler Galleries show that 46% of these paintings contain an organic red colorant, or *yan zhi*. While FORS can identify both *enji* and *yan zhi* as an insect dye, it does not differentiate between the various insect dyes.

Insect dyes in East Asian paintings

Scale insect dyes have been used for thousands of years worldwide. Although there are over 200 scale insects in the Coccoidea superfamily, only a few are important commercially as sources of dyestuff. The most economically important dyestuffs in the past were American cochineal, lac dye, kermes, Polish cochineal and Armenian cochineal. Only American cochineal and lac dye remain commercially important in modern times.

Insects of the *Kerria* genus produce lac dye. These insects are found throughout southern Asia. Lac dye was used as a dye in China as early as the 2nd century BC.[6] A sample of sticklac dating from the 8th century has been identified in the Shōsōin repository in Nara, Japan.[7] Lac dye is a by-product in the processing of sticklac into shellac. The water-soluble lac dye is washed out of crushed sticklac, concentrated and absorbed onto pieces of cotton.[8] The cotton saturated with the lac dyestuff, similar to medieval clothlets, was traded and referred to in Japan as *enji wata*.[9]

American cochineal (*Dactylopius coccus* Costa), is native to Latin America, but was imported into India as early as 1612 and known in China as early as 1700.[10] By 1800 it was known in Japan.[11] Kermes (*Kermes vermilio* Planchon), Polish cochineal (*Porphyrophora polonica* L.) and Armenian cochineal (*Poryphyrophora hamelii* Brandt) are found around the Mediterranean basin and the Middle East, northern Europe and the area around modern-day Armenia, respectively. While they are no longer commercially important, it is possible that they could have been traded along the Silk Route.

The insect dyestuffs are all composed of closely related anthraquinones. The similarity of the anthraquinone structures of insect dyes, each having between five and seven substituent groups with multiple hydroxyl end groups, results in almost identical spectroscopic and chromatographic properties. Lac dye is composed of at least five laccaic acids. The principal component of lac dye is laccaic acid A, which generally constitutes 71–96% of the dyestuff. The second component of lac dye is laccaic acid B, constituting 0–20% of the dye. Other laccaic acids rarely contribute more than 10% of lac dye. Kermes dye contains 75–100% kermesic acid, the remainder being mostly flavokermesic acid, also known as laccaic acid D. The dye extracted from American cochineal is composed almost entirely (94–98%) of carminic acid. Carminic acid is also the main dyestuff present in two other, less commercially important insect dyes: Polish cochineal (62–88%) and Armenian cochineal (approximately 95–99%). The similarity of these proportions can make them difficult to differentiate from each other and from American cochineal in small quantities. Studies by Wouters and Verhecken show that the various insect dyes can be separated based on the quantities of the anthraquinones present in the dyes; unfortunately differentiating between Armenian and American cochineals requires a sample size larger than that which can safely be obtained from typical East Asian paintings.[12]

Studies of modern *wata enji* in our laboratory by HPLC identified the presence of carminic acid.[13] An 1878 study of *wata enji* (referred to as *shoyenji* in the English translation) determined the presence of a carminic acid lake, although the same chemical properties would have been found in a lac dye lake.[14]

Identification of insect dyes

While FORS can determine the presence of insect dyes non-invasively, present techniques do not allow differentiation between the different insect dyes. After evaluating the various analytical options both frequently and infrequently used to examine insect dyes and paint samples, two techniques became the focus of our investigation: three-dimensional fluorimetry and reverse phase high-performance liquid chromatography (HPLC).

Three-dimensional fluorimetry, where the fluorescence is measured versus both the excitation and emission wavelengths, has been used with fibre optics to identify many dyestuffs, although not insect dyes, on *ukiyo-e* school prints and on textiles non-invasively.[15] Some investigators have used fluorescence to differentiate between insect dyes using solution methods that require removing a sample from the object.[16] Other methods are non-invasive, but require a powerful excitation method: laser-induced fluorimetry.[17]

The most common technique used to analyse insect dyes is reverse phase HPLC. A number of methods obtained from the literature for HPLC of insect dyes were examined.[18] Some methods focus on differentiating between the various insect dyes and sources; others aim for differentiating insect dyes from other red dyes or quantifying the amount of carminic acid present.

Experimental method

Fluorimetry was carried out using a Perkin-Elmer LS50B fluorimeter. Work in solution was carried out in quartz cuvettes; work with fibre optic probes used a probe custom-made by CIC Photonics Inc. (Albuquerque, NM, USA) of the six-around-one tapered TAPROBE design with 0.8 mm thick quartz fibres. A BG-24 filter was used on the excitation side and a 500 nm cut-off filter was used on the emission side of the instrument.

HPLC work was performed using both a Hewlett-Packard 1090M HPLC with diode array detection at the National Gallery of Art, Washington DC and an Agilent 1100 with a binary pump, diode array and fluorescence detection at the Freer Gallery of Art. All work was carried out on a Brownlee RP C1-18 narrow bore column (Spheri-5 ODS 5 μm, 100 × 2.1 mm) using a RP-18 Newguard precolumn and HPLC grade solvents.

Eluent systems involving methanol and acetonitrile gradients were examined with eight different additives, including ion pairing agents, a buffer and chelating agents at two concentrations (see Table 1). Four sample preparation methods, involving both methylating and hydrolyzing the insect dyes, were simultaneously evaluated.[19] To eliminate other factors, all experiments were run with the same gradient profile used by Halpine, changing the solvent, additive or additive concentration to that specified by the statistical experimental matrix.[20]

The statistical experimental matrix for the HPLC method determination was a modified fractional factorial design of the type 2^{9-3}_{IV}.[21] This type of matrix, when carefully designed, allows for the evaluation of a number of experimental factors and some of their interactions with a limited number of experiments. In this design, 64 HPLC runs were carried out, varying the additive, additive concentration, solvent and sample preparation method as determined by the experimental matrix. In all work, a signal-to-noise detection limit of 5 was required for a peak to be considered definite.

Results and discussion

Fluorimetry

The fluorescence level of carminic acid, like other anthraquinones, is known to vary with the pH and the chelation state of the molecule.[22] Any low levels of fluorescence emitted from carminic acid activated in its native state could not be monitored. However, when carminic acid was chelated with Al^{3+} in solution, the fluorescence intensity could be seen to increase with the amount of chelating ions present (Fig. 1). Fluorescence could also be measured non-invasively using the fibre optic probe when samples were painted out on alum-sized paper, or from an alum/insect dye solution. As Figure 2 shows, carminic acid and lac dye can be differentiated when in a suitable form chelated by aluminium. However, non-invasive evaluation of the *enji* on nine paintings, previously determined to be an insect dye by FORS, showed no observable fluorescence, either by spectrometer or with the naked eye.

Table 1 Lists of additives, additive concentrations, solvents and sample preparation methods for statistically designed HPLC experiments.

Additives	Additive concentration	Solvent (in gradient with water)	Sample preparation
None	0.1% by volume	Acetonitrile	None
Alum	0.25% by volume	Methanol	Hydrolysation
Potassium sodium tartrate			Methylation
Tartaric acid			Hydrolysate dried and followed by methylation
Potassium phosphate (monobasic) buffer			
Phosphoric acid			
Oxalic acid			
Trifluoroacetic acid (TFA)			

HPLC

Preliminary HPLC work was carried out using only diode array detection and the elution system developed by Halpine.[23] However, our experiments were not sensitive enough to detect the amount of colorant in samples from East Asian paintings consistently. As little as 2.5 ng of carminic acid could be detected, but 10 ng of lac dye was required for a similar detector response. When *enji* samples from four paintings were tested with HPLC, cochineal was found in two samples, but the remaining two samples showed no peaks. The very thin washes of *enji*, painted over a thick white calcium carbonate layer or directly onto the silk or paper support, contain significantly less colorant than typical samples available from textiles or oil paintings.

If the analyte fluoresces, or can be made to fluoresce, an HPLC fluorescence detector often becomes more sensitive than a diode array detector. However, using the Halpine method, the insect dyes do not fluoresce enough to use a fluorescence detector (see Fig. 2). Information from HPLC methods described in the literature was combined with previous experiments in fluorimetry to determine a new HPLC method. In order to evaluate the HPLC methods being considered, a statistically designed matrix of 64 experiments was planned, as described above.

Analysis of the experimental results is performed by averaging the results and calculating the error for each individual variable,

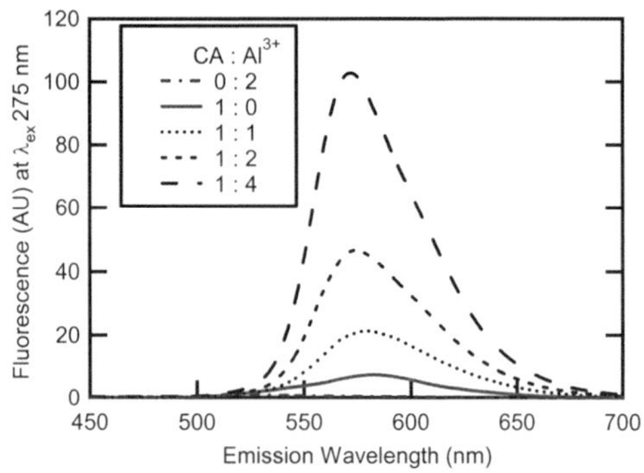

Figure 1 Fluorescence intensity spectra plotted against emission wavelength for carminic acid in alum solutions. The solutions were excited at 275 nm. Carminic acid concentration is 25 μM with varying concentrations of alum ($KAl(SO_4)_2$).

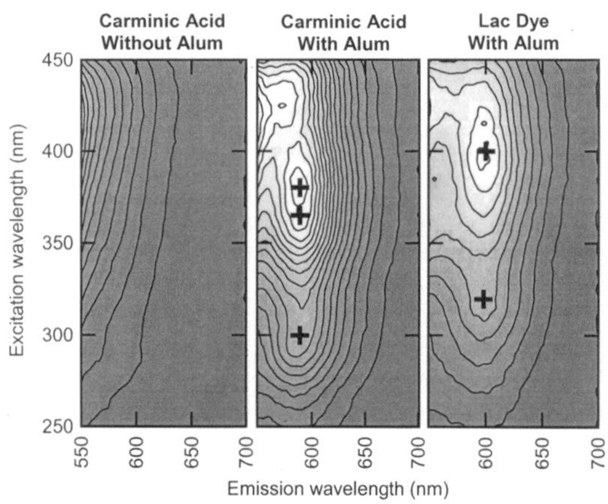

Figure 2 Three-dimensional fluorescence contour plots of insect dye solutions painted out on paper.

independent of the other variables. Interactions can be monitored by separate calculations. The results showed that neither the additive concentration nor the sample preparation methods affected fluorescence detection. However, the solution additives gave interesting results, as shown in Figure 3. As expected, all of the acids worked as ion pairing agents. While the phosphate buffer also acted as an ion pairing agent, neither of the chelating additives, alum and potassium sodium tartrate, were acidic enough to enable separation of the insect dyes. Only phosphoric acid gave a significant improvement in fluorescence detector response while maintaining adequate photodiode array detection.

Although for fluorescence detection, methanol appeared to be the solvent of choice (Fig. 3), the results between the two solvents methanol and acetonitrile were statistically insignificant ($p = 0.269$). At this gradient, the elution time is much shorter in acetonitrile gradients than in methanol gradients (12 minutes as opposed to 25 minutes retention times for carminic acid), which allows shorter overall runs. Acetonitrile was chosen for further experiments to allow for a shorter overall run time while maintaining a gradient consistent with previous experiments.

Trial HPLC runs compared the Halpine method with a new eluent system using acetonitrile with both phosphoric acid and alum additives (Fig. 4). The addition of alum to the acid/acetonitrile eluent systems gave prominent fluorescence peaks for the first time. Although the fluorescent peak for carminic acid in this figure shows significant improvement over previous experimental runs, it should be noted that the diode array detector was still the more sensitive of the two HPLC detectors used. It was observed in more recent experiments with HPLC that using the alum and potassium sodium tartrate salts in the HPLC pump would occasionally cause blockages of the HPLC system. Recent experiments in the Freer Gallery laboratory avoided this problem by adding these salts using a Pickering PCX-5200 post column derivatizer between the diode array and fluorescence detectors.

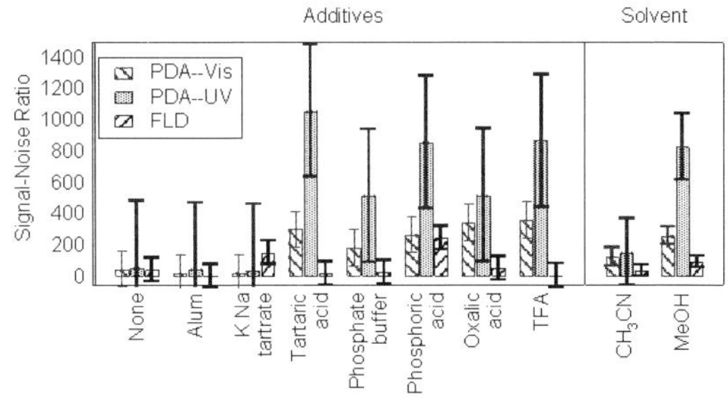

Figure 3 Signal-to-noise detector response for fluorescence detector and UV-visible diode array detector for solution additive and solvent factors.

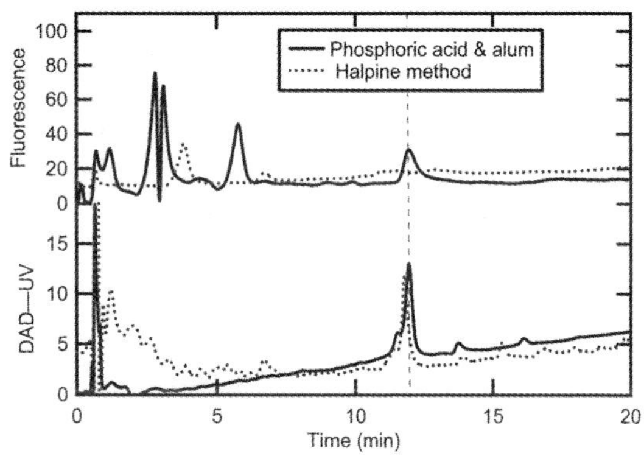

Figure 4 Chromatograms showing fluorescence detector and diode array detector (at 275 nm) response for carminic acid eluted following the Halpine method and using 0.1% phosphoric acid and 0.1% alum in water with an acetonitrile gradient.

Conclusion

Fibre optic fluorescence spectroscopy was unable to provide any information about *enji* used in the East Asian paintings examined. The requirements for an HPLC method specialising in insect dyes are different than for an HPLC method developed for distinguishing a larger general group of dyes. The desire to keep the sample sizes as small as possible drives the search for a more sensitive HPLC method. In this study fluorescence detection in the HPLC was investigated. The insect dye must be in the proper chemical state for fluorescence detection to have the required sensitivity, yet the needs for fluorescence detection of insect dyes differ from what is required to elute and separate the insect dyes on the HPLC column. This set of experiments has not been able to provide an HPLC method in which the fluorescence sensitivity is greater than that of the diode array detector. However, the improvement of fluorescence sensitivity by changing a few of the HPLC method parameters shows promise and future work is intended.

Editor's note

This paper was presented as a poster at the Meeting. Address for correspondence: Walters Art Museum, 600 North Charles Street, Baltimore, MD 21201, USA (jgiaccai@thewalters.org). Paper received 11 April 2003; revised version received 24 April 2004. When the research was carried out, the author was employed at the Department of Conservation and Scientific Research, Freer Gallery of Art and Arthur M. Sackler Gallery, Smithsonian Institution, Washington DC, USA.

Notes and references

1. FitzHugh, E.W. (1979) 'A pigment census of *ukiyo-e* paintings in the Freer Gallery of Art', *Ars Orientalis* 11, pp. 27–38.
2. Yamasaki, K. and Emoto, Y. (1979) 'Pigments used on Japanese paintings from the protohistoric period through the 17th century', *Ars Orientalis* 11, pp. 1–14.
3. Taguchi, E.-I. and Taguchi, M. (1977) 'Spectrophotometric identification of color materials, especially of organic pigments and combined mixed colors. With reference to the pigments used in the scroll, "Genji-monogatari 54-jo Emaki" of the Edo period', *Kobunkazai no Kagaku* 20/21, pp. 18-37.
4. FitzHugh 1979 (see note 1 above); Oguchi, H. (1969) 'A scientific study on the coloring materials in Japanese paintings', *Bulletin of Faculty of Fine Arts, Tokyo University of Arts* 5, pp. 27–82; John Winter, Department of Conservation and Scientific Research, Freer Gallery of Art, pers. comm.
5. Leona, M. and Winter, J. (2001) 'Fiber optics reflectance spectroscopy: a unique tool for the investigation of Japanese paintings', *Studies in Conservation* 46, pp. 153–62.
6. Donkin, R.A. (1977) 'The insect dyes of western and west-central Asia', *Anthropos* 72, pp. 847–80.
7. Mahdihassan, S. (1952) 'The oldest specimens of lac in the world', *Indian Journal of Entomology* 14, pp. 39–43.
8. Mahdihassan, S. (1954) 'The natural history of lac as known to the Chinese: Li-shih-chen's contribution to our knowledge of lac', *Indian Journal of Entomology* 16, pp. 309–26.
9. *Enji wata* is the Japanese term for the dye-saturated cotton. When referring specifically to the dye that comes from the cotton when it is soaked in water, the term *wata enji* is used.
10. Donkin 1977 (see note 6 above); Donkin, R.A. (1977) 'Spanish red: an ethnogeographical study of cochineal and the Opuntia cactus', *Transactions of the American Philosophical Society* NS, 67, 5, pp. 5–84.
11. FitzHugh 1979 (see note 1 above).
12. Wouters, J. and Verhecken, A. (1989) 'The scale insect dyes (Homoptera: Coccoidea). Species recognition by HPLC and diode-array analysis of the dyestuffs', *Annales de la Société Entomologique de France* NS, 25, 4, pp. 393–410.
13. *Wata enji* was obtained by Betty Fiske in 1986 from Kukodo in Kyoto, Japan.
14. Takamatsu, T. (1878) 'On Japanese pigments', Department of Science, Tokyo Daigaku, Tokyo, pp. 21–2.
15. Shimoyama, S., Noda, Y. and Katsuhara, S. (1997) 'Non-destructive analysis of *ukiyo-e* prints', *Dyes in History and Archaeology* 15, pp. 27–42; Shimoyama, S. and Noda, Y. (1997) 'Non-destructive analysis of dyes in a Chinese brocade: determination of plant dyes in a 16th/17th-century textile by a three-dimensional fluorescence spectrum technique with fibre optics', *Dyes in History and Archaeology* 15, pp. 70–84; Shimoyama, S., Noda, Y. and Katsuhara, S. (1998) 'Non-destructive determination of colorants used for traditional Japanese *ukiyo-e* woodblock prints by the three-dimensional fluorescence spectrum using fibre optics', *Bunseki Kagaku* 47, 2, pp. 93–100 and references within.
16. Wallert, A. (1991) '"Cimatura de grana": identification of natural organic colorants and binding media in mediaeval manuscript illumination',

Zeitschrift für Kunsttechnologie und Konservierung 5, 1, pp. 74–83; Wallert, A. (1986) 'Fluorescent assay of quinone, lichen, and redwood dyestuffs', *Studies in Conservation* 31, pp. 146–55.

17. Miyoshi, T. and Matsuda, Y. (1987) 'Laser-induced fluorescence and reflection spectra of red and purple natural dyes on silk cloths', *Kobunkazai no Kagaku* 32, pp. 47–53; Miyoshi, T. (1988) 'Fluorescence from colours used for Japanese painting under N_2 laser excitation', *Japanese Journal of Applied Physics* 27, 4, pp. 627–30.
18. Wouters and Verhecken 1989 (see note 12 above); Carvalho, P.R.N. and Collins, C.H. (1997) 'HPLC determination of carminic acid in foodstuffs and beverages using diode array and fluorescence detection', *Chromatographia* 45, pp. 63–6; Halpine, S.M. (1996) 'An improved dye and lake pigment analysis method for high-performance liquid chromatography and diode-array detector', *Studies in Conservation* 41, pp. 76–94; Hayashi, A. and Saito, M. (2001) 'Characterization of red natural dyes on the fabric by HPLC (I) cochineal, lac, kermes and madder', *Bunzaki Hozon-Syuhuku Gakkaisi* 45, pp. 27–43; Kirby, J. and White, R. (1996) 'The identification of red lake pigment dyestuffs and a discussion of their use'. *National Gallery Technical Bulletin* 17, pp. 56–80; Koren, Z.C. (1994) 'HPLC analysis of the natural scale insect, madder and indigoid dyes', *Journal of the Society of Dyers and Colourists* 110, pp. 273–7; Novotná, P., Pacáková, V., Bosáková, Z. and Stulík, K. (1999) 'High-performance liquid chromatographic determination of some anthraquinone and naphthaquinone dyes occurring in historical textiles', *Journal of Chromatography A* 863, pp. 235–41; Shibayama, N., Yamaoka, R., Sato, M. and Iida, J. (1991) 'Investigation of natural dyes from fabrics using thermospray liquid chromatography/mass spectrometry', *Mass Spectroscopy* 39, 3, pp. 123–31.
19. Hydrolyzation was performed as described by Halpine 1996 (see note 18 above), methylation was performed as described by Kirby and White 1996 (see note 18 above). Hydrolyzation and methylation were performed by hydrolyzing the sample, drying it and then methylating it.
20. Halpine 1996 (see note 18 above).
21. A further explanation of statistical experimental design is beyond the scope of this paper. The reference used by the author is Box, G.E.P., Hunter, W.G. and Hunter, J.S. (1978) *Statistics for Experimenters: An Introduction to Design, Data Analysis, and Model Building*, New York. The matrix used here is described on pp. 374–418.
22. Wallert 1986 (see note 16 above); Favaro, G., Miliani, C., Romani, A. and Vagnini, M. (2002) 'Role of protolytic interactions in photo-aging processes of carminic acid and carminic lake in solution and painted layers', *Journal of the Chemical Society, Perkin Transactions 2* (2002), pp. 192–7.
23. Halpine 1996 (see note 18 above).

A Study of Model Carminic Acid Lakes Prepared Using Cochineal, Collagen and Different Mordant Salts

Valery Golikov* and Zoë Zharikova

Abstract Cochineal lakes with some associated protein have been found in late Russian icons of the 17th to the 19th centuries. Following from work presented earlier,[1] model cochineal lakes prepared with collagen protein are the subject of the present study. The types of complex formed between carminic acid, different mordant cations and collagen are investigated, examining the influences of dyestuff and mordant cation concentration and the presence or absence of cream of tartar in the mixture.

Introduction

Lake pigments used in traditional painting up to the late 19th century were made using natural organic dyes of both vegetable and animal (scale insect) origin. They were used as final layers of translucent glaze, as coloured layers beneath metal leaf and also as coloured varnishes. Lakes containing cochineal dyestuff have, for example, been identified in late Russian icons of the 17th to the 19th centuries and in these cases there seems to be some protein associated with the pigment (Plate 27).[2] In previous work, an account of the use of organic pigments in Russian art has been given.[3] A classification of the pigments and the paint layers in which they are found, based on their visual appearance under the microscope, has also been suggested.[4] According to this scheme, the lake pigment layers seen in the Russian icons are described as single-phase systems, since they consist of a homogeneous coloured film.

The manufacture of lake pigments is quite varied and complicated, and also a very early technology: recipes for such pigments date back to antiquity. This technology has, however, been little studied by currently available scientific methods, notably in the case of the cochineal-containing pigments. Cochineal dyestuff is easily extracted from the insect raw material in neutral or slightly acidic aqueous solution. As a result of the complex formed between the dyestuff and the mordant cations, with the addition of collagen protein in this case, extremely brightly coloured compounds are formed. The dried, powdered material may be used as a pigment and the use of such pigments has been critical in the achievement of subtle effects of light and shade in traditional Russian religious painting.

Taking these pigment layers as a starting point, the present paper is concerned with the behaviour of model systems prepared using the dyestuff extracted from the cochineal insect, *Dactylopius coccus* Costa (essentially carminic acid), collagen and different mordant salts.

The aims of the work were to establish if a cochineal lake containing protein could be obtained from the starting materials of collagen, carminic acid and mordant salts; to study the properties of the lakes prepared; and to estimate the influence of certain technological factors on these properties. These factors were the type of single mordant cations; binary combinations of the trivalent aluminium (Al) cation and bivalent cations; the concentration of the mordant cation salts; the concentration of the cochineal extract; and finally, the presence or absence of cream of tartar (potassium hydrogen tartrate, the mono potassium salt of (+)-2,3,-dihydroxybutane-dioic acid).

distorting the absorption spectra of the coloured protein lake films. Secondly, only those forms of dye bonded to the collagen in the film were desired.

The lake gel films on slides, which are illustrated in Plates 28 and 29, have been relatively stable for several years. In addition, absorption or transmission spectra of the dry cochineal pigments can easily be recorded. Spectrophotometry was a very effective method for the analysis of the coloured compounds in the dry films directly.

Materials and methods

Methods of making the model pigments

Two different techniques were used to make the pigments. The first corresponded roughly to the technology of lake pigment-making described in Russian icon painting formulae.[5] Aqueous solutions of the dyestuff extracted from dried cochineal insects with a pH of 5.5–7 were prepared. At the same time, aqueous solutions of collagen containing the mordant cation salts and, occasionally, cream of tartar were prepared. The appropriate collagen and dye solutions were then mixed. The absorption spectra of examples of these aqueous solutions are shown in Figure 1. The rather stringy liquid cochineal pigments formed were deposited on a glass slide in the shape of a gel strip. After drying, a transparent coloured film of dry coloured proteinaceous lake was formed on the slide.

The second technique consisted of preparing the two solutions as described above, but in this case they were not mixed. The collagen-containing solution was deposited on slides as strips of gel as before. The slides were then immersed in the appropriate dyestuff solutions until the intensity of colour of the dyed gel strips ceased to change. The strips of coloured gel were then dried. This technique was used for two reasons: firstly, to avoid the formation of insoluble complex compounds that can appear upon mixing the collagen, mordant salts and dye solutions. Such compounds are capable of

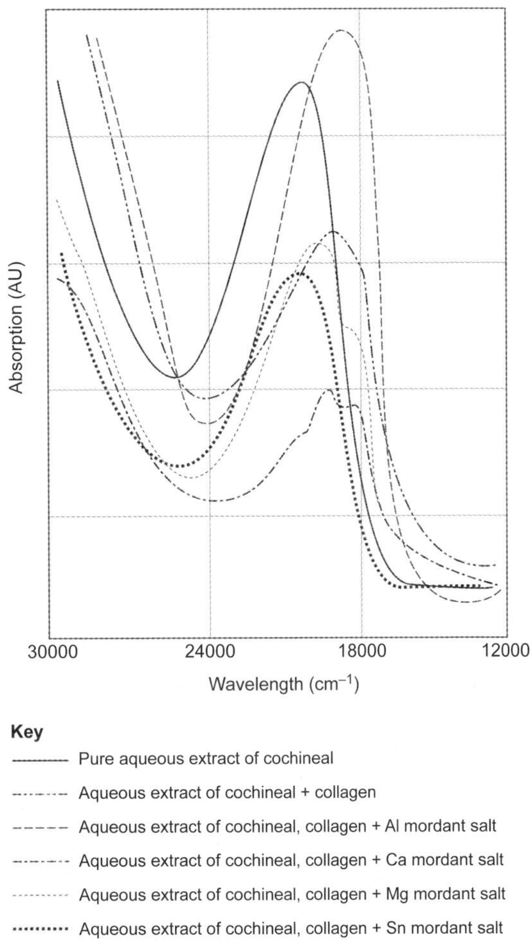

Figure 1 Absorption spectra of aqueous cochineal extracts with additions of collagen and different mordant salts.

Variation of technological factors

In the present experiment, the factors that might influence the properties of the model lakes included the source of collagen; the concentration of the dye solution; the pH; the mordant cations present; and the presence or absence of cream of tartar.

- In the majority of the modelling experiments, edible gelatin and hare glue, that is, collagen extracted from hare skin, were used as the main sources of collagen. The collagen concentration in the coloured protein lake solution amounted to 2 or 3%.
- Three concentrations of carminic acid – 0.26%, 0.13% and 0.06% – were used for preparing the standard samples.
- The pH value of the cochineal extract was 5.5–6.
- Five different mordant cations, either identified in microsamples taken from works of art or mentioned in traditional formulae,[6] were used, in the form of soluble salts. These were the trivalent Al^{3+} cation and four divalent cations: Ca^{2+}, Mg^{2+}, Zn^{2+} and Sn^{2+}; mixtures of Al^{3+} with these four divalent ions were also used. The salts used were potassium aluminium sulfate ($AlK(SO_4)_2 \cdot 12H_2O$), calcium chloride ($CaCl_2$), magnesium chloride ($MgCl_2$), zinc chloride ($ZnCl_2$) and tin(II) chloride ($SnCl_2$). Four concentrations of the salts or salt mixtures, varying at a twofold rate, were used for preparing the standard samples. The salt ratio with respect to the collagen concentration reached its maximum at 20%.
- In certain experiments standard cochineal-collagen mixtures with and without the addition of cream of tartar were compared.

Methods of investigating the model lakes

The properties of the model lakes examined included the colour, the homogeneity of the films, the carminic acid content and the method by which carminic acid was thought to be bonded to the collagen (for example, direct hydrogen bonding or by complex formation). Colour characteristics were estimated visually and the homogeneity of the dry lake films illuminated by transmitted polarised light was estimated visually under the microscope.

The total content of carminic acid combined with collagen in the pigment films was evaluated in relative units according to the absorption maxima (in the visible region of the spectrum) of the films (see, for example, the spectra illustrated in Figs 2–4). The percentage carminic acid concentration present in extracts from the dried films (described below) was evaluated similarly.

In order to estimate separately the concentrations of the different carminic acid–collagen

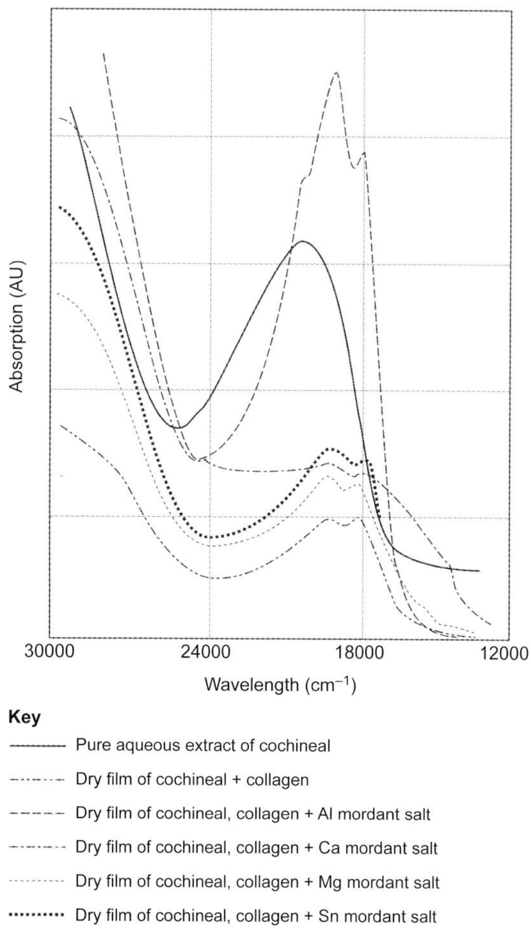

Figure 2 Absorption spectra of dry cochineal/collagen lakes with additions of different mordant salts.

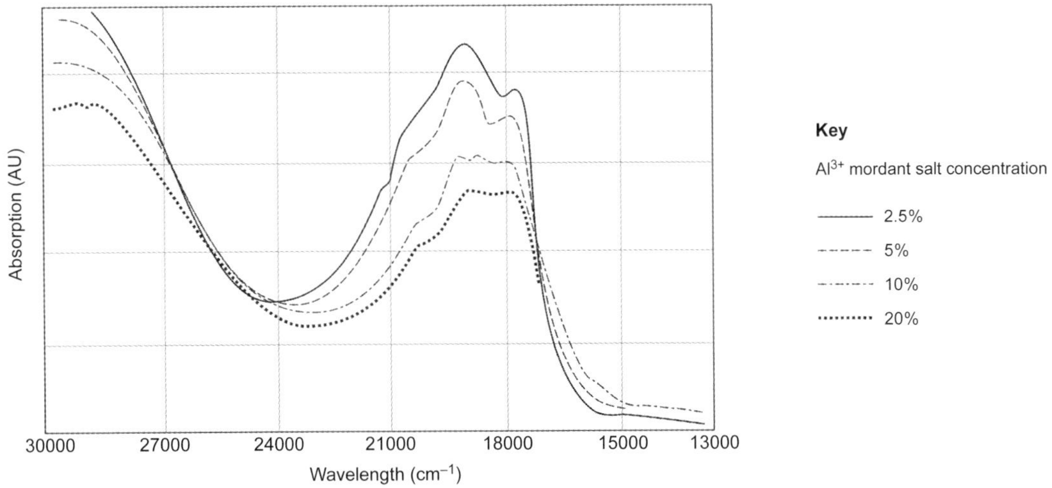

Figure 3 Absorption spectra of dry Al-containing cochineal/collagen lakes showing the effect of the addition of different percentage concentrations of Al^{3+} mordant salts. A 0.26% aqueous extract of cochineal was used in each case.

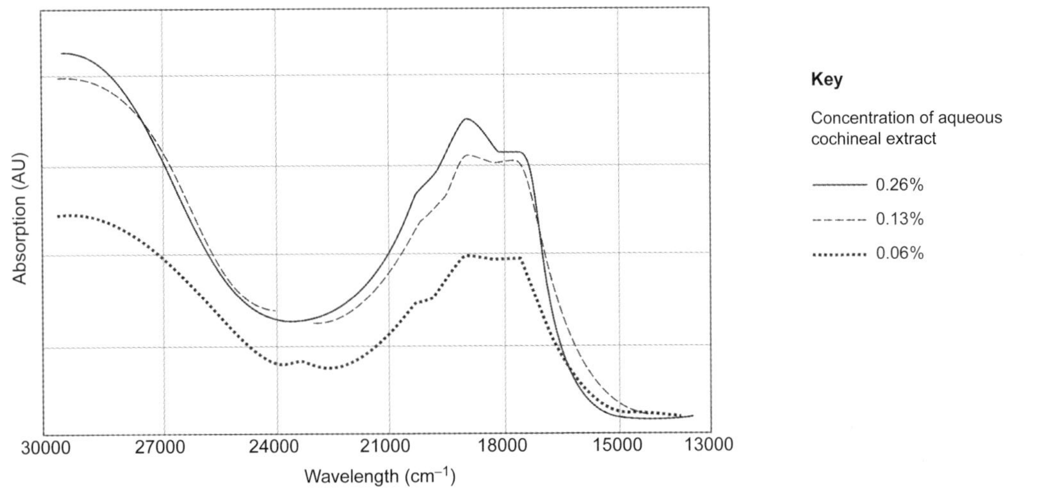

Figure 4 Absorption spectra of dry Al-containing cochineal/collagen lakes showing the effect of the use of a different percentage concentration of aqueous cochineal extracts. A 5% Al^{3+} mordant salt solution was used for each preparation.

combinations in the dry lakes, the films were treated successively with reagents, each of which destroys a different variety of dye–collagen bond and extracts dye bonded in that particular manner into solution. Samples were taken of each film under test and the mass of each sample was measured; these masses were about 2 mg. The sample was then treated with 2 mL of one of three extracting agents.

Reagent 1 was pure dimethylformamide (DMF). Treatment with DMF at about 100 °C for 90 minutes was used to destroy hydrogen bonds between the dye and collagen. Some other reagents are also able to do this, but DMF was chosen because both hydrophilic and hydrophobic colouring matters are easily dissolved in it while collagen itself remains practically intact. Dye extracted from the dyed film samples by

DMF can be related to a direct dyeing in which the dye is bound to the protein only by hydrogen bonds.

Samples from which any cochineal dye colouring the gel by direct dyeing had been removed with DMF were then treated for 90 minutes at about 100 °C with Reagent 2: DMF with one drop of a saturated (2–3 %) aqueous solution of the chelating agent DTPA (diethylenetriaminepentaacetic acid; [(carboxymethyl)imino]bis(ethylenenitrilo)]-tetra-acetic acid) at neutral pH. The concentration of DTPA in the DMF was about 0.04%. This mixture destroys the least stable complexes between the dye and the collagen.[7]

If bound dye still remained in the sample, it was treated with Reagent 3: the mixture used for Reagent 2 to which one drop of 3% hydrochloric acid (HCl) was added (about 0.04% DTPA in DMF; concentration of HCl in the mixture was 0.05%). The pH of this reagent was about 3.0. This reagent destroys the most fast complexes between collagen, the carminic acid dye and the mordant cations.[8]

At present the authors do not intend to establish a line of demarcation between mechanisms of interaction with collagen shown by the two latter forms of dye–mordant cation complex formation. However, even a phenomenological estimate of the types of dye–mordant cation–collagen complex present in the experimental gels is very useful. Coupled with the results of spectrophotometric studies, such an estimate enables the establishment of more precise correlations between the variability of the different technological factors (such as dyestuff concentration) and the properties of the dry coloured films, in particular to compare binding mechanisms between the dye and different mordant cations.

The properties of the dry lake films

About 200 samples of model cochineal protein lake gels were made. The majority of samples yielded red cochineal films of various tints and intensity (see Plates 28 and 29).

The optical homogeneity of the model films

By examining the samples under the microscope under transmitted polarised light, it was possible to see how many coloured 'phases' were present in the dry film. In the majority of samples, a film of uniformly coloured, transparent dried gel was present: a single phase system. In some cases, insoluble black, grey or red carminic acid derivatives were present. Usually this second 'phase' was present in addition to the first phase, thus these samples could be described as 'two-phase' lakes.[9]

The absorption spectra of dry model lake films

Absorption spectra of the aqueous cochineal extracts were examined first (Fig. 1). Without admixture of other reagents, these spectra show a long-wave maximum with some indication of the presence of a triplet of components, although very poorly resolved. If salts of mordant cations are added to the water extract, a shift of the maximum to a longer wavelength (bathochromic shift) takes place and the resolution of its triplet shape is improved. This is indicative of the emergence of complex compounds of carminic acid with the mordant cations.

Intensive triplet long-wave maxima can also be observed in the absorption spectra of dry lake films (Figs 2–4). These maxima are shifted to a longer wavelength by comparison with those in the absorption spectra of the pure water cochineal extracts without mordant salt admixtures (as shown in Figs 1–4). The values for the long-wave maxima for the dry films with mordant cations present and for water extracts of cochineal with admixtures of the same cations are relatively close (compare Figs 1 and 2). This suggests that collagen exerts a negligible influence on the optical properties of complexes formed between carminic acid and mordant cations in the lake gel films, even if it is participating in these chelate complexes as a ligand.

The resolution of the triplet components in the long-wave maximum of the absorption spectra of the aluminium-containing lakes diminishes with the increase in concentration of mordant salts (Fig. 3), but is improved by an increase in dye concentration (Fig. 4). The value of the bathochromic shift of the long-wave maximum of these lakes has a negligible correlation with the concentrations of dye and mordant salts.

The main forms of carminic acid–collagen combination in the model lake films

Successive treatments with the three different extraction reagents permitted the isolation of the principal carminic acid–collagen combinations. Possible structures of some forms are represented in Figure 5.

As a direct dye, carminic acid could combine with collagen through relatively weak intermolecular interactions: for example, through hydrogen bonds. A direct form could be, for example, a water-soluble salt of carminic acid combined with the Al^{3+} cation, as shown in Figure 5. After drying the coloured gel, a direct form could participate in the formation of a transparent, single-phase coloured lake film.

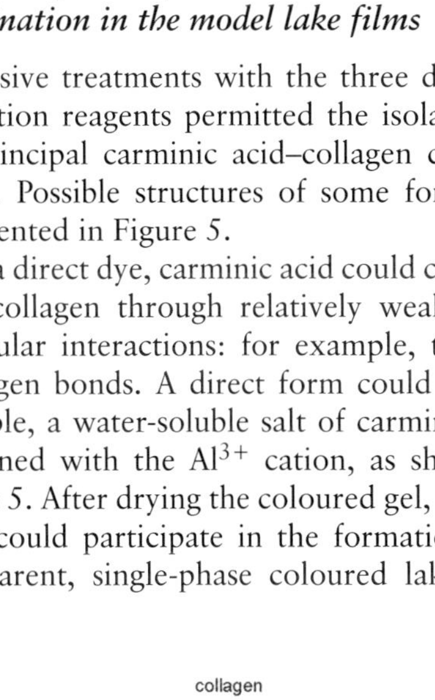

Figure 5 The principal possible types of interaction between carminic acid, mordant cations and collagen in a protein-containing cochineal lake.

Such a form could be eliminated from the dry film using DMF.

As a mordant dye, carminic acid could interact with collagen as a complex through coordination bonds; for this the participation of mordant cations is required. The mordant form could, for example, produce some kind of water-soluble ternary chelate complex 'carminic acid–mordant cations–collagen'. In such a case, carminic acid and collagen play the roles of specific ligands. After drying, the ternary chelate complex also forms a transparent dry single-phase lake film. Mordanted forms of carminic acid, of any type, could only be eliminated from the dry films with Reagents 2 and 3, an approximately 3% aqueous solution of the chelating agents EDTA (ethylenediaminetetraacetic acid) or DTPA as described above, with or without hydrochloric acid. The exact type of mordanted form of carminic acid has not usually been characterised further in the results that follow. Many samples of the cochineal–protein lakes were found to contain about 80–100% mordant forms of carminic acid.

At least two main kinds of insoluble compound can be formed. Firstly, two-component insoluble complexes of carminic acid and certain cations may occur. In earlier work, insoluble compounds with similar colour characteristics to those observed in the films were obtained during reactions between aqueous solutions of carminic acid water solutions and solutions of Ca^{2+} and Sn^{2+} mordant cations. Secondly, it is possible that collagen also participates in the formation of insoluble carminic acid compounds with Ca^{2+} and Sn^{2+} mordant cations. It has been shown in earlier work that under certain conditions 3-component insoluble complexes of the type 'carminic acid–Ca/Sn/Mg cations–collagen' are formed. It is evident that the formation of such insoluble coloured compounds (with the participation of different combinations of an organic dye, mordant cations and collagen) could be the basis for the technology of the making of insoluble organic pigments.

The relationship between carminic acid and the mordant cations in the dry coloured collagen lake films

For any one gel sample, the actual carminic acid and mordant cation contents could be calculated using spectrophotometry. For example, in the dry aluminium-containing lake prepared using a 5% concentration of the aluminium mordant salt solution (potassium aluminium sulfate) and 0.26% cochineal extract, the actual content of carminic acid and mordant cations could be calculated. A carminic acid sample prepared from an extract of cochineal insects was used to prepare a spectrophotometric standard. Cations from the cochineal were removed from the extract using a Dowex® cation exchange resin and cochineal biopolymers were removed using Sephadex®-G25. It can thus be calculated that 100 g dry lake gel contains:

- About 1×10^{-2} moles of Al;
- About 6.8×10^{-2} moles of carminic acid in total;
- About 5.4×10^{-2} moles of carminic acid in a mordanted form (80% of the total carminic acid content) combined with collagen in the form of different chelate complexes.

Thus the ratio Al^{3+}/ carminic acid in this particular sample is 1/5.4.

If after purification the cochineal extract was still not sufficiently pure and contained other substances besides carminic acid, the ratio given above could be less: 1/4, for example. Naturally, for the other dry lake samples the ratios of collagen, Al^{3+} and carminic acid were different.

Influence of specific technological factors on the characteristics of the dry carminic acid lake gels

Influence of mordant cation type

The influence of four tri- and bivalent cations was examined – Al^{3+}, Ca^{2+}, Mg^{2+} and Sn^{2+} – as well as mixtures of Al^{3+} with these bivalent cations. It is possible to divide the cations studied into two main groups according to their influence on the model systems: Al^{3+}; and the bivalent cations (Ca^{2+}, Mg^{2+}, Sn^{2+}) as a whole.

Colour characteristics

The colour characteristics varied for different concentrations of Al^3 and carminic acids (see Plate 28). It should be noted that the colour characteristics of Al-containing cochineal lakes and 18th-century Al-mordanted silk fabrics are closely similar (potassium aluminium sulfate was commonly used as a mordant during this period).[10]

Dry films of lake gels prepared with bivalent cations are a black or grey-rose colour (Plate 28). The colour characteristics of cochineal pigment films prepared using bivalent cations thus differ from those of silk fibres mordanted with Ca^{2+}, Mg^{2+} or Sn^{2+} solutions and dyed with cochineal.[11] The ratio of black, grey and rose colours in the films depends on the cation type: Ca^{2+} pigment films are mainly black or grey, while Mg^{2+} and Sn^{2+} lake films are grey-rose of different tints.

Optical homogeneity

The Al-containing lakes consisted of films of uniformly coloured and transparent crimson single-phase collagen lakes: no other coloured particulate phases were present. The dry films prepared with bivalent cations contained two phases: a uniformly coloured and transparent rose phase, and a second phase consisting of insoluble black, grey or red carminic acid derivatives.

Absorption spectra

The triplet form of the long wavelength maxima of aluminium complexes of carminic acid can be clearly seen in the absorption spectra of Al-containing lake gels (Figs 3 and 4). The absorption spectra of lake gels containing bivalent cations have two components: the triplet form long-wavelength maxima given by ternary complexes of carminic acid and the broad spectra of the black and grey insoluble products.

The influence of mordant cation concentrations

This depends on the type of mordant cation present.

Optical density

The triplet form long-wave optical density maximum is more pronounced in Al-containing lake gels than in the case of the corresponding lake gels made with bivalent cations, thus variations in Al^{3+} concentrations have a more marked effect. For high carminic acid concentrations, the long-wave optical density maximum of the dry pigments decreases as the Al^{3+} concentration increases. For minimal concentrations of carminic acid, however, the dependence of optical density on Al^{3+} concentration is negligible. The results of varying the mordant salt and its concentration and the carminic acid concentration on the optical densities of the samples are given in Table 1.

The triplet long-wave optical density maxima of lakes made with bivalent cations is very low and remains constant, or shows a negligible increase, with an increase in bivalent cation concentration for the majority of dry lake gels. The exception is the Ca-containing lake with a maximum concentration of carminic acid, in which the optical density is very high and constant for all concentrations of the calcium mordant salt. This effect is connected with the formation of insoluble calcium-containing derivatives.

Table 1 Intensity of absorption of dry cochineal lake films, with and without cream of tartar, as a function of the concentrations of different mordant salts and cochineal extract.

Mordant cation	Cochineal concentration %	Mordant salt concentration, %							
		Optical density, relative units: no cream of tartar				Optical density, relative units: cream of tartar present			
		2.5	5	10	20	2.5	5	10	20
Al	0.26	1.38	1.17	0.9	0.8	1.38	1.4	1.14	0.98
	0.13	0.7	0.94	0.76	0.28	0.76	0.8	0.8	0.71
	0.06	0.42	0.56	0.5	0.56	0.34	0.46	0.54	0.55
Al + Ca	0.26	0.74	0.75	0.75	0.8	0.64	1.04	1.07	1.22
	0.13	0.46	0.68	0.6	0.72	0.3	0.38	0.68	0.84
	0.06	0.36	0.4	0.45	0.52	0.16	0.34	0.52	0.52
Al + Mg	0.26	1.24	1.2	1.2	1.4	0.53	1	1	1.28
	0.13	0.66	1.16	1	1.04	0.52	0.84	0.84	0.84
	0.06	0.44	0.68	0.64	0.7	0.31	0.42	0.48	0.64
Al + Sn	0.26	1.4	1.4	1.4	1.4	0.52	0.76	0.83	0.69
	0.13	0.86	0.77	0.8	0.86	0.52	0.76	0.83	0.9
	0.06	0.56	0.5	0.5	0.6	0.34	0.58	0.52	0.64
Ca	0.26	1.4	1.4	1.4	1.4	0.26	0.46	0.4	0.26
	0.13	0.24	0.46	0.42	0.5	0.15	0.2	0.16	0.32
	0.06	0.1	0.16	0.11	0.16	0.1	0.1	0.16	0.24
Mg	0.26	0.2	0.28	0.3	0.48	0.18	0.28	0.38	0.5
	0.13	0.14	0.18	0.24	0.3	0.2	0.36	0.4	0.41
	0.06	0.13	0.23	0.24	0.2	0.1	0.16	0.18	0.24
Sn	0.26	0.24	0.36	0.26	0.52	0.17	0.3	0.46	0.39
	0.13	0.14	0.22	0.16	0.2	0.1	0.32	0.44	0.36
	0.06	0.16	0.18	0.18	0.18	0.08	0.16	0.18	0.17

Total concentration of carminic acid combined with collagen

The total concentration of carminic acid in the dry Al-containing pigment gels is greater than in the corresponding pigments made with bivalent cations (Fig. 6). The dependence of the total concentration of carminic acid on the Al^{3+} concentration is not equal for Al-containing lakes with different carminic acid concentrations, however (Figs 7 and 8): this dependence is at a maximum when the concentration of carminic acid is greatest and, for lower dye concentrations, it increases uniformly with the growth of concentration of carminic acid.

The total concentration of carminic acid in the Mg- and Sn-containing pigment gels is very low and the increase with an increase in bivalent cation concentration is negligible.

The percentage of mordant forms of carminic acid

Table 2 shows how the concentration of mordant forms of carminic acid in the pigment films varies with the concentration of the various mordant salts and carminic acid. As described above, the results of treatment of the dry Al-containing pigments with the three extraction reagents suggest that the percentage of carminic acid present in a mordant form could be as much as 80–100%. The percentage of mordanted forms of the dyestuff in the Al-containing lakes where the concentration of Al^{3+} cations is not so great shows a negligible increase or is constant as Al^{3+} concentration increases, but it diminishes at the maximum Al^{3+} concentration. The percentage of mordanted forms of carminic acid in dry pigment gels prepared with bivalent cations is very complicated: it depends on the increase in the concentration of bivalent cations. This is connected with the formation of insoluble products in the samples prepared using bivalent cations.

The results of researches into the influences of cation types and their concentrations lead us to conclude that only Al^{3+} forms very intensely and uniformly coloured lake films with collagen and cochineal extract. Bivalent cations form non-uniform, less strongly coloured films, with the additional presence of insoluble derivatives.

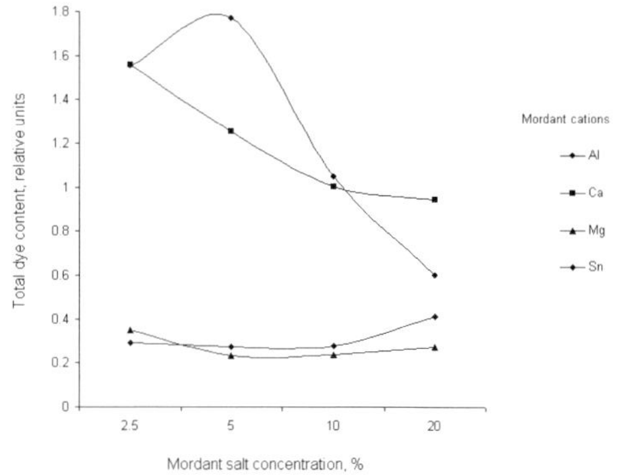

Figure 6 The total carminic acid content in dry lake films, prepared without cream of tartar, as a function of the concentration of different mordant salts.

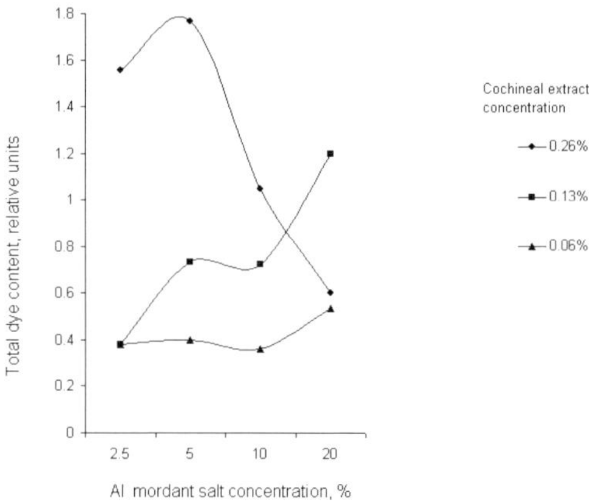

Figure 7 The total carminic acid content in dry lake films, prepared without cream of tartar, as a function of Al^{3+} mordant salt concentration.

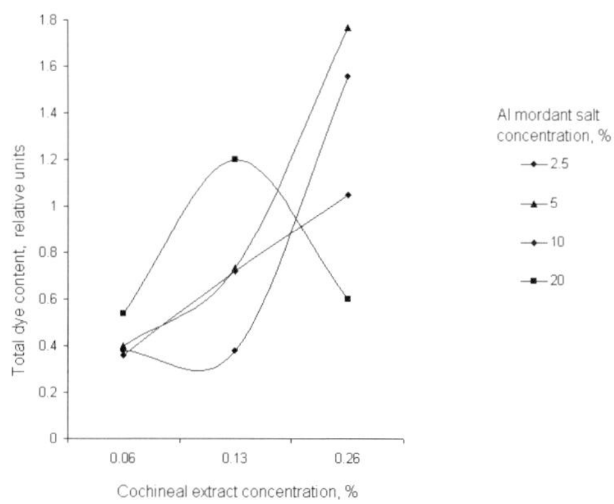

Figure 8 The total carminic acid content in dry aluminium-containing lake films, prepared without cream of tartar, as a function of cochineal extract concentration.

Table 2 The concentration of mordant forms of carminic acid in dry cochineal lake films, with and without cream of tartar, as a function of the concentrations of different mordant salts and cochineal extract.

		Mordant salt concentration, %							
		Carminic acid in mordant form, % concentration: no cream of tartar				Carminic acid in mordant form, % concentration: cream of tartar present			
Mordant cation	Cochineal concentration %	2.5	5	10	20	2.5	5	10	20
Al	0.26	67.67	78.21	89.65	87.78	36.85	46.85	51.75	46.43
	0.13	92.6	92.2	92.16	80.63	64.62	45.34	70.71	66.38
	0.06	100	100	100	88.89	100	100	100	100
Al + Ca	0.26	76.42	87.36	80.31	56.91	36.88	48.34	47.61	47.85
	0.13	76.71	79.21	70.23	73.63	59.32	57.04	60.17	57.67
	0.06	100	100	100	84.54	100	100	100	100
Al + Mg	0.26	66.65	71.43	64.72	60.76	36.53	44.3	46.94	50.3
	0.13	89.84	91.64	84.78	83.76	67.41	66.81	55.43	60.8
	0.06	100	100	100	93.95	100	100	92.87	87.03
Al + Sn	0.26	55.26	58.79	49.79	42.17	36.81	55.25	46.76	82.59
	0.13	95.4	80.94	71.65	90.92	72.65	59.43	60.53	69.97
	0.06	100	100	100	91.36	100	100	74.26	69.42
Ca	0.26	90.69	82.99	88.79	54.09	46.29	56.7	49.14	28.69
	0.13	70.28	77.01	57.5	57.49	51.58	44	44.1	41.12
	0.06	100	91.91	100	40	100	0	65.79	100
Mg	0.26	57	62.67	60.17	47.6	51.25	28.86	44.08	37.1
	0.13	87	69.74	37.7	46.51	39.59	64.71	74.58	67.86
	0.06	100	100	62.71	100	0	100	100	100
Sn	0.26	88.7	56.72	35.4	30.46	34.02	34.05	30.27	18.83
	0.13	65.16	60.88	47.91	46	48.67	66.67	65.57	51.06
	0.06	100	62.89	100	46	100	100	100	100

The influence of carminic acid concentration

From the figures given in Table 1, it can be seen that, for the cochineal extract (carminic acid) concentrations studied, the long-wave maximum optical density does not depend – or has only a very slight dependence – on the type of mordant cation. For the majority of the dry pigment films the optical density increases with an increase in carminic acid concentration. For the Al-containing pigment gels, the total concentration of carminic acid combined with collagen increases as the carminic acid concentration increases (Fig. 8). The exception is the pigment containing the maximum concentration of the Al^{3+} mordant salt – 20%, where the lake made with a carminic acid concentration of 0.13% contains more combined carminic acid than the pigment made with the more concentrated, 0.26% solution. (This figure should be compared with Fig. 9, where cream of tartar, discussed below, is also present.)

For the Al-containing lakes, it can be seen from Table 2 that the proportion of mordanted forms of carminic acid diminishes with an increase in carminic acid concentration. Thus the combination of dye with the metal cation and collagen in the mordanted form is at its greatest for the lowest concentrations of carminic acid in the Al-containing lakes. The situation in the lake gels made with bivalent cations is very complicated

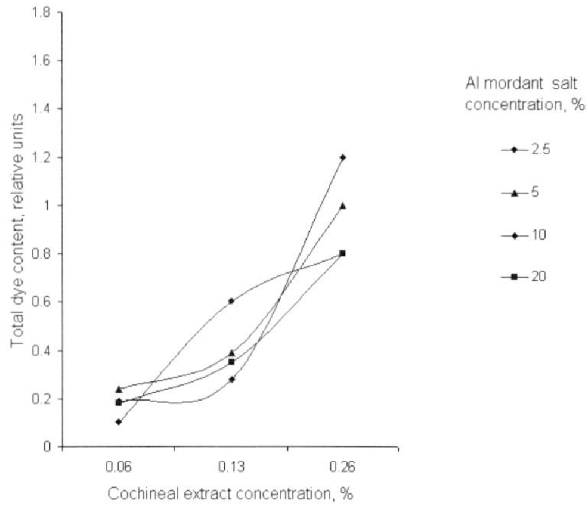

Figure 9 The total carminic acid content in dry aluminium-containing lake films, in the presence of cream of tartar, as a function of cochineal extract concentration.

The triplet long-wave optical density maxima for the lakes containing bivalent mordant cations in mixture with Al^{3+} has a comparatively small dependence on the concentration of the mordant cations. For the pure aluminium-containing pigments, however, the optical density decreases with an increase in the Al^{3+} concentration, as shown in Table 1. The table also shows that, in the presence of Mg^{2+} and Sn^{2+} mixed with Al^{3+}, the optical density maximum for lakes with maximum cation concentration increases markedly with increasing carminic acid concentration. There is a quite close dependence of the presence of mordant forms of carminic acid and depends on the increase in carminic acid concentration. It is connected with the formation of insoluble products in these particular pigment systems.

The influence of additions of bivalent cations to trivalent Al^{3+}

Pigment gels were made containing mixtures of the Al^{3+} cation and the bivalent mordant cations Ca^{2+}, Mg^{2+} and Sn^{2+}, as described above and illustrated in Plates 28 and 29. Mordant mixtures of this type (Al^{3+}/Ca^{2+}, for example) have been found in 18th-century silks dyed with cochineal.[12] The influence of additions of bivalent mordant cations on the characteristics of the model Al^{3+}-containing lakes can thus be assessed.

Plate 28 and Figure 10 show that the colour characteristics and the absorption spectra of the dry films containing mixtures of mordant cations are very similar to those of the films containing Al^{3+} alone. The addition of bivalent mordant cations to the Al^{3+}-containing systems does not lead to the formation of insoluble forms of carminic acids: on the contrary, these insoluble forms are typical only for the pigments prepared with pure bivalent mordant cations.

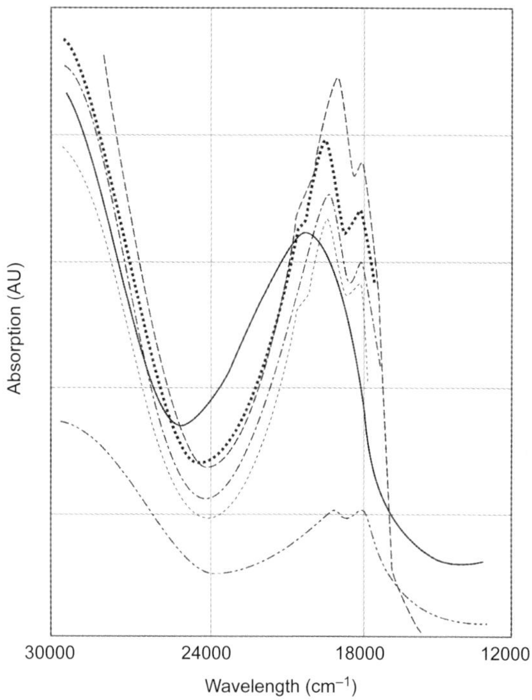

Key

- Pure aqueous extract of cochineal
- Dry film of cochineal pigment + collagen
- Dry film of cochineal lake + collagen + Al mordant salt
- Dry film of cochineal lake + collagen + Al and Ca mordant salts
- Dry film of cochineal lake + collagen + Al + Mg mordant salts
- Dry film of cochineal lake + collagen + Al + Sn mordant salts

Figure 10 Absorption spectra of dry cochineal lakes containing Al^{3+} and mixtures of Al^{3+} and bivalent mordant cations.

on the concentrations of Al^{3+} and of carminic acid in pure Al^{3+}-containing lakes and in lakes containing a mixture of mordant cations.

Influence of additions of cream of tartar

Cream of tartar (potassium hydrogen tartrate), the mono potassium salt of (+)-2,3,-dihydroxybutanedioic acid, is able to form bidentate complexes with metal cations.[13] It is thus possible that formation of poly-ligand complexes of metal cations with cream of tartar and carminic acid could occur, and also those between metal cations, cream of tartar, carminic acid and collagen.

For this part of the study, a set of model cochineal lakes containing 10% cream of tartar, with respect to the mordant salt content, was made (Plate 29). The same factors were varied in these pigment systems as in those without cream of tartar: for example, the concentration of the dye solution, the pH and the mordant cations present. The analysis of the experimental results, given in Tables 1–3, allowed the influence of cream of tartar on the properties of the model lakes to be evaluated.

The most important effect is that cream of tartar effectively prevents the formation of insoluble black, grey and greyish-rose forms of carminic acid typically present in lake systems prepared with bivalent mordant cations. This can be seen if Plates 28 and 29 are compared.

The presence of cream of tartar causes a slight increase in the triplet long-wave optical density maxima in the absorption spectra of the dry lakes prepared with the higher concentrations of the Al^{3+} mordant salt and carminic acid solutions. Its effects are more varied in the pigment systems prepared with other mordant salts or mixtures.

The effects of the presence of cream of tartar on the total content of carminic acid in the lake films, depending on the concentrations of the different mordant salts and the cochineal extract, are shown in Table 3 and, as an example, illustrated in the case of the aluminium mordant salt in Figure 9, which can be compared with Figure 8, discussed above. Cream of tartar diminishes the total content of carminic acid in lakes prepared with the greatest concentration of Al^{3+} mordant salt (20%) and carminic acid, particularly at lower concentrations of carminic acid; the position is more complicated for the highest concentration of carminic acid.

It can be seen from Table 2 that the presence of cream of tartar increases the combination of carminic acid with collagen in a mordant dye form for aluminium-containing lakes with the greatest concentration of Al^{3+} and carminic acid. The combination of dye with the metal cation and collagen in a mordant dye form is greatest – 100% – for the lowest concentrations of carminic acid in the Al^{3+} and cream of tartar lakes, and also in the pigment systems containing Al^{3+} with bivalent cations, notably at lower concentrations of the mordant salt solutions.

Table 3 The total content of carminic acid in dry cochineal lake films, in the presence of cream of tartar, as a function of cochineal extract and mordant salt concentrations. The values for the aluminium-containing lake film are shown graphically in Figure 9.

Mordant cation	Mordant salt concentration, %	Cochineal concentration, % — Total content of carminic acid, relative units		
		0.06	0.13	0.26
Al	2.5	0.1	0.6	0.76
	5	0.24	0.39	1.1
	10	0.19	0.28	1.2
	20	0.18	0.35	0.8
Al + Ca	2.5	0.177	0.13	0.3
	5	0.27	0.102	0.7
	10	0.241	0.097	1.1
	20	0.3	0.114	0.6
Al + Mg	2.5	0.061	0.49	0.44
	5	0.137	0.47	0.78
	10	0.136	0.43	1.24
	20	0.293	0.42	0.79
Al + Sn	2.5	0.062	0.49	0.83
	5	0.18	0.31	1.26
	10	0.2	0.34	1.58
	20	0.3	0.39	0.52
Ca	2.5	0.025	0.29	0.5
	5	0.02	0.23	0.74
	10	0.038	0.07	0.5
	20	0.05	0.107	0.24
Mg	2.5	0	0.14	0.165
	5	0.04	0.1	0.15
	10	0.055	0.059	0.211
	20	0.045	0.08	0.19
Sn	2.5	0.03	0.15	0.17
	5	0.028	0.08	0.14
	10	0.032	0.061	0.19
	20	0.009	0.05	0.154

Conclusion

The experiments show that it is possible to obtain intensely coloured lake pigments using an aqueous extract of cochineal, solutions of mordant salts and the protein collagen. These are similar to the pigments found in Russian icons dating from the late 17th to the 19th century. In these pigments, carminic acid can combine with collagen in two main forms: as a direct dye and as a mordant dye.

The properties of the dry lakes depend on the mordant cation type and its concentration, and also on the carminic acid concentration. Of all the mordant salts studied, the Al^{3+} cation was found to form the most durable and intensely coloured cochineal collagen pigments. Additions of bivalent cations – Ca^{2+}, Mg^{2+}, Sn^{2+} – to Al^{3+} increase the combination of carminic acid with collagen.

For aluminium lakes, the proportion of carminic acid present in a mordant dye form decreases for maximal Al^{3+} and total carminic acid concentrations. Thus the combination of dye with metal cation and collagen in a mordant form is greatest for minimal concentrations of Al^{3+} and carminic acid in aluminium-containing lakes.

Cream of tartar is an effective bidentate complexing agent and can interact as an additional ligand with mordant cations and collagen by means of coordination bonds. In lakes, therefore, cream of tartar prevents the formation of insoluble derivatives of carminic acid with bivalent cations and collagen. It increases the combination of carminic acid with collagen for lakes containing maximum concentrations of Al^{3+} and carminic acid. Here it is playing the same role as in the dyeing of silk with cochineal.[14]

Editor's note

*Author for correspondence: Russian Research Institute for Cultural and Natural Heritage, 2 Kosmonaut Street, 129366 Moscow, Russia (heritage@mtu-net.ru). Paper received 23 April 2003; revised version received 2 March 2004.

References

1. Golikov, V. (2002) 'The classification of coloured organic materials in painting and applied arts on the basis of natural dyes', *Dyes in History and Archaeology* 18, pp. 87–100; Golikov, V. and Zharikova, Z. (2002) 'The use of coloured organic materials based on natural dyestuffs in Russian icon painting of the 15th–19th centuries', *Dyes in History and Archaeology* 18, pp. 101–16.
2. Lantratova, O. and Golikov, V. (2003) 'Organic colorants in the 17th-century *Parsuna* of Patriarch Nikon from the State Historical Museum, Moscow', *Dyes in History and Archaeology* 19, pp. 60–66.
3. Golikov and Zharikova 2002 (see note 1 above).
4. Golikov 2002 (see note 1 above); Golikov, V. (2000) 'The technology of colour protein lacquers', paper presented at the 19th Meeting of Dyes in History and Archaeology, Royal Museum of Scotland (now National Museums Scotland), Edinburgh, 19–20 October 2000, unpublished.
5. Golikov and Zharikova 2002 (see note 1 above).
6. Traditional sources consulted include Agentov, M.A. (1769) *Otkrytie sokrovennykh khudozhectv* [A development of mysterious arts], vol. 2, Moscow [in Russian]; Reshetnikov, A. (1791) *Lyubopytnyĭ khudozhnik i remeslennik* [A curious artist and craftsman], vols 1 and 2, Moscow [in Russian]; Regan, P. (1795) *Nauka sostavlyat' dvenadtsat' sortov krasil'nykh tush'* [A new book for dyers], Moscow [in Russian]; Levshin, V. (1818–19) *Krasochnyĭ fabrikant, ili nastavlenie dlya sostavleniya vsyakogo roda krasok* [The complete dyer], vols I–IV, Moscow [in Russian]; Lavrov, N.I. (1869) *Opisanie krasok, upotreblyaemykh i predlozhennykh dlya upotrebleniya na zhidkostyakh v kartinnoĭ i dekoratsionnoĭ zhivopisi i pri okraske zdaniĭ* [The description of liquid paints used for applications in easel and decorative painting and for the decoration of buildings], St Petersburg [in Russian]; Grenberg, Yu.I. (1995) *Svod pis'mennykh istochnikov po tekhnike drevnerusskoĭ zhivopisi, knizhnogo dela i khudozhestvennogo remesla v spiskakh XV–XIX vv.* [Collection of written sources on the technique of old Russian paintings, manuscripts and artistic crafts in the 15th–19th centuries], 2 vols, St Petersburg [in Russian].

7. Golikov 2002 (see note 1 above).
8. Golikov 2002 (see note 1 above).
9. Golikov 2002 (see note 1 above).
10. Golikov, V. (2001a) 'The technology of silk dyeing by cochineal. II. The experimental investigation of the influences of types and concentrations of cations', *Dyes in History and Archaeology* 16/17, pp. 10–20.
11. Golikov 2001a (see note 10 above).
12. Golikov, V. (2001b) 'The technology of silk dyeing by cochineal. III. The experimental investigation of the influences of pH, water quality, cream of tartar and oak galls', *Dyes in History and Archaeology* 16/17, pp. 21–33.
13. Dyatlova, N.M., Temkina, V.Ya. and Popov, K.I. (1988) *Kompleksony i kompleksonaty metallov* [The complexing agents for and the complexes of metals], Moscow, Khimiya [in Russian].
14. Golikov 2001b (see note 12 above).

Natural Dyes Used in Polish Workshops in the 17th and 18th Centuries

Katarzyna Schmidt-Przewoźna

Abstract The Institute of Natural Fibres, Poznań, has studied 17th–18th-century 'patterned weaves' produced by the so-called 'Persjarnia' factories from the towns of Słuck, Kobyłki, Grodno and Gdańsk. This paper presents a history of the foundation of Polish dyeing workshops in the 17th and 18th centuries. The most famous and traditional patterned weaves from this time were Polish silk waist sashes. The Western and Eastern cultural influences and their role in the creation of colour and design of the Polish style of waist sashes are described. This paper also presents the plants used for textile applications in those periods.

Introduction

The 17th and 18th centuries were periods of very rapid development in the Polish textile industry, especially in spinning and weaving technologies and in design. The dyeing technologies remained more conservative and continued to follow earlier experiences and traditions.

The 18th century was characterised by a baroque style in art and architecture. The inherent trend in this style towards richness and ostentation resulted in the development of a palette of colours underlining the attractiveness of fabrics and encouraging a lively reaction to changes of fashion and taste.

During these two centuries, Poland capitalised on the opportunity to improve economic potential and independence, based on domestic raw materials and the internal market. One major development was that of the textile industry: raw materials that had previously been exported, such as wool and raw flax, began to be processed internally. Weaving manufacturing workshops were organised by the initiative of the Polish nobility. The founding of these workshops was assigned to foreigners and after a few years the first results were visible.

The search for colours in these times was not restrained by any regulations, unlike past centuries when particular colours were reserved for specific classes of society: for instance, red for noblemen. From this time onwards, the colour of clothes was determined by frequently changing fashion, which was highly influenced by western Europe, principally France, and by Persian and Turkish oriental cultures.

The main buyers of clothes were noblemen, who wore characteristic Polish national costumes: the *kontusz* and the *zupan*.[1] The *zupan* was similar to a long dress, with narrow sleeves and folds gathered into a belt fastened with a clasp. The *kontusz* dates from the end of the 16th century; this element of costume was based on the *zupan* and was tied by the traditional waist sash.

The factories and dye houses of Polish waist sashes

The main element of these traditional clothes was a special waist sash, which is also associated with Polish culture. Originally waist sashes were imported from Persia and Turkey by merchants from Armenia. The waist sashes were characterised by a subtle colour composition and their popularity resulted in the foundation of workshops for their manufacture: the 'Persjarnia' workshops, some of which soon became famous throughout the country. The five most important workshops were located in the towns of Słuck, Kobyłki, Grodno, Gdańsk and Lipków. The most important factory in Słuck, which was founded by the Radziwił family, was managed from 1758 by two Armenians, Jan and Leon Madżarski.

A characteristic feature of Polish waist sashes is the fact that they were double-sided. The waist sash effectively had four different faces as it was folded lengthways into two halves (Plates 30 and 31). Four-sided waist sashes of this sort were, in fact, a Polish invention.

Because of the different colours and ornaments used in the different parts of the waist sash (see, for example, Plates 30–35 and Fig. 1), it could be folded and worn with different apparel depending on the wearer's requirements. The colours of a waist sash symbolised different emotions: white, for instance, symbolised joy, while black indicated sadness.

The design of a waist sash is shown in Figure 1.[2] A floral-patterned border (a) surrounds the principal decorative fields. One end of the central main part of the sash is divided into a series of small linear fields (b) running across the width of the sash and consisting of alternate geometrical and floral motifs. Each end of the sash (c) is decorated by an ornament, the 'head' (a bouquet of flowers), while the background of the three principal decorative fields (d) is divided lengthwise into two colours on each side. Comparing Polish waist sashes with those from Istanbul, an evolution of the colours used can be observed. The Istanbul waist sashes were characterised by three colours of delicate shades; in those of the Madżarskis far more highly contrasting colours can be found.[3] Thanks to the archival documents from the factory in Nieświerz, belonging to the Radziwił family,[4] we know the specification of silk fabric colours used in the factory in Słuck (Plates 36 and 37): green, violet, pale blue, orange, dark and light walnut, navy blue, cinnamon, sapphire blue, grey, black, white, lilac, pink, lemon, yellow and red. Dyeing accounted for 50% of the costs of silk yarn production.

Natural dyestuffs used in manufacturing workshops in the 17th and 18th centuries

The natural dyestuffs used in the 17th and 18th centuries fall into three groups: dye-woods, cultivated plants and animal (insect) dyestuffs. It can be seen that, apart from the insect dyes, all the dyestuffs used in the workshops were of plant origin. The dye-woods and other plant

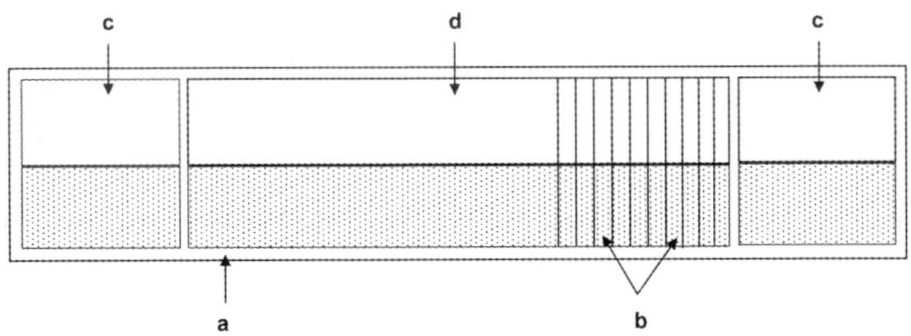

Figure 1 Diagram showing the structure of a waist sash: a: floral-patterned border; b: smaller fields – stripes – of alternating geometrical and floral motifs; c: an ornament called the 'head' – a bouquet of flowers; d: background of two colours.

dyes include those cultivated in Europe and those imported from further afield.[5] Dye-woods were imported from South America, the Pacific Islands and India in the form of whole trunks or as wood chips: red sandalwood (*Pterocarpus santalinus* L. f.), brazilwood (*Caesalpina* spp., C. *echinata* Lam. being the most important) and cutch (*Acacia catechu* Willd), for dyeing red and brown; and logwood (*Haematoxylon campechianum* L.), for dyeing blue and violet. The rhizome of turmeric (*Curcuma longa* L.) was imported from China and India, allowing yellow and orange colours to be obtained. Annatto (*Bixa orellana* L.) and indigo (*Indigofera tinctoria* L.) were also known and used in small amounts.

To secure easier access to the dyestuffs, plantations of dye plants were established, a good example being the plantation and botanical garden established in Grodno by Frenchman Vincent Dupiney,[6] who imported plants and seedlings for the garden from France. Dyeing plants thus cultivated included madder (*Rubia tinctorum* L.; see also Table 2), woad (*Isatis tinctoria* L.; see also Table 3) and weld (*Reseda luteola* L.).

Obtaining a dyestuff from a plant was not an easy task: it required a great deal of experience. Madder, for example, yielding a red dyestuff, required a fertile soil and care during its prolonged growing period before the root could be harvested and dried, and fertilisation. The dyestuff could be obtained no sooner than after three years of cultivation. Woad, yielding blue indigotin, was planted in March. The leaves were harvested in July and September, spread out on the ground and crushed with a stone wheel. The pulp obtained was made into small balls and dried. Safflower (*Carthamus tinctorius* L.; Table 2) was used to obtain a red colour.[7] The colouring matter is contained in the petals, which had to be collected in July and August. Safflower contains both yellow and red dyestuffs. The yellow dyestuff is soluble in water and very impermanent; the red dyestuff, on the other hand, is an excellent dyeing agent. Before the red dye could be used, the yellow dye was removed by sprinkling the safflower petals with salted water.

One of the Polish plants used in the dyeing workshops was sawwort (*Serratula tinctoria* L.; Table 1), which was collected in the woods.[8] It gives a yellow dye.

Besides the plant dyestuffs, insect dyestuffs were also used. For a considerable period of time, Poland was famous for the red dyestuff obtained from the Polish cochineal insect (*Porphyropora polonica* L.), which was exported to both western and eastern Europe.[9] The insect thrives on a plant called the perennial knawel (*Scleranthus perennis* L.), which can be found across northern Europe, including Poland.[10] The precious dyestuff was extracted from the insect larvae. The insect is a root pathogen, therefore it was necessary to dig up the plant to obtain the encysted larvae. The cysts were collected by children, who separated them from the roots using little sticks, then boiling water was poured onto the cysts to kill them and they were dried in the sun. The Polish cochineal dyestuff was in great demand until the discovery of America. The cochineal insect imported from there (the Mexican cochineal, *Dactylopius coccus* Costa) was richer in dyestuff and more economical to use. It replaced Polish cochineal entirely in the 17th and 18th centuries, although royal privileges were granted during this period in order to re-establish the raising of the indigenous Polish cochineal.

Today, Polish cochineal is described in the literature as a representative of a family related to the huge aphid family and is considered to be only a plant pest. According to the Plant Protection Institute in Poznań, no locations of Polish cochineal have been identified recently, although there are confirmed reports of its occurrence, for instance in the flooding area of the Warta River in the eastern part of the Wielkopolska region (central-western Poland). In fact searches for the insect conducted from 1960 to 1995 led to the conclusion that it is a species on its way to extinction. However, only quite recently, Polish cochineal has been reported again in central-southern and south-eastern Poland as well as in the eastern Wielkopolska region. Attempts have been scheduled in cooperation with the Plant Protection

Table 1 Wild plants used to obtain yellow and brown colours.[11]

Family	Scientific name	Common name
Asteraceae*	*Achillea millefolium* L.	Yarrow
Asteraceae	*Anthemis arvensis* L.	Corn chamomile
Asteraceae	*Anthemis tinctoria* L.	Yellow chamomile
Asteraceae	*Arctium lappa* L.	Greater burdock
Asteraceae	*Bidens tripartita* L.	Burr marigold, trifid bur-marigold
Asteraceae	*Calendula arvensis* L.	Field marigold
Asteraceae	*Calendula officinalis* L.	Pot marigold
Ericaceae	*Calluna vulgaris* (L.) Hull	Heather, ling
Asteraceae	*Carthamus tinctorius* L.	Safflower
Equisetaceae	*Equisetum arvense* L.	Field horsetail
Papaveraceae	*Fumaria officinalis* L.	Common fumitory
Fabaceae*	*Genista tinctoria* L.	Dyer's greenweed
Geraniaceae	*Geranium robertianum* L.	Herb Robert
Clusiaceae*	*Hypericum perforatum* L.	Perforate St John's wort
Asteraceae	*Matricaria chamomilla* L.	German chamomile, scented mayweed
Polygonaceae	*Polygonum aviculare* L	Knotgrass
Polygonaceae	*Polygonum persicaria* L.	Redshank
Polygonaceae	*Rumex acetosa* L.	Common sorrel
Polygonaceae	*Rumex acetosella* L.	Sheep's sorrel
Caryophyllaceae (Illecebraceae)	*Scleranthus annuus* L.	Annual knawel
Caryophyllaceae (Illecebraceae)	*Scleranthus perennis* L.	Perennial knawel
Asteraceae	*Serratula tinctoria* L.	Sawwort
Asteraceae	*Solidago serotina* Ait.	November goldenrod
Asteraceae	*Tanacetum vulgare* L.	Tansy
Asteraceae	*Taraxacum officinale* Weber	Dandelion

* Note: certain plant family names have been revised in order to give the consistent and recognisable ending '-aceae'. The preferred name for the daisy family is thus Asteraceae, rather than the Compositae seen in older publications. Fabaceae is preferred to Leguminosae, Clusiaceae to Guttiferae and Brassicaceae to Cruciferae.

Table 2 Wild plants used to obtained red colours.[12]

Family	Scientific name	Common name
Boraginaceae	*Alkanna tinctoria* (L.) Tausch (= *Anchusa tinctoria* L.)	Alkanet
Asteraceae	*Carthamus tinctorius* L.	Safflower
Rubiaceae	*Galium aparine* L.	Cleavers, goosegrass
Rubiaceae	*Galium mollugo* L.	Hedge bedstraw
Rubiaceae	*Galium verum* L.	Lady's bedstraw
Malvaceae	*Hibiscus trionum* L.	Flower-of-an-hour
Clusiaceae	*Hypericum perforatum* L.	Perforate St John's wort
Rubiaceae	*Rubia tinctorum* L.	Madder
Ericaceae	*Vaccinium myrtillus* L.	Bilberry

Table 3 Wild plants used to obtain blue colours.[13]

Family	Scientific name	Common name
Asteraceae	*Cirsium arvense* (L.) Scop.	Creeping thistle
Brassicaceae*	*Isatis tinctoria* L.	Woad
Oleaceae	*Ligustrum vulgare* L.	Privet
Polygonaceae	*Polygonum aviculare* L.	Knotgrass
Rosaceae	*Prunus spinosa* L.	Blackthorn, sloe
Ericaceae	*Vaccinium myrtillus* L.	Bilberry

* Note: certain plant family names have been revised in order to give the consistent and recognisable ending '-aceae'. The preferred name for the daisy family is thus Asteraceae, rather than the Compositae seen in older publications. Fabaceae is preferred to Leguminosae, Clusiaceae to Guttiferae and Brassicaceae to Cruciferae.

Institute to find these once famous insects in June 2004 or 2005.

Plates 30–37 show the range of colours that could be obtained with these dyestuffs used in the dyeing factory workshops associated with the weaving workshops. However, dyeing workshops were also located at the palaces of noblemen; these were associated with the cultivation and production of flax and wool. For dyeing, low-cost plants growing in the wild were used. Examples of the plants that could have been available for use are listed in Table 1 (yellow and brown), Table 2 (red) and Table 3 (blue).

Conclusion

The 17th and 18th centuries were a period of great expansion in the Polish textile industry, seen in the foundation of the 'Persjarnia' workshops for the production of the characteristic Polish waist sashes. The natural dyestuffs used in their production included imported dyes and those available locally, among them the famous Polish cochineal insect. Dye plants were cultivated for use by the dyeing factory workshops, but dyeing workshops at the palaces of noblemen, which concentrated on dyeing linen and wool rather than silk, made use of a wide selection of wild plants which were more economical to use.

Editor's note

Address for correspondence: Instytut Włókien Naturalnych (Institute of Natural Fibres), ul. Wojska Polskiego 71b, 60-630 Poznań, Poland (kasia@inf.poznan.pl). Paper received 7 May 2003; revised version received 5 January 2004.

Notes and references

1. Bartkiewicz, M. (1979) *Polski ubiór do 1864 roku* [Polish costume to 1864], Wrocław (Ossolineum), pp. 20–33 and 78–90.
2. Mańkowski, T (1976) *Mecenat artystyczny Stanisława Augusta* [The artistic patronage of Stanislaus Augustus], Warsaw, pp. 244–55 and 279–87.
3. Mańkowski 1976 (see note 2 above).
4. Mańkowski 1976 (see note 2 above); archive documents in the Archive Nieświerz.
5. Mańkowski 1976 (see note 2 above); Kowecka, E. (1963) *Farbiarstwo tekstylne na ziemiach Polskich (1750–1870)* [Textile dyeing in Poland 1750–1870], 2 vols, Wrocław/Warsaw/ Kraków, vol. 2, pp. 20–30 and 46–9; Dean, J. and Diadick Casselman, K. (1999) *Wild Colour*, London; Sobisiak W. (1968) *Wiejskie włókiennictwo w Wielkopolsce: porównawcze studium historyczno-etnograficzne* [The textiles industry in the Wielkopolska region: comparative historical-ethnographical studies], Poznań, p. 155.
6. Mańkowski 1976 (see note 2 above).
7. Kowecka 1963 (see note 5 above).
8. Szafer, W., Kulczyński, S. and Pawłowski, B.

(1976) *Rośliny polskie* [Polish plants], Warsaw (first published 1953).

9. Kowecka 1963 (see note 5 above).
10. Szafer *et al.* 1976 (see note 8 above).
11. Reinfuss R. (n.d.) 'Contribution for research on folk dyeing and printing in the Białostockie district', *Polish Folk Art*, 5, pp. 315–17; Tuszyńska, W. (1986) *Farbowanie barwnikami naturalnymi* [Dyeing with natural dyes], Warsaw, pp. 90–93; Dean and Diadick Casselman 1999 (see note 5 above); Sobisiak 1968 (see note 5 above); Szafer *et al.* 1976 (see note 8 above).
12. Reinfuss (see note 11 above); Tuszyńska 1986 (see note 11 above); Dean and Diadick Casselman 1999 (see note 5 above); Sobisiak 1968 (see note 5 above); Szafer *et al.* 1976 (see note 8 above).
13. Reinfuss (see note 11 above); Tuszyńska 1986 (see note 11 above); Dean and Diadick Casselman 1999 (see note 5 above); Sobisiak 1968 (see note 5 above); Szafer *et al.* 1976 (see note 8 above).

Archaeology and Dyeing Traditions in West Africa

Marie Marquet

Abstract Studying the present-day uses of vegetable dyes in Mali may provide a better understanding of West African history through its textiles. Three main traditions using plants for dyeing textiles can be described: *bogolan* of the Malinke; indigo dyeing among the Soninke; and Fulani dyeing techniques on wool. A comparison of those dyes with the colorants of the archaeological textiles from Tellem funeral caves has been made to assist with the identification of ancient dyestuffs.

Studying the past through natural dyes

In ancient times, dyes were mainly extracted from plants, with a minority from animals. Plants and animals used in dyeing are related to different ecosystems, therefore the identification of the plant source of a colour, combined with a study of the fibres used, may give us some indication of the area of fabrication of a textile. Through recipes, it is even possible in some cases to find precise cultural traditions within which a textile was made as some recipes are used only by one specific group. This, together with the technical analyses of the textiles, can help to define their cultural origin.

In most ancient societies, textiles were not only important from a cultural and social point of view, but also had great economic value, sometimes even being used as money.[1] Their study, combined with a study of the history of dyeing, may thus offer an insight into long-distance trade, cultural influences, or religious attitudes. This is particularly true for sub-Saharan Africa, where textiles have played an essential role from ancient times until now and where traditional textile techniques are still alive, in spite of the competition from manufactured products.

The present study mainly deals with Mali, where archaeological textiles from the 11th and 12th centuries, known as Tellem textiles, have been discovered. The aim of my degree research (DEA) in 2002–03 was to design an experimental methodology, comparing dyes in modern traditions with the colorants of these ancient textiles.[2] Discovered in the 1960s by the archaeological team of Utrecht University in funeral caves of the Bandiagara cliff in the Dogon area,[3] Tellem textiles were studied by Rita Bolland of the Amsterdam Tropenmuseum, together with Rogier Bedaux. The results were published first in articles and then in complete form in the book *Tellem Textiles* in 1991.[4] Most of the collection is now kept in the Mali National Museum in Bamako. Some dyestuff analyses by thin-layer chromatography (TLC) were carried out by Judith Hofenk de Graaff, the results of which were reported in the 1991 publication.[5] She found the following vegetable dyestuffs: indigo; a type of redwood she called 'brazil wood'; varieties of tannin; varieties of madder; alizarin; and a red dye similar to that of cochineal.[6] Hofenk de Graaff identified the latter as extracted from *Coccus lacca*, more correctly known as *Kerria lacca* Kerr, the lac insect. Apart from this dye, it was rather difficult to identify the plant sources of the colorants in the medieval West African context and some colorants could not be identified. It is important to mention here that our knowledge of African plant chemistry is far from complete, even though some important publications deal with medicinal aspects. Much work has still to be done, particularly concerning colorants.

To answer some of the questions raised by the analyses and to study the vegetable dyes of Mali, a large collection of dye plants and recipes was needed. The study followed an ethno-archaeological methodology with several strands:

- A precise study of the ethno-botanical publications on uses of West African plants;
- A survey in Mali to collect plants and transcribe traditional recipes using natural dyes, still known by craftsmen;
- A historical study of textiles in West Africa, focusing on the Tellem corpus, but also encompassing a review of the archaeological finds linked to textile production and of the written sources, mainly descriptions by Arabian travellers;
- Analyses of new samples from some of the Tellem textiles stored in Bamako.

As only eight months were available for this research, it does not claim to be exhaustive or complete. It should be understood rather as an overall synthesis, suggesting directions for further research.

A living tradition: natural dyes in West Africa

As a start, a database was constructed based on a study of the ethno-botanical literature on West African plants. All the plants known for their colouring properties were sorted by geographical area, cultural context of use, recipes (when given by the author) and local names. While there are few publications on dyeing plants as such, it was possible to compile many different publications with some relevance, including:

- Ancient inventories of useful plants, dating from the time of the exploration of Africa and from the first period of colonisation, written by missionaries, administrators and travellers;
- Recent ethnographical inventories of plant uses, concerned with very small areas or groups from a cultural perspective;
- Recent botanical inventories covering large areas, some with a particular topic, usually medicinal plants. Among these recent works, the inventory of the dye plants of Burkina Faso by Narco and Millogo-Rasolodimbi (1993) is a major contribution to the subject.[7] Kerharo's work on Senegal and the Ivory Coast is also very useful (Kerharo and Adam 1974), although more accurate on medicinal aspects.[8]

The problem was then to link botanical information with technical knowledge, including examples of recipes and textiles. Thanks to the database, the result of this bibliographical research is an exhaustive list of plants that grow in the Mali climate and which are known for their dyeing properties, or could be used for this purpose.[9]

When travelling in Mali, the first step of my survey was to meet people and to ask them about dye plants. The list was then used to translate local names into scientific names, and to check whether bibliographical data were confirmed or not by contemporary Malian craftsmen. The identification of plants was confirmed thanks to the expertise of Michel Arbonnier, botanist at CIRAD (Centre de coopération internationale en recherche agronomique pour le développement) and author of a flora on the woody plants of the dried zones of West Africa.[10]

I had previously stayed in Mali in 1998, working on archaeological research. When starting the 2003 study, I went back to places where I knew people with whom I had learned the use of dye plants that are part of the Bogolan tradition.[11] The main work then involved meeting different craftsmen with a wide knowledge of natural dyes. Enlarging the inventory of dyeing plants and recording recipes was a first goal.[12] As the time available for research in Mali was relatively short, much time was spent collecting as many dye plants as possible to be used for the interpretation of archaeological sample analyses.

Three main dyeing traditions still extant in Mali are discussed here, based on personal observations and bibliographical information. Each uses different plants and produces different kinds of textiles, characteristic of cultural groups. Plants that have colouring properties, but which are not used in any of these main traditions, will be mentioned briefly.

Indigo blue

It seems difficult not to say a word about indigo when writing on African dyes, since it is the most famous and most studied of all the West African vegetable dyes. Several important studies on indigo blue have already been published, from the cultural, historical and technical points of view. As Renée Boser-Sarivaxevanis (1969) notes, as far as West Africa is concerned, it seems that a clear borderline separates dye traditions linked to indigo plants and those linked to the other colorants.[13] This separation can be observed in the specific technical knowledge needed for the creation of the vat,[14] and also in the practical context: indigo blue is very commonly used alone, whereas the other colorants are less often used this way.[15]

In Mali, the Marka (or Soninké or Maraka) women are the holders of the indigo vat recipe. Nowadays, indigo dyeing is also carried out by other groups, in particular Dogons and Bambara, but the origins of this practice still remain in the Soninke tradition. These people live together, which explains the fact that there are ancient borrowings of techniques. The indigo colorant can be extracted from several plants: the Yoruba indigo (*Lonchocarpus cyanescens* Benth.) is mainly used in the wooded savannah and forest area, whereas the genus *Indigofera* is present in dry savannah.[16] This grows profusely in many areas of Mali. *Indigofera* seeds have been found on medieval archaeological sites on the inner Niger delta.[17]

Indigo plants are not the subject of this present work, because it is still practically impossible to determine which indigo plant was the source of the colouring matter on a blue textile. For this reason, they are an unreliable basis for research on ancient textiles, despite the fact that many Tellem cotton textiles show beautiful indigo dyes, such as the well-known cap with tie-and-dye patterns.

Bogolan: fermented clay and tannins

The *bogolanfini* (or *bogolan*) is a specific Mande technique,[18] of great importance in Mali. The name is applied to cotton cloth, dyed and decorated with clay used to draw patterns: '*fini*' means 'cloth', '*bogo*' means 'clay' and '*lan*' means 'with' in the Bamana (or Bambara) language. The reaction between tannins, present in the dye, and iron, present in the fermented clay, is used to create black designs. In the past, this technique was women's work and the drawings were highly significant signs.[19] Today, there are many *bogolan* workshops in Bamako, where this technique is used to create modern-style cloth for tourists or export, or for modern Malians. Some designers and creators, such as the Bogolan Kasobané group, are using this technique as a way of artistic expression (Plate 38).[20]

The thesis of Pauline Duponchel-du Barry is the most complete approach to this subject; her botanical identification of the plants she observed can be used as a solid base. *Bogolan* is far from an unknown topic: publications on this subject exist, but until now no comparison with ancient textiles has been carried out.[21]

My aim was to collect as many of the *bogolan* plants as possible as samples that could help in the identification of colorants on Tellem textiles. When meeting craftsmen, I asked them if they knew of other dyeing practices or plants. It was mainly through discussions with Augustin Sagara and Issa Niambélé, artists in Bogolan, in Bamako, that I was first able to compare the plant list derived from the ethno-botanical publications with their knowledge of plants.[22] I travelled also to Ségou, where I met another artist, Boubakar Doumbia, from the Bogolan Kasobané group.[23]

In order to prepare the dyes, the plants are covered with water in a big pot. They are left to simmer on the fire for at least one hour, in the case of leaves, or two hours when barks are used. When the dye is ready, it is poured into a calabash or a plastic bowl. Once the preparation is removed from the fire, but still hot, the cloth is steeped in it. The dye is used up completely: for the second and third soakings it has cooled down, but this is of no importance; the dye solution can be kept for a few days. Sometimes the decoction ferments and is used as it is.

After soaking the cloth in the dye, it is spread in the sun to dry; the shade also thus becomes darker. The design is then drawn, traditionally using a small stick or, nowadays, a brush, to apply the clay solution. The parts of the cloth covered with clay will become black, whereas the cloth left free from clay will keep the original colour, from yellow ochre to brown and brownish red, depending on the dye plant used. Once the clay is dry, it is taken off the cloth by shaking and rubbing it, followed by rinsing. To obtain a deep colour and a deep black, all the operations need to be repeated several times: soaking the cloth in the dye, drying it in the sun (an important step) and drawing with clay on the black parts. In some works, the part of the cloth left without clay is finally made white using a potash solution made from ashes, soap or even chlorine to obtain a black and white design.

Traditional *bogolanfini* cloths have white designs on a black background – the designer covered the entire cloth with the clay solution, avoiding the design intended to be white. The ferruginous mud used for drawing generally comes from the bottom of the river or the marshland. In Africa, the climate leads to iron concentration in the soils: an example is the red colour of laterite soils, very common in West Africa. This clay is collected and then left in a jar for a year. During this time, it must ferment. Some dyers add rusty nails to improve the quality of the solution. When it has fermented and become blackish, a small amount is diluted in a pot with water. It can also be diluted with an addition of potash to obtain brown shades instead of black.

The main plants used for dyeing *bogolan* are still very well known by the Malinke: *Anogeissus leiocarpa* Guill. & Perr. (called '*n'galama*' in Bambara) is the foundation of *bogolan* dyes because of the strong reaction of its tannin with iron; it gives the characteristic yellow-ochre colour. Two species of *Lannea* – *Lannea acida* L. Rich. and *Lannea microcarpa* Engl. & K. Krause (both named '*n'pekuba*' in Bambara)[24] – are used: their bark gives a brownish red. Some craftsmen also employ various species of *Combretum*, *Terminalia* and *Ficus*. A decoction of fresh leaves of *Combretum glutinosum* Perr. ex DC. is used cold to dye yellow; dry leaves mashed and macerated also give a yellow dye and the roots give an orange colour. *Combretum nigricans* Leprieur ex Guill. & Perr. gives a reddish-orange colour and *Combretum micranthum* G. Don a red. A decoction of the bark of *Terminalia laxiflora* Engl. (Plate 39) gives a brown; its roots give a reddish brown, several soakings being necessary. The mashed bark of *Terminalia macroptera* Guill. & Perr. is boiled to give an orange colour. A decoction of the bark of *Ficus gnaphalocarpa* Steud. ex Miq. gives a brown, while that of *Ficus platyphylla* Delile gives a red. Other plants, such as *Piliostigma reticulatum* (DC.) Hochst. and *Hexalobus monopetalus* Engl. & Diels, are known locally, but scarcely used nowadays.

In the cities, dye plants are bought in the market, together with the medicinal plants commonly used by Malians (Plate 40). Barks and roots are sold in large pieces or in powder, in the case of the most popular plants; leaves are tied up in balls. In the countryside, they are cut by the villagers in the nearby bush. The availability of these resources for dyeing partly explains why they are still in use in Mali: it costs much more to obtain chemical dyes.

Fulani dyes on wool

The Fulani are great animal breeders and they possess sheep with long wool, which are scarce in sub-Saharan Africa: here, sheep are generally of a shorthaired breed. Fulani weavers called Mabbube developed a specific textile craft using wool to make the famous Kaasa and Kerka blankets from Macina.[25] These blankets are highly valued in other parts of West Africa and have been traded to other regions for centuries. Nowadays these blankets are still commonly woven with natural dyed yarns. The colours are white, yellow, red and black; indigo blue can be found, but it is borrowed from other traditions (Plates 41 and 42).[26]

Gardi (1985) and Imperato (1973) are, to my knowledge, the main authors on the subject of Fulani textiles in general.[27] They both mentioned the use of dye plants, but a mushroom they observed being used was not then identified. While staying in the village of Banguineda I met Mariam Kampo, a Fulani dyer who is very interested in natural dyes, even though she uses synthetic dyestuffs for her work. There I first saw this mushroom, collected in Macina, and the other plants used in Fulani recipes. Later, I met Colado Cissé, a Fulani merchant of traditional textiles in Bamako, who helped me to collect some of these mushrooms. Back in France, they were identified by Alain Gerault, a mycologist of Brest University, as *Podaxis pistillaris* (L.:Pers.) Fr., from the family of Gasteromycetes.[28]

Using spores of this fungus to prepare a dye-bath is a unique Fulani recipe, for it seems that no other West African group knows this technique.[29] The fungus is collected in millet fields and a yellow dye is obtained from its spores by decoction. Two or three small cups of the spores are necessary to make the dye-bath, in which the fibres are left to boil at the same time, for 15 minutes. After the mushroom bath it is then possible to dye the threads red, using a decoction of *Hibiscus sabdariffa* L. flowers. Alone, this dye is unstable and soon vanishes, but in association with the *Podaxis pistillaris* bath it gives a strong, pleasant red.

Another shade of red can be obtained with the sorghum stem. This dye does not need the preliminary mushroom dye-bath treatment and can be applied onto white hanks, but the red obtained is dull. Pods and leaves of *Tamarindus indica* L. are also used for reds and browns, again in hot dye-baths. For black, pods of *Acacia nilotica* (L.) Willd. ex Del. are combined with fermented mud containing iron, as in Bogolan (Plate 43). The recipe is more complex, requiring maceration and fermentation for a few days before it is ready.

The Fulani still dye and weave Kaasa in the traditional manner, but it seems that synthetic dyes are already starting to invade the practice. There is much that could be said about the Fulani textile tradition, whose origin, linked to the coming of long-haired sheep in sub-Saharan Africa, has still to be explored.[30]

Other plants

Beside these three traditions of dyeing techniques, certain other plants are also used for dyeing, but knowledge of their use is fast disappearing. From the survey the plants described in the following section were brought back for analysis. Although one can envisage carrying out many other experiments, these are good examples of the plants growing in Mali commonly mentioned in the literature.

A few years ago, cola nuts, *Cola nitida* (Vent.) Schott & Endl., were in fashion among the women dyers in Bamako (Plate 44). The nuts must be mashed and macerated in water to produce a beautiful orange dye. This is still quite a lively tradition, compared to the others, and very often cola nut dyes are made by women indigo dyers.[31]

The leaves and bark of several species of *Ficus* giving green or red colours are mentioned in the ethno-botanical literature. I did observe the use of some *Ficus* barks among the dyers making *bogolan* cloth (for example, *Ficus platyphylla* and *gnaphalocarpa*) but never the decoction of *Ficus* leaves. *Parkia biglobosa* (Jacq.) R. Br. ex G. Don and kaïlcedrat (*Khaya senegalensis* A.

Juss) also have colouring properties. There are many of these trees in the bush, largely used in medicine, magic, and as a condiment; dyeing is a very marginal possibility, but mentioned in publications. In the area of Segou, the pods of *Parkia biglobosa* are included in the building of clay houses to obtain a red-coloured cob.

The use of the roots of *Cochlospermum tinctorium* Per.[32] and *Adansonia digitata* L.[33] is still a matter for further research: some authors describe them as dye plants, but their use seems to be lost in contemporary Mali. More surprisingly, the analyses of the dye-bath made with these plants failed to identify any precise colorants!

Dye fixation

Dye fixation is, in my opinion, the most interesting aspect of this research: it is the key to the recipes and one of the less studied points. When uses for a dye plant are mentioned in publications, they are very seldom accompanied by a detailed recipe and even less often by information concerning fixation or colour. Unlike the European tradition, mordant dyes using metallic salts are not very common in traditional dyeing practices in Mali. The technique of mordanting is replaced by other techniques that help to achieve a stable colour.

The first question is how people think of the colour, its meaning, its use and its stability. In many cases dyeing is not necessarily related to the idea of keeping the cloth as coloured for a long period of use. Sometimes, dyes are 'ingredients' for special fabrics, for example in ceremonial costume or circumcision wear. The colours, or even the plant source of colour, play an important role in the global symbolism of the ritual. Once the event is over, there is no need and no desire to keep the costume in the same state, or to reuse it, because it is now socially and emotionally very 'charged'. Consequently, concern for fixing the colour is in fact irrelevant.

During my fieldwork, I observed three techniques of fixation:

- The association of several plants;
- The action of the sun;
- The reaction with iron.

The minor use of additives will also be examined. These techniques are nowadays applied less strictly, because it is always possible to draw on chemical colours. Generally speaking, there is probably a loss of know-how in the practice of the craftsmen, who transfer recipes known for one plant to another new one, and do not precisely remember which technique is used for which plant.

The association of plants

Some plants giving fast dyes are used to fix the less stable colorants of other plants. These may – but not necessarily – be tannin plants. They can be called impregnation dyes or plant-association dyes. For fixing a great number of dyes in *bogolan*, such as bark from *Lannea* and *Ficus* species, or from *Piliostigma reticulata*, the plant *Anogeissus leiocarpa* is used for a fixing bath.

A first dye-bath 'for colouring' is made with the decoction of the chosen plant, such as *Lannea microcarpa*. The cloth is soaked in this dye and then dried in the sun. This is followed by the 'fixing dye', a bath with *Anogeissus leiocarpa*. The cloth is again left to dry in the sun. This operation is repeated three times at least. Finally, the cloth is soaked again in the first, coloured, dye-bath and dried in the sun. The fixing properties of *A. leiocarpa* are probably due to the presence of tannins in the leaves, also used for their reaction with iron when applying clay. According to the dyers, it gives the other plant dye more stability.

The same kind of technique can be seen in Fulani recipes, in which white wool is prepared in a bath of *Podaxis pistillaris* before going into the *Hibiscus sabdariffa* bath. In the first bath, sheep droppings are often added, which seem to increase the stability of the colour. There are no tannins here, but the efficiency of this technique is proved by the results: the red obtained is indeed strong and stable, whereas dye with hibiscus alone is unstable. This association of several plants is quite a common technique.

The sun

For most dyers, drying the cloth in the sun is vital to fix the colour of many dyes. It can be associated with the other two techniques, or used alone. The dye-bath is obtained by decoction or maceration. After the cloth has been soaked, it is taken out and spread in the sun on a flat horizontal surface, generally the ground: there must be no creases. Once it is dry, the cloth is soaked again and exposed to the sun once more. This operation may be repeated four or five times, or even more.

According to the craftsmen, the darker and faster the desired colour, the more the cloth must be soaked and dried. Indeed, the side of the cloth facing the sun turns darker, whereas the other side remains lighter. Dyes with *Anogeissus leiocarpa* show the most striking change with the sun's action. All the dyers give this as a rule for a stable dye with most of the plants.

A systematic study of the plants for which this technique is used should be undertaken for a better understanding of their chemical reaction under exposure to light, heat and air. The final point is to define the cultural link that may exist between this technique and the indigo-dyeing method, in which exposure to the air reveals the colour. Could this technique come from the practices of indigo dyes, applied to the other dyes?

Iron from the earth

Using fermented mud containing iron sulfates is a method of obtaining a black colour. In Bambara mythology, the colour black originates in the earth: this is how the clay from the marshland dyes loincloths black.[34] This technique of colouring is also a fixation process: iron sulfate is a mordant, and the black obtained is strong. When used on wool, iron mordants may corrode the fibres. This is a means of identifying this technique on ancient textiles, where, quite often, dark fibres are weakened or destroyed.

Additives

Potash is often used as ash lye, added to the dye-baths to darken the colour;[35] or for extracting and fixing the colorant (e.g. with indigo); or used as a paste to whiten the cloth after dyeing (e.g. in *bogolan*). Potash plants are generally well known; villagers commonly use them for making soap and washing clothes. For the indigo vat, great attention is paid to the choice of the potash plant sources, and to the manner of filtering the ashes and preparing the liquor,[36] but for some other recipes, common ashes from any kind of wood, or from the fire under the dye-bath pot, are added to the decoction. The ashes of several bush plants are used to fix dyes; they are mentioned in ethno-botanical literature by Arbonnier (2002) and Abbiw (1990).[37] Abbiw also mentions the use of salt and natron (crystallised sodium carbonate). Alum is found in many mines in the Saharan area, known from the first millennium AD at least, and described by trade travellers. I have never observed the use of these additives during my fieldwork, but maybe a more thorough search on this subject could elucidate this point.

The use of acidic plants is also known in West Africa and needed for some dyes. Lemon is used nowadays. *Citrus limon* (L.) Burm. f. was introduced in sub-Saharan Africa with the Saharan trade, and probably replaced the more ancient use of local acidic plants. We find an example of this in the report of an Arab traveller, Al-Umari, writing in the first half of the 14th century AD, describing a recipe mixing henna with the dry flesh of the baobab fruit to obtain a black dye. This fruit is both acidic and sweet.[38]

The recipes aiding the fixation of colorants are especially interesting, because all the knowledge of the dyers lies in these techniques. As a first conclusion, these techniques seem to be rather different from the traditions of the North African and Mediterranean world, especially with respect to the importance of plant association and the significance given to dyeing in the sun and the absence of the use of alum. These are the 'unknown side' of West African dyeing techniques: chemical studies on this subject would be very useful to answer the remaining questions about colour fixation, the recipes and the additives.

More generally, the topic of the relationship between dyeing traditions and the choice of fibres to be dyed, and between medicinal and dyeing plants, is a fascinating field to be explored. From an ethnological point of view, the meaning of the colour and the choice of plants used for dyeing 'special' cloth or ritual costumes are important fields for further research.

The Tellem collection

This unique group of particularly well-preserved archaeological textiles was found in caves used as tombs by the Tellem, together with skeletons, basketry and wooden objects. The Tellem people occupied the surroundings of the Bandiagara cliff from the 10th to the 15th century AD, until they were displaced by the famous Dogon people. The Dogon are now the inhabitants of this region.

During the period of the Tellem occupation, cotton cultivation and textile production developed significantly in sub-Saharan West Africa, and trade in luxury goods, such as gold, salt, metal, ceramics, and also textiles, took place between African regions and through the Sahara to the north.[39] Our knowledge of the Tellem civilisation is based entirely on discoveries in the funerary context; their daily life, and especially textile fabrication, is still unknown so there are no obvious proofs of whether these textiles were produced locally or imported.

The textiles can be separated into two types: cotton and woollen. The cotton cloths are quite fine, and a great number are dyed with indigo. Other colours, brownish red and ochre, may have been used, but for most of them, the colour conservation is not really good. The woollen fabrics have very fresh and varied colours, and tapestry patterns. This second category occurs only in the more ancient caves, dating from the 11th and 12th centuries. The cultural context of the use of the textiles and their technique and place of fabrication are unknown. Through the existing traditions and their study, we can, however, try to find indications as to their origin.

Experimental archaeology and dye analyses

The dyeing plants collected in Mali served as a reference collection for understanding the results of HPLC analyses on small textile samples (Table 1). In Mali, 63 samples were taken from 29 textiles kept in the National Museum of Mali in Bamako, thanks to the kind permission of Mr Samuel Sidibé, its director, and with the help of Mr Drahmane Berthé, curator. They were analysed by Witold Nowik from the Research Laboratory of Historical Monuments in Champs-sur-Marne (France).[40]

Dyes containing laccaic acid associated with another anthraquinone, indicating the use of lac insects (*Kerria lacca*), were the most unusual results. This accords with the results of Judith Hofenk de Graaff (1991), who found 'cochineal insect-type' dyes that she identified as *Coccus lacca* or *Coccus ficus*.[41] All these insects are, for the time being, unknown in this Malian area, but some lac insects have been identified in Nubia and in Central Sahara (Tassili N'Ajjer).[42] The cloths or threads could thus come from quite a distance away. The possibility of long-trade exchanges is confirmed by technical studies of the weaving techniques for some other cloths, which also show a Nubian origin.[43] However, we must not exclude the possibility of a future discovery of some 'unknown' African lac insects (Plate 45).

Results of the analyses of black threads reveal an association of iron and ellagic acid. This can be compared either to *bogolan* techniques using *Anogeissus leiocarpa*, or to the use of *Acacia nilotica* in Fulani work. The latter comparison is closer as it also concerns woollen textiles (Plate 46).

The presence of apigenin on other samples (red and yellow) could be explained by the use of a sorghum dye, from *Sorghum caudatum* Stapf. or *Sorghum vulgare* Pers., also known as

Table 1 Reference collection of plant material and other sources of dye.

Scientific name of source of dye	Parts used for dyeing
Acacia nilotica (L.)Willd. ex Del. subsp. *adstringens* (Schum. & Thonn.)Roberty	fruits
Adansonia digitata L.	foots (powder)
Anacardium occidentale L.	bark, seeds
Caesalpinia sp.	food
Cochlospermum tinctorium Perr	roots (powder)
Cola nitida (Vent.) Schott & Endl.	fruits (nuts)
Combretum glutinosum Perr. ex DC	leaves
Combretum micranthum G. Don.	leaves
Ficus platyphylla Delile	bark
Haematoxylum sp.	wood
Hexalobus monopetalus Engl. & Diels	leaves
Hibiscus sabdariffa L.	flowers
Indigofera sp.	leaves
Kerria lacca (Kerr)	insects
Khaya senegalensis A. Juss.	bark (powder)
Lannea acida A. Rich.	bark
Lawsonia inermis L.	leaves
Parkia biglobosa (Jacq.) R. Br. ex G. Don	pods
Piliostigma reticulatum (DC.) Hochst.	bark, *frets??*
Podaxis pistillaris (L.: Pers.) Fr.	stalk and spores
Pterocarpus sp.	bark
Sorghum bicolor (L.) Moench	stem
Sorghum caudatum Stapf	stem
Sorghum vulgare Pers.	stem
Tamarindus indica L.	fruits (pods)
Terminalia laxiflora Engl.	leaves, bark

Royal Botanic Gardens, Kew (2002). Electronic Plant Information Centre. Published on the Internet; http://www.kew.org/epic/ [accessed 3 January 2007; 22:00 GMT]; The International Plant Names Index (2004). Published on the Internet http://www.ipni.org [accessed 3 January 2007].

a Fulani technique (Plate 47). Other colorants giving a yellow colour were detected by the analyses; these were distinctive, but still unidentified. Their identification might be possible once the plant source is included in the reference collection (Plate 48). Unfortunately, some samples gave no results. Sometimes the amount of dye present on the threads was too small; in other cases, the threads may not have been dyed (the dyeing of 12 samples was doubtful). Finally, in some cases the interpretation of the very small peaks visible on the chromatogram was impossible.

Enlarging the plant inventory and assembling a collection of coloured threads dyed using all the traditional recipes could encourage progress in making comparisons with archaeological textiles. These results have already shown two different kinds of textiles within the corpus of woollen fabrics: those imported from remote places and those either imported from a nearby region, or produced locally. Future research will throw more light on sub-Saharan exchanges, and on long-distance trade between the cultures surrounding the Sahara.

Conclusion

This work is the result of a first field mission. It gives an insight into questions about ancient textiles and traditional dyes in Mali. Its aim is to define prospects for future research. The inventory of the recipes and factual knowledge still alive in Africa should be made urgently: these are fast disappearing or changing. The development of studies on the African textile heritage will refine our knowledge of ancient African history, still imperfectly known. From an ecological point of view, it seems necessary to consider the exploitation of these vegetable resources and to start plantations and cultivation for those widely used.

For all these reasons, an inventory of data on plant uses, recipes and colorants is a fundamental requirement from the cultural, ethnological, historical and chemical points of view. Consequently in 2003, as an experiment, I started to build a database on African dyes for my research. This database model, first used for West Africa, will be extended to integrate new research and to include the dyes used in other civilisations and continents.

Acknowledgements

This survey of West African dye plants was possible thanks to the help of many people. I would like to thank them all warmly and in particular: Mr Samuel Sidibé for the authorisation to study and analyse the Tellem textiles and for his confidence, and the National Museum team for their welcome and help; Issa Niambélé, Augustin Sagara, Mariam Kampo and Colado Cissé for their help in collecting plants and recipes in Mali; Bernhard Gardi for his valuable advice; Michel Arbonnier and Alain Gerault for their botanical identifications and their availability; Witold Nowick for HPLC analysis, advice and support; Ann Deck and Marinette Stern for their help with the English translation; and Dominique Cardon, who led me with precise bibliographical and methodological advice, who found time to edit this paper and, above all, who has supported me during this work until now and hopefully will continue to do so for new projects to come.

Editor's note

This paper was presented as a poster at the 22nd Meeting of Dyes in History and Archaeology, Riggisberg, 2003. Address for correspondence: CNRS, UMR 5648, Louis Lumière – Lyon II University, 18, quai Claude Bernard, F-69365, Lyon Cédex 07, France (marquet.marie@yahoo.fr). Paper received 18 May 2004.

All plant names checked: Royal Botanic Gardens, Kew (2002). Electronic Plant Information Centre. Published on the Internet: http://www.kew.org/epic/ [accessed 3 January 2007; 22:00 GMT]; The International Plant Names Index (2004). Published on the Internet: http://www.ipni.org [accessed 3 January 2007].

Note and references

1. Johnson, M. (1980) 'Cloth as money, the cloth strip currencies of Africa', in D. Idiens and K.G. Ponting (eds), *Textiles of Africa*, Bath (Pasold Research Fund), pp. 193–202; Johnson, M. (1976) 'The calico caravans: the Tripoli-Kano trade after 1880', *Journal of African History* XVII, pp. ??.
2. Marquet, M. (2003) *Plantes tinctoriales d'Afrique de l'Ouest, l'exemple des textiles Tellem du Mali médiéval*, DEA thesis, Université Louis Lumière – Lyon II.
3. Research by the Institute of Human Biology.
4. Bedaux, R.M.A. and Bolland, R. (1980) 'Tellem, reconnaissance archéologique d'une culture de l'Ouest africain au Moyen-âge: les textiles', *Journal des africanistes* 50, pp. 9–23; Bedaux, R.M.A. and Bolland, R. (1989) 'Vêtements médiévaux du Mali: les cache-sexe de fibres des Tellem', in B. Engelbrecht and B. Gardi (eds), *Man does not go Naked: Textilien und Handwerk aus afrikanischen und anderen Länder*, Basel (Basler Beiträge zur Ethnologie, vol. 30), p. 15; Bolland, R. (1991) *Tellem Textiles: Archaeological Finds from Burial Caves in Mali's Bandiagara Cliff*, with contributions by R.M.A. Bedaux and R. Boser-Sarivaxévanis, Amsterdam (Royal Tropical Institute), p. 294.
5. Bolland 1991 (see note 4 above).
6. Method described in Hofenk De Graaff, J. (1969) *Natural Dyestuffs: Origin, Chemical Constitution, Identification: ICOM Committee for Conservation, Plenary Meeting, Amsterdam, September 15–19 1969*, Amsterdam.
7. Narco, M. and Millogo-Rasolodimbi, J. (1993) *Plantes tinctoriales et plantes à tanins du Burkina Faso*, Amiens.
8. Keharo, J. and Adam, J.G. (1974) *La pharmacopée sénégalaise traditionnelle, plantes médicinales et toxiques*, Paris; Kerharo, J. and Bouquet, A. (1950) *Plantes médicinales et toxiques de la Côte d'Ivoire- Haute-Volta*, Paris.
9. This database, used as a personal tool for the research described, might serve as a prototype for a larger database, including more data, for collective use.
10. Arbonnier, M. (2002) *Arbres, arbustes et lianes des zones sèches d'Afrique de l'Ouest*, 2nd edn, CIRAD-MNHN, Montpellier.
11. See below for description of this technique.
12. Indeed, a survey with the aim of discovering completely new uses for a plant requires a long time: I stayed in Mali for exactly four weeks.
13. Boser-Sarivaxévanis, R. (1969) 'Aperçus sur la teinture à l'indigo en Afrique occidentale', *Verhandlungen der Naturforschenden Gesellschaft in Basel* 80, 1, pp. 151–315.

14. Balfour-Paul, J. (1998) *Indigo*, London.
15. *Afrique bleue, les routes de l'indigo* (2000) exh. cat., Clermont-Ferrand, Musée du tapis et des arts textiles, Aix-en-Provence (Edisud).
16. Cardon, D. (2003) *Le monde des teintures naturelles*, Paris.
17. *Vallées du Niger* (1993) exh. cat., Musée National des Arts d'Afrique et d'Oceanie, Paris 12 October 1993–10 January 1994, p. 456.
18. Including Malinke, Bamana, Dogon, Bobo Sénoufo, Minianka and others.
19. For more information on the meaning of the drawings, see Muteba Luntumbue, T. (1998) *Bogolan, un art textile du Mali*, Brussels.
20. For more information on Bogolan's tradition and modernity, see Duponchel-Du Barry, P. (1997) *Textile de coton, Bogolan: teinture à la terre*, PhD thesis, École Pratique des Hautes Études, Paris.
21. See also on this subject Balfour-Paul, J. (1998) 'Dyeing with indigo and mud in Senegal and Mali', *Dyes in History and Archaeology* 16/17, pp. 63–71.
22. Issa Niambélé and Augustin Sagara are artists from the INA (National Institute of Art) in Bamako. They created a workshop where young people in social difficulties can learn *bogolan* techniques and drawing, and start to work with these techniques. Augustin Sagara is Dogon, and also knows about the uses of plants in his native place.
23. Boubakar Doumbia has created Conservatoire of Mandé traditions in Ségou. He works with *bogolan* and indigo and, with the aim of social inclusion, employs young people in his workshop.
24. It is a difficult exercise to establish a connection between the different Malian local names (whether they are in Bambara, Fulani, Soninké or other languages) and scientific names: some plants are different from a scientific point of view, but have similar names in local languages; or the same plant can have several names. In the case of 'n'pekuba', the local name corresponds to two different kind of *Lannea*.
25. The Macina is an area of the inner delta of the Niger, where Fulani are found in great numbers.
26. S. Frojat in *Afrique bleue, les routes de l'indigo* (2000) (see note 15 above), p. 78.
27. Gardi, B. (1985) *Ein Markt wie Mopti, Handwerkerkasten und traditionelle Techniken in Mali* (Basler Beiträge zur Ethnologie, vol. 25), Basel; Imperato, P. J. (1973) 'Wool blankets of the Peul of Mali', *African Arts* VI, 3, Los Angeles.
28. The use of this species for dyeing is also known in Australia and Central Asia, according to Alain Gerault.
29. For other uses of this mushroom in Senegal, mostly medicinal, see Dahourou, A.G. (1989) *Les champignons supérieurs dans la pharmacopée traditionnelle en Afrique de l'Ouest*, thesis, Faculté de Médecine et Pharmacie, Université Cheikh Anta Diop, Dakar, pp. 54–64.
30. Muzzolini, A. (1993) 'L'origine des chèvres et des moutons domestiques en Afrique. Reconsidération de la thèse diffusionniste traditionnelle', *Empúries*, Barcelona, 48–50, 2, pp. 160–71.
31. Jenny Balfour-Paul observed dyes from cola nuts in Gambia, also associated with indigo in tie-dye techniques: Balfour-Paul 1998 (see note 21 above).
32. *Cochlospermum tinctorium,* named 'N'dilibara' in Bambara, is mentioned for its yellow dye by Picton, J. and Mack, J. (1979) *African Textiles, Looms, Weaving and Design*, London; Narco and Millogo-Rasolodimbi 1993 (see note 7 above); Kerharo and Bouquet 1950 (see note 8 above).
33. *Adansonia digitata*, the baobab tree, is mentioned for the red dye extracted from its roots in Narco and Millogo-Rasolodimbi 1993 (see note 7 above) and Kerharo and Bouquet 1950 (see note 8 above).
34. Zahan, D. (1951) 'Les couleurs chez les bambaras du Soudan français', in *Notes Africaines* 50, Paris, April 1951, pp. 52–6.
35. For example, with *Lannea* sp., the bath begins to lather and become darker after the addition of ashes half an hour before the end of the decoction.
36. Balfour-Paul 1998 (see note 21 above) p. 65.
37. Arbonnier 2002 (see note 10 above); Abbiw, K.D. (1990) *Useful Plants of Ghana: West African Uses of Wild and Cultivated Plants*, London (Intermediate Technology and Royal Botanic Gardens, Kew).
38. Marquet 2003 (see note 2 above), p. 31; Ahmad Ibn Yahyā, called Ibn Fadl Allāh, al-'Umarī (1927) *Masalik el absar fi mamalik al-amsar*, tr. M. Gaudefroy-Demombynes, Paris, ch. 9.
39. The development of the production of cotton and textiles can be observed by the increase in the presence of cotton pollen or seeds and by the great number of spindle whorls found on archaeological sites. The principal publications giving information about textiles in archaeology are: Chavannes, B.A. (1985) *Villages de l'ancien*

Tekrour, recherches archéologiques dans la moyenne vallée du Sénégal, Paris; Thilmans, G., Descamps, C. and Khayat, B. (1980) *Protohistoire du Sénégal: les sites megalithiques,* Mémoires de l'IFAN, Dakar, no. 91, vol. I; Shaw, T. (1970) *Igbo-Ukwu: An Account of Archaeological Discoveries in Eastern Nigeria*, London; Paris, F. (1992) *Les sépultures du Sahara nigérien du Néolithique à l'islamisation, coutumes funéraires, chronologie, civilisations*, PhD thesis, University of Paris I, Panthéon-Sorbonne.

40. Witold Nowik, Laboratoire de Recherche des Monuments Historiques de Champs sur Marne.
41. Hofenk de Graaff, J. (1991) 'Dyestuff and fibres analysis', in Bolland 1991 (see note 4 above) p. 295.
42. On the existence of these insects in the Sahara, see: Balachowsky, A. (1950) 'Sur deux Tachardina CKLL. (Coccoidea-Lacciferinae) nouveaux du Sahara central' in *'EOS', Revista Espanola de Entomologia* (Instituto Espanol de Entomologia, Madrid) XXVI, Jan–Jun 1950, pp. 7–17. As medieval trade routes were passing through the Tassili n'Ajjer area, a Central Saharan origin has to be considered, but according to the author, the species identified in this area originates in a tropical zone; its presence in the southern part of Africa and the possibility of its exploitation in dyeing are still to be studied.
43. Technical analyses performed by Rita Bolland, 1991.

Bibliography

Aubreville, A. (1950) *Flore forestière soudano-guinéenne*, Paris (Société d'Editions Géographiques, Maritimes et Coloniales).

Bailleul, *Père* Charles (1998) *Dictionnaire Français Bambara*, Bamako (Donniya).

Barth, H. (1857–88) *Travels and Discoveries in North and Central Africa,* 5 vols, London.

Bensaï, S. (1994) 'Plantes tinctoriales et teinture indigène au Soudan', in *Notes Africaines* 23, Paris, June 1994, pp. 17–19.

Bensaï, S. (1994) 'Plantes à tanins: tannage et teintures des cuirs au Soudan', in *Notes Africaines* 24, Paris, Juin 1994, pp. 20–22.

Binger, L.G. (1892) *Du Niger au golfe de Guinée par le pays de Kong et le Mossi (1887–1889)*, 2 vols, Paris (Hachette).

Boser-Sarivaxévanis, R. (1991) 'An introduction to weavers and dyers in West Africa: some notes on textiles archaeology in West Africa', in Bolland, R. (1991) *Tellem Textiles: Archaeological Finds from Burial Caves in Mali's Bandiagara Cliff*, with contributions by R.M.A. Bedaux and R. Boser-Sarivaxévanis, Amsterdam (Royal Tropical Institute).

Boser-Sarivaxévanis, R. (1972) *Les tissus de l'Afrique occidentale* (Basler Beiträge zur Ethnologie, vol. 13), Basel.

Burkill, H.M. (1985–2004) *The Useful Plants of West Tropical Africa*, 2nd edn (1st edn by J.M. Dalziel, 1937), London (Royal Botanic Gardens, Kew), 6 vols.

Cardon, D. (1990) *Guide des teintures naturelles*, Paris-Lausanne.

Connah, G. (1975) *The Archaeology of Benin, Excavations and other Researches in and around Benin City, Nigeria*, Oxford.

Dapper, O. (1686) *Description de l'Afrique*, Amsterdam.

Finin, étoffes et motifs du Musée National du Mali (2002) Exposition du Pavillon des Arts, Paris.

Gardi B. and Seydou, C. (1988) 'Arkilla Kerka, la tenture de mariage des Peuls du Mali' in *Man does not go Naked: Textilien und Handwerk aus afrikanischen und anderen Länder* (Basler Beiträge zur Ethnologie, vol. 30), Basel.

Imperato, P.J. (1979) 'Blankets and covers from the Niger Bend', *African Arts* XII, 4, Los Angeles, pp. 38–47.

Jacques-Felix, H. (1963) Contribution de René Caillé à 1'Ethnobotanique Africaine Muséum d'Histoire naturelle, Paris.

Lamb, V. (1975) *West African Weaving*, London.

Lamb, V. and Lamb, A. (1980) 'The classification and distribution of horizontal treadle looms in sub-Saharan Africa', *Textile History* 11, pp. 22–62.

Lombard, M. (1978) *Études d'ecmomie médiévale, 3. Les textiles dans le monde musulman au VIIe au XIIe siècle*, Paris-New York.

Mauny, R. (1961) *Tableau géographique de l'Ouest Africain au Moyen-âge, d'après les sources écrites, la tradition et l'archéologie*, Mémoire de l'Ifan, no. 61, Dakar.

Mauny, R., Monteil, V., Djenidi, A., Robert, S. and Devisse, J. (trs) (1966) 'Textes et documents relatifs à l'histoire de l'Afrique, extraits tirés des voyages d'Ibn Battuta', *Histoire* (Publications de la Faculté des Lettres et Sciences Humaines) 9, University of Dakar.

Maydell, H.J. von (1983) *Arbres et arbustes du Sahel, leurs caractéristiques et leurs utilisations*, Weiskersheim.

Miège, J. (1991–92) 'Couleurs, teintures et plantes tinctoriales en Afrique Occidentale', *Bulletin du Centre genevois d'Anthropologie* 3, pp. 115–31.

Origins of African Plant Domestication (1976), eds J.R. Harlan, J.M.J. de Wet and A.B.L. Stemler, The Hague.

Park, M. (1800) *Voyage dans l'intérieur de l'Afrique*, tr. J.-B. Duvoisin, Hamburg-Brunswick.

Portères, R. (1962) 'Plantes tinctoriales de Guinée' and 'Plantes tannantes et plantes tinctoriales chez les cordonniers de Dalata' in *Ethnobotanique Guinéenne*, Centre d'étude ethnobotanique guinéen.

Recueil des sources arabes concernant l'Afrique occidentale du VIIIème au XVIème siècle (*Bilal-al-Sudan*) (1985) tr. J.M. Cuoq, Paris (CNRS).

Sidiyene, E., Le Floch', E. and Bernus, E. (1996) *Des arbres et des arbustes spontanés de l'Adrar des Iforas, (Mali) Une étude ethnobotanique et linguistique*, Paris (ORSTOM/CIRAD).

Chinese Green: Experimental Investigations

Christopher J. Cooksey,* Alan T. Dronsfield and Jo Kirby

Abstract Chinese green (*Lo-kao*) was manufactured in China in the mid-19th century by extraction of the bark of *Rhamnus dahurica* Pall. and *R. tinctoria* Waldst. & Kit. (formerly *R. utilis* Decne. and *R. chlorophorus* Decne.). Samples of the raw blue *Lo-kao* dyestuff supported on cotton and alum-mordanted silk dyed green with *Lo-kao* were analysed by high-performance liquid chromatography (HPLC) following release of the dye using 4% boron trifluoride-methanol reagent with a 16–18 hour reaction time; this showed that the majority of components were different. Samples of wool which had been dyed with an extract of the stems of an unknown *Rhamnus* sp. to give green (alum mordant) and purple (no mordant) colours were treated in the same way, giving a different result again. A sample of sap green, a watercolour pigment derived from the ripe berries of *Rhamnus cathartica* L., revealed a different set of components when examined by HPLC. None of the components obtained at this stage from any of the substrates was coloured green. However, when the *Lo-kao* samples were treated with the boron trifluoride-methanol reagent for a brief period (5 minutes), new HPLC peaks were detected which had visible absorption maxima at *c*.680 nm.

Introduction

The brief heyday of *Lo-kao*, Chinese green, was the mid-19th century. *Lo-kao* was first introduced into France from China in the mid-19th century. Derived from the bark of *Rhamnus dahurica* Pall. and *Rhamnus tinctoria* Waldst. & Kit., it is unusual among green colours obtained from natural dyes in that it can be used in a single direct dye-bath, giving a fast green colour with an alum mordant. The discovery of *Lo-kao* was announced in 1852 and its use declined abruptly in the 1860s with the advent of aniline greens. The early chemical investigations of *Lo-kao* by Cloez and Guignet (1872), Kayser (1885) and Rüdiger (1914) suggested that it may be a metal complex of a flavone of partially known structure. The similar green dye from the bark of *Rhamnus cathartica* L. (buckthorn), formerly known incorrectly as *Rhamnus catharticus*, which was more accessible in the West, was promoted as a substitute for Chinese green by Michel (1858) and Rommier (1860). Chemical investigations of this dye suggested the origin of the green lay in either anthraquinone derivatives (Bridel and Charaux 1925) or a mixture of sorigenin derivatives (Rauwald 1983). These investigations have been described in an earlier article.[1]

We are most fortunate in having available samples of both *Lo-kao* and of a product of the bark of *Rhamnus* species, probably *Rhamnus cathartica*. This paper describes the first investigations of these two green colours since those of Rüdiger in 1914, and also compares them to another green colour derived from *Rhamnus cathartica*, sap green.

Experimental: materials and methods

The samples

The samples of *Lo-kao* were originally part of the Leeds collection of natural dyestuffs and dyestuff constituents (the latter assembled by Professor A.G. Perkin), then in the Department of Colour Chemistry, University of Leeds, now partly dispersed. The original dyestuff sample, from *Rhamnus dahurica*, Pall., Leeds collection No. 253, consisted of deep blue flakes of the extracted and dried colorant supported on cotton. Following the recipes of Perkin and Everest (1918), three specimens of dyed yarns were produced.[2] Cotton in weakly alkaline solution gave a bright malachite blue-green, alum-mordanted silk gave celadon green and tin(II) (stannous) chloride-mordanted wool gave lilac (Plate 49).

Two samples of wool dyed with an extract of the bark of a species of *Rhamnus* (possibly R. *cathartica*), one using no mordant (purple) and the other using an alum mordant (green), had been supplied to one of the authors (Plate 50).[3] In addition, two samples of sap green[4], one prepared from the juice of ripe *Rhamnus cathartica* berries as a lake on hydrated alumina, the other consisting simply of the expressed juice of the ripe berries, were also available.[5]

A summary of the sample data is shown in Table 1.

Reflectance spectra

Reflectance spectra of samples 1, 3, 5 and 6 were measured in two ways: firstly with a Monolight Optical Spectrum Analyser (OSA), Model 6100, and secondly with a Minolta CM-2600d spectrophotometer. The first technique (Monolight OSA) records the UV-visible (UV-VIS) reflectance spectrum in the region 380–780 nm. The second technique (Minolta spectrophotometer) measures colour and renders the results numerically. It records the spectrum (400–700 nm) and converts the result to numerical colour coordinates.

Equipment used for visible reflectance spectra

- Monolight OSA: measurements were made with the Monolight OSA, Model 6100, with a photomultiplier tube detector, Model 6117, using a tungsten halogen light source and silica fibre optics. 45°/0

Table 1 The samples used in this study.

Sample number	Dye	Substrate/support	Mordant	Colour
1	*Lo-kao* Leeds collection No. 253	raw dyestuff supported on cotton	none	deep blue
2	*Lo-kao* Leeds collection, dyed by S. Grierson	cotton	none	bright malachite blue-green
3	*Lo-kao* Leeds collection, dyed by S. Grierson	silk	alum	celadon green
4	*Lo-kao* Leeds collection, dyed by S. Grierson	wool	tin(II) chloride	lilac
5	*Rhamnus* sp. (possibly *R. cathartica* L.) bark, dyed by R. Hill	wool	none	purple
6	*Rhamnus* sp. (possibly *R. cathartica* L.) bark, dyed by R. Hill	wool	alum	green
7	*Rhamnus cathartica* L. ripe berries (sap green) via National Gallery, 1990	none	none	green
8	*Rhamnus cathartica* L. ripe berries (sap green) via National Gallery, 1989	hydrated alumina		green

geometry was used: that is, the sample is illuminated at 45° and the reflected light is measured perpendicular to the surface, thus excluding the specular, or gloss, component from the colour measurement. A 5 mm diameter measuring area was used and the monochromator slit size was set to 0.89 mm. Spectra were measured in the region 380–780 nm, with respect to a Spectralon white standard, dark current subtracted. Wavelength calibration was carried out against a helium/neon laser before any measurement was carried out.

- Minolta CM-2600d: this spectrophotometer has a xenon light source. The measurement area was set to 3 mm, specular component and UV radiation excluded, and the observer setting was 10°. The meter was calibrated against the white standard supplied with the equipment before any measurement was carried out. Measurements were obtained as reflectance spectra from 400 to 700 nm at 10 nm intervals and converted by the meter to CIE (Commission Internationale de l'Eclairage) L*, a* and b* coordinates with respect to the CIE standard illuminant D65 (daylight containing the ultraviolet component, colour temperature 6504 K); values for C (chroma) and h (hue) were also recorded. L* corresponds to the lightness of the colour (from 0 – black – to 100 – pure white); a* is the red–green coordinate, positive values indicating redness, negative values greenness; b* is the yellow–blue coordinate, positive values indicate yellowness, negative values blueness.

Solubility tests

Small pieces, about 2 mm^2, of the raw dyestuff on cotton (sample 1) were tested with a range of solvents: water, methanol, acetone, ethyl acetate, propan-2-ol, tertiary butyl methyl ether, tetrahydrofuran, 1-methyl-2-pyrrolidinone, dioxane, xylene, cyclohexane, dichloromethane, chloroform (trichloromethane) and aqueous acetic (ethanoic) acid (about 30%).

HPLC methods

High-performance liquid chromatography (HPLC) is a powerful technique for the separation and identification of the components of complex mixtures. Often this technique is hyphenated with diode array detection (HPLC–PDA), mass spectrometry (HPLC–MS) or even NMR (HPLC–NMR). Here, HPLC–PDA has been used which gives the opportunity to characterise and identify HPLC-separated components by their retention times and by UV-VIS spectra. But in all cases involving coloured materials such as textiles or pigments, the constituents need to be dissolved in a compatible solvent. Quite often this requires the release of chemically bound molecules such as are found in lakes. A common technique is heating the sample with hydrochloric acid in methanol, but this is quite aggressive and leads, partially or completely, to hydrolysis of glycosides and methylation of phenolic –OH groups. (It was also impractical as the dyestuff was found to be insoluble in this reagent.) A milder reaction, used here, which has been successfully applied to lakes in paint films is the room temperature cleavage using diluted, 4%, boron trifluoride in methanol.[6]

Analysis was carried out using Hewlett-Packard (now Agilent) HP1100 series binary pumps and vacuum degasser, modified for use with a capillary microbore column by the incorporation of an LC Packings Acurate™ microflow processor which splits off a calibrated proportion (20%) of the solvent stream, thus reducing a flow rate of 100 $\mu l\ min^{-1}$ at the pump to 20 $\mu l\ min^{-1}$ at the injector. The column used, a 25 cm 800 μm i.d. Zorbax ODS (C18) column, 5 μm packing, was supplied by Presearch. Detection was performed using the HP1100 diode array detector, fitted with a 0.5 μL flow cell, path length 10 mm, 4 nm slit width, monitoring signals at 254, 275, 330, 491 and 540 nm.

- Eluents: (A) 99.9% water/0.1% trifluoroacetic acid; (B) 94.9% acetonitrile/5% methanol/0.1% trifluoroacetic acid; initial flow rate 20 $\mu l\ min^{-1}$.
- Gradient: 95% A: 5% B for 5 min; 5–30% B in 100 min; 30–45% B in 45 min; 45–

6% B, reduce flow rate to 16 μl min^{-1} in 4 min, then 46–75% B in 136 min; maintain at 75% B for 30 min, 75–95% B and increase flow rate to 20 μl min^{-1} in 10 min, 95% B 20 min, 95–5% B 10 min; post run equilibration time 15 min.

HP Chemstation software was used to process the data. The HPLC equipment (including the microflow processor), computer and software have been most generously lent to the National Gallery Scientific Department by Hewlett-Packard Ltd.

First set of analyses

1. A sample of the silk yarn dyed green with *Lo-kao* (sample 3), approximately 20 mm long, was treated with 50 μl 4% BF_3/MeOH reagent (2 parts Fluka 10% boron trifluoride-methanol: 3 parts methanol; nominal concentration), causing the silk to turn a greyish colour almost immediately, while the solution became pinkish. The sample was left overnight. The next day, the supernatant solution was colourless and the silk yellowish grey. The sample was centrifuged and a 10 μl aliquot was injected directly onto the column.
2. Some threads of the cotton dyed deep blue with *Lo-kao* (sample 1) were treated with 30 μl 4% BF_3/MeOH reagent as before. The solution rapidly became navy blue in colour, but part of this coloration appeared to be due to a suspension of dyestuff particles in the solution. It was left overnight. The following morning the solution was yellowish and the fibre greyish in colour. After centrifuging the sample, 10 μl was injected onto the column.
3. A 5–10 mm length of the pale green wool yarn dyed with *Rhamnus* sp. bark extract (sample 6) was treated with 10 μl 4% BF_3/MeOH reagent and left overnight as before. The following day the supernatant liquid had no marked colour while the wool was a pale pinkish beige. The solution was diluted with 10 μl methanol, centrifuged and 5.4 μl was injected onto the column.

Second set of analyses

As the reagent was affecting the dyestuff rapidly, the analyses were repeated as follows, using the analytical method and conditions described above:

1. Sample 3 (see Table 1), *Lo-kao* on silk, 1 cm sample treated with 30 μl 4% BF_3/MeOH reagent for 40 min only; the solution was faintly pinkish.
2. A further 10 μl of sample 3 was injected onto the column 24 hours later.
3. Sample 1, *Lo-kao* raw dyestuff on cotton, 1 cm sample treated with 30 μl 4% BF_3/MeOH reagent for 30 min only, by which time the solution had changed colour from blue to grey to yellow, as seen in the earlier run.
4. Sample 3, *Lo-kao* on silk, 1 cm sample treated with 30 μl 4% BF_3/MeOH reagent for 5 min only.
5. Sample 1, *Lo-kao* raw dyestuff on cotton, 1 cm sample treated with 30 μl 4% BF_3/MeOH reagent for at most 5 min, by which time the solution was blue-grey in colour. This change is rapid: the change towards grey is apparent in about 30 seconds.
6. Sample 6, pale green wool,[7] *c*.5 mm treated with 20 μl BF_3/MeOH reagent for about one hour. Initially the wool appeared a little bluer, but this may have been the effect of wetting; there was no coloration of the solution. After one hour the wool was a slightly yellow green, the solution still had no apparent colour.

Subsequently 10 μl aliquots of the pale green solution of the dyestuff (*Lo-kao*, sample 1) in 1-methyl-2-pyrrolidinone was examined after treatment lasting for 5–10 minutes, 45 minutes and 18 hours, using the same chromatographic conditions.

Post acquisitions processing

Images of the HPLC and UV-VIS spectra traces were scanned at 300 dots per inch. The appropriate line was digitised using WinDIG 2.5 (lovy@scsun.unige.ch) giving an x-y file of typically 2700 values. The data were converted into graphs using Microsoft Excel 2000. Comparisons of two curves can be conveniently made by plotting one as positive data and the other as negative. This shows small peaks close together without overlap.

Results and discussion

Reflectance spectra

- Monolight OSA: the four reflectance spectra curves obtained, one for each sample, are broad featureless curves showing maximum absorption at 545 nm (sample 1), 640 nm (sample 3), 540 nm (sample 5) and 630 nm (sample 6). These are illustrated in Figure 1.
- Minolta CM-2600d spectrophotometer: five curves were obtained for each sample and averaged. The results are shown in Table 2.

Table 2 Averaged spectrophotometric data for *Lo-kao* and *Rhamnus* sp. dyed samples.

Sample	L*	a*	b*	C*	h
6 Green on wool	47.99	-9.90	2.39	10.20	166.38
5 Purple on wool	38.16	15.54	-9.76	18.35	327.87
3 Green on silk	65.09	-7.21	4.19	8.37	149.87
2 Blue on cotton	27.06	4.33	-7.35	8.53	300.49

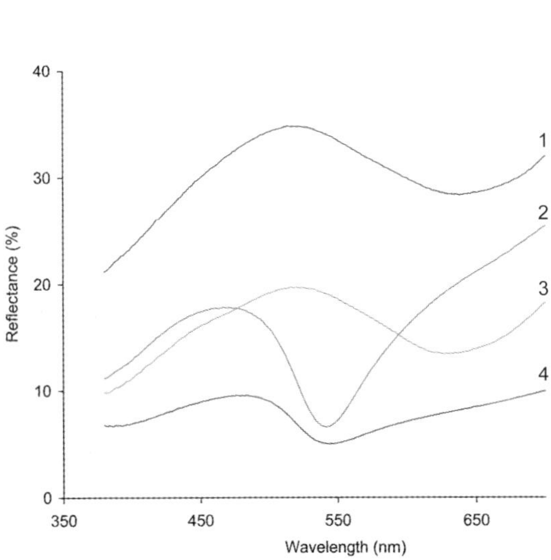

Figure 1 Reflectance spectra of *Lo-kao* and *Rhamnus cathartica* (?) samples and their absorption maxima. 1: sample 3, λ_{max} 640 nm; 2: sample 5, λ_{max} 540 nm; 3: sample 6, λ_{max} 630 nm; 4: sample 1, λ_{max} 545 nm.

Solubility tests

The colour changed from dark blue to a reddish purple in aqueous acetic acid, but the colouring matter did not dissolve. It was insoluble in all the solvents except 1-methyl-2-pyrrolidinone, with which the colour of the dye changed to a dark blue green; it then dissolved to give a bluish-green solution. This colour gradually faded; after 24 hours the solution was colourless.

Simple tests with 3M HCl and 2M NaOH (for the *Lo-kao*; Table 1, samples 1 and 3) or *c*.7M KOH (for the wool; Table 1, samples 5 and 6) revealed marked differences between the two sets of samples, suggesting that the buckthorn dyestuff used to dye the wool was not *Lo-kao*. In the case of the wool samples, a constituent of the dyestuff showed characteristics reminiscent of an anthocyanidin, becoming red with acid and blue with alkali; this change was reversible. There seemed also to be a yellowish constituent present, which did not change colour. The *Lo-kao* did not show this behaviour: the raw, navy blue dyestuff became a slightly greener blue with the acid and possibly a more hyacinth blue at most with the alkali. There was no marked effect on the green silk (sample 3) with acid; in alkali the silk turned turquoise blue at once. In neither case did the dyestuff dissolve.

HPLC

When applied to the two samples of wool dyed with *Rhamnus cathartica* bark extract using no mordant (purple) and an alum mordant (green), samples 5 and 6, HPLC–DAD showed that the samples were very similar. Both consisted of a variety of widely separated components which could not be identified with known database dyestuff constituents (Fig. 2). In contrast, when treated in the same way as above, the raw *Lo-*

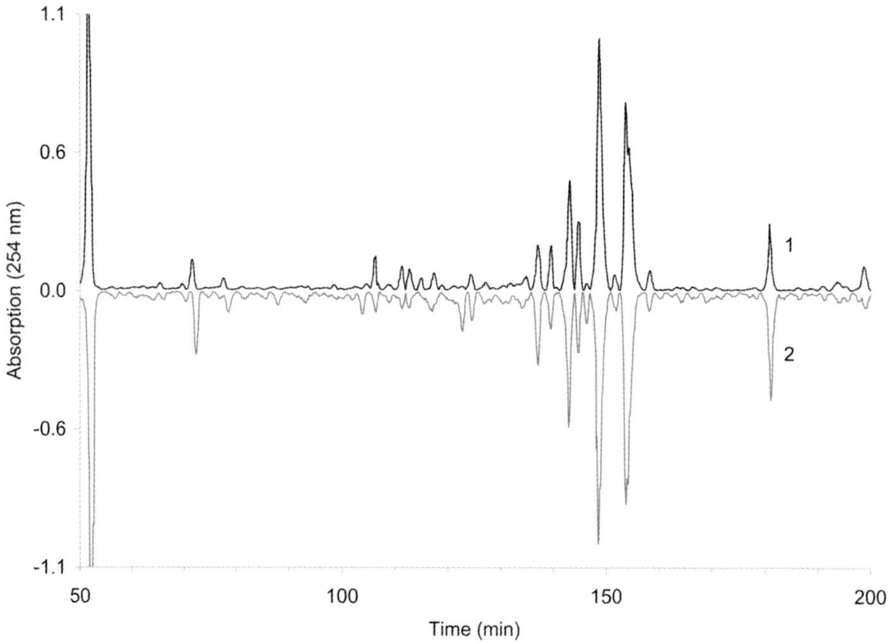

Figure 2 HPLC of samples of wool dyed with *Rhamnus* sp. (?*R. cathartica*) bark with no mordant (1: sample 5) and alum mordant (2: sample 6) after reaction with BF_3/methanol, plotted inverted.

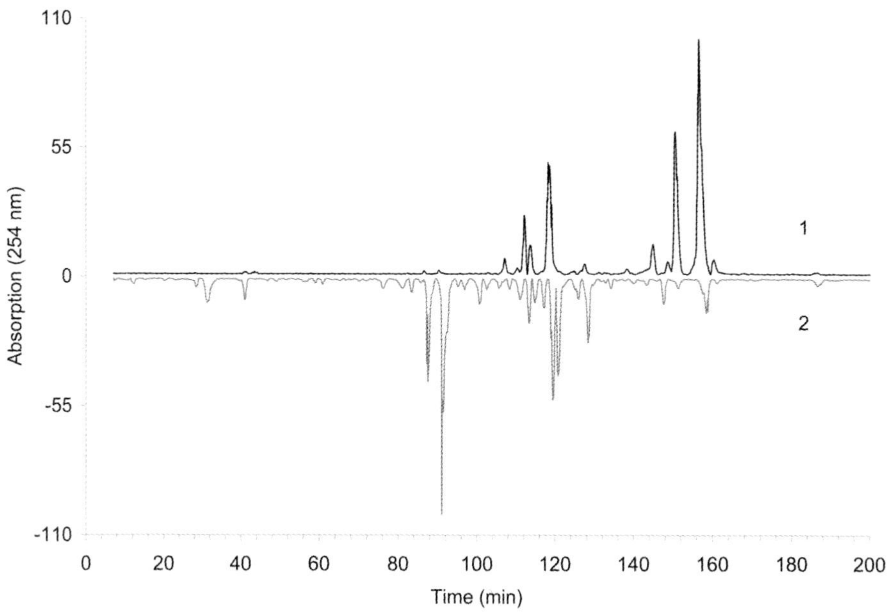

Figure 3 HPLC of *Lo-kao* samples after reaction with BF_3/methanol. 1: sample 1; 2: sample 3, plotted inverted.

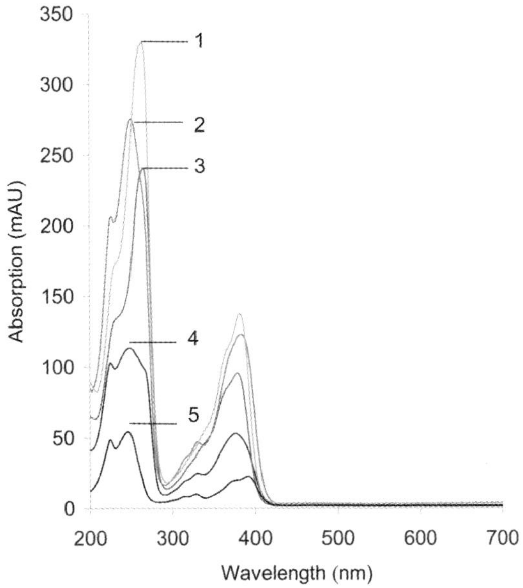

Figure 4 UV-VIS spectra of sample 1 HPLC peaks after reaction with BF_3/methanol. 1: 156.12 min; 2: 154.47 min; 3: 118.37 min; 4: 112.07 min; 5: 144.62 min.

kao dyestuff (sample 1) and the green-coloured silk sample (sample 3) gave distinctly different results. Some of the peaks have identical retention times (Fig. 3) but the composition is markedly different. An unusual feature of all the data obtained up to this point is that the UV-VIS spectra of the components suggest that none of them is green or blue in colour, but a transient blue colour was seen in a solution of sample 1 which faded after 5–10 minutes. The UV-VIS spectra obtained for the principal HPLC peaks after a long reaction time (24 h) for sample 1 are shown in Figure 4.

Prompted by the observation of a transient blue colour following the reaction of samples 1 and 3 with BF_3/MeOH, a further HPLC analysis was performed after a short (5 minute) exposure of the samples to the BF_3/MeOH reagent. This revealed a quite similar set of peaks, but with some components – those with shorter retention times (85–135 min) – showing a visible absorption at about 670–680 nm. Peaks with retention times 135–165 min appear to be identical for both short and long reaction times and are not green. The comparison between the

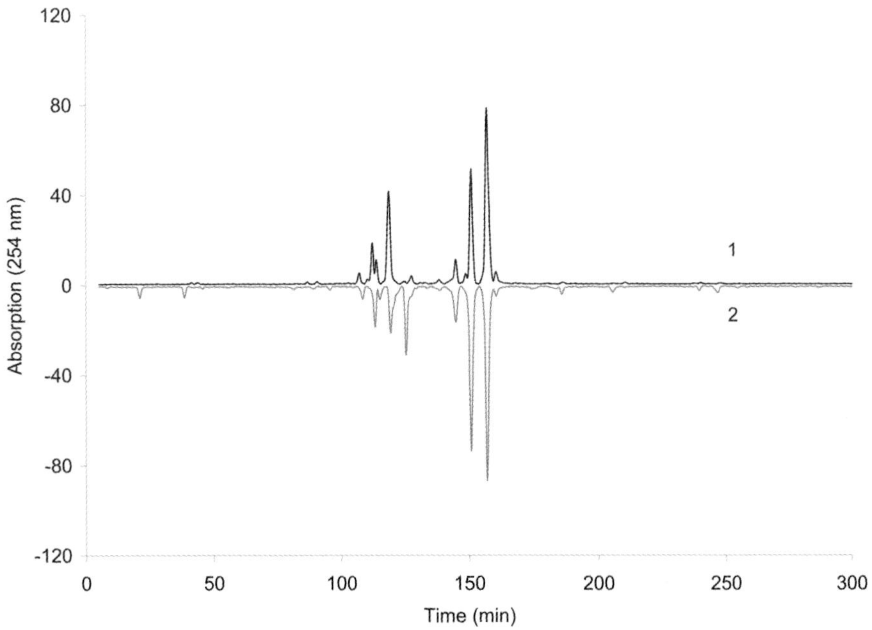

Figure 5 HPLC of sample 1: *Lo-kao*, after short (2: plotted inverted) and long (1) reaction times with BF_3/methanol.

results obtained for short/long reaction times for sample 1 is shown in Figure 5 and the UV-VIS spectra for the 'green' peaks are shown in Figure 6. Similar, but not identical, results were obtained for sample 3. The corresponding chromatograms for sample 3 are shown in Figure 7; the absorption spectra of the 'green' HPLC peaks can be seen in Figure 8.

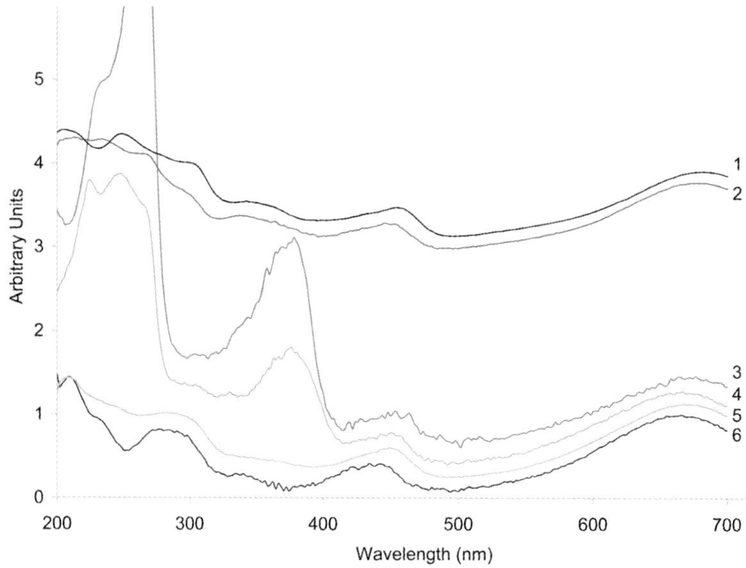

Figure 6 Absorption spectra of HPLC peaks of sample 1 (5 min reaction time), rearranged to minimise overlap, consequently the absorption scale is arbitrary. Retention times: 1: 124.8 min; 2: 120.2 min; 3: 119.0 min; 4: 113.7 min; 5: 111.7 min; 6: 106.1 min.

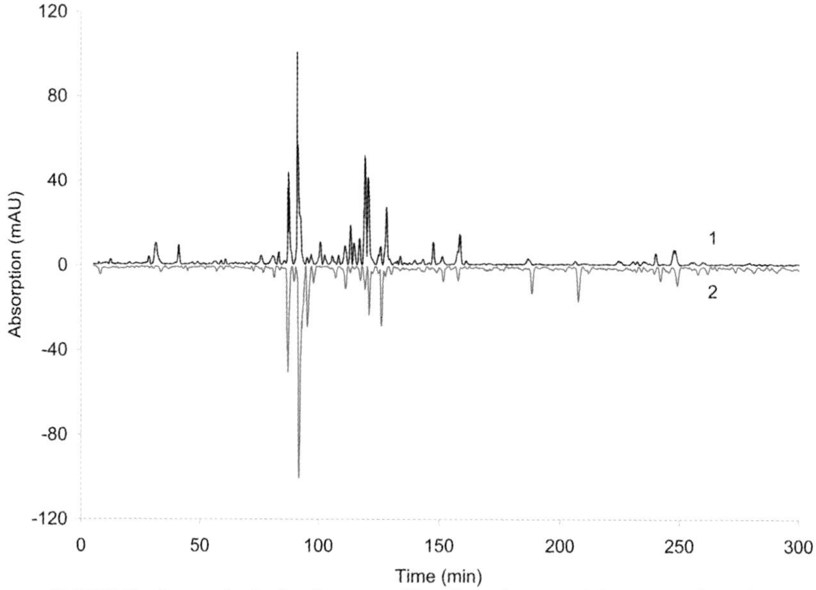

Figure 7 HPLC of sample 3: *Lo-kao* on silk, after short and long reaction times with BF_3/methanol. 1: long reaction time; 2: short reaction time, plotted inverted.

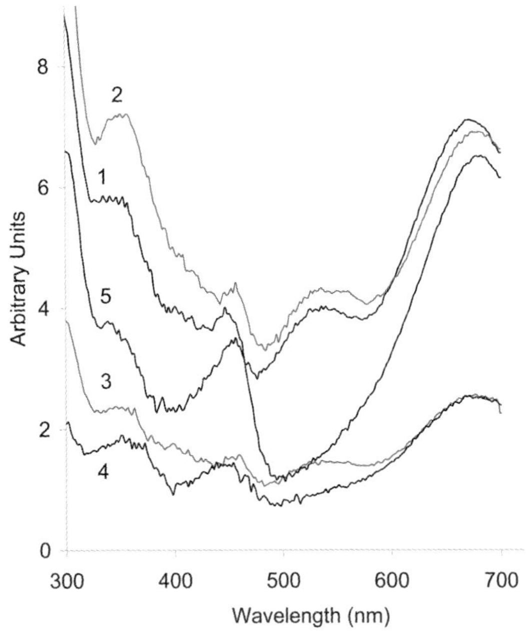

Figure 8 Absorption spectra of HPLC peaks of sample 3, *Lo-kao* on silk (5 min reaction time). Retention times: 1: 87 min; 2: 91.7 min; 3: 95 min; 4: 120.5 min; 5: 125.8 min.

When a solution of sap green, sample 7 (expressed berry juice), was subjected to HPLC without the BF_3/MeOH treatment, multiple peaks were obtained, mainly with a shorter retention time than those derived from *Lo-kao* (Fig. 9). The minor early peaks at 30–40 min can be assigned to anthocyanidin glycosides (not yet identified) and the major peaks at 65–103 min to flavonoid glycosides (including a trace of quercitrin, Rt = 81 min). The aglycones which have been found in alum-mordanted wool dyed with unripe *Rhamnus cathartica* berries,[8] and in buckthorn lakes,[9] quercetin (Rt = 104 min), kaempferol (Rt = 125 min), rhamnetin (Rt = 141 min) and possibly rhamnocitrin (Rt = 160 min), shown in Figure 10,[10] can be detected as very minor components. Similar results were obtained for the sap green 'laked' on hydrated alumina (sample 8), derived from a different batch of berries, after treatment with BF_3/MeOH. In this case anthocyanidin peaks were undetectable at 30–40 min but flavonoid glycosides were clearly identifiable around 60 min and rhamnocitrin appeared to be absent. The

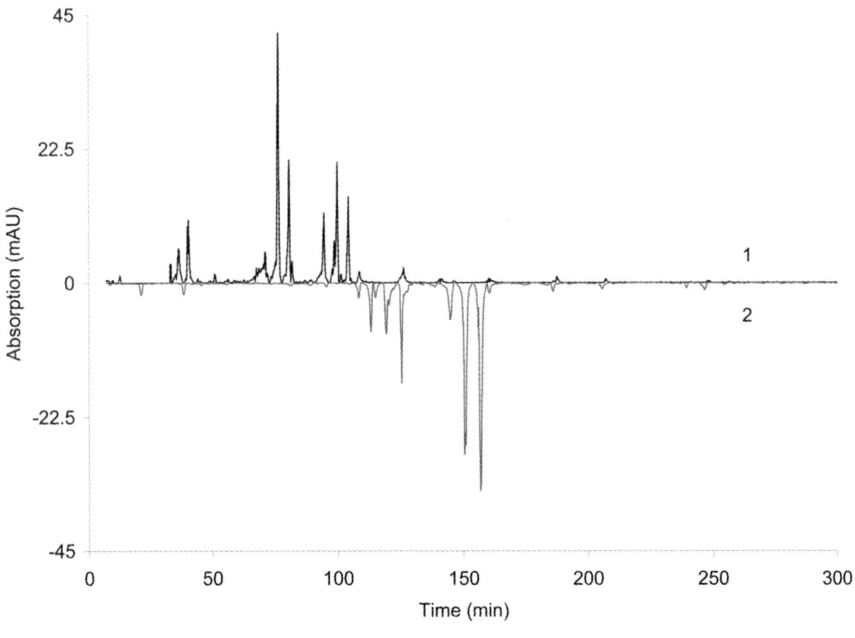

Figure 9 HPLCs of sap green (sample 7; not reacted with BF_3/methanol) compared with *Lo-kao* (sample 1) after a short reaction time with BF_3/methanol. 1: sap green; 2: *Lo-kao*, plotted inverted.

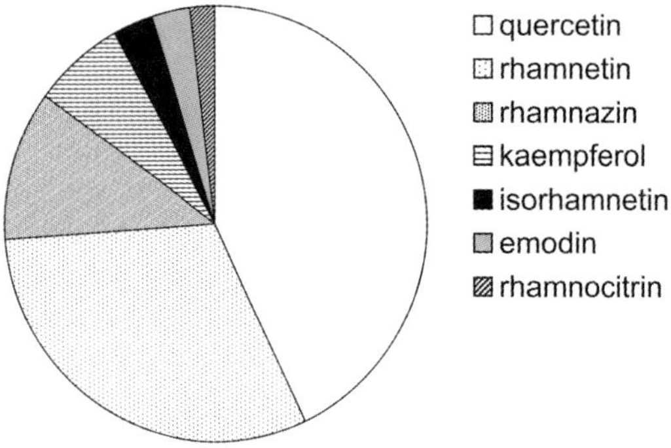

Figure 10 Composition of the dye from Persian berries (unripe), *Rhamnus cathartica*, as given by Ferreira *et al.* 2003 (see note 8).

HPLC of this sample, compared with that of *Lo-kao*, is shown in Figure 11.

First set of analyses

There appeared to be very little similarity at all between the chromatograms obtained from *Lo-kao* dyestuff, either raw or as a dye on silk (samples 1 and 3), and the dyestuff on wool (sample 6), possibly extracted from *R. cathartica* bark: the dyestuffs appear to be different. In all three cases, the constituents present eluted in the earlier part of the run where flavonoid glycosides or certain flavonoids might be expected, but none of these constituents were recognisable from previous analyses of buckthorn lakes or extracts, or were identifiable. The anthraquinones emodin and chrysophanol, present in *Rhamnus cathartica* bark,[11] which elute at about 184 and 214 min respectively, could not be detected on the wool sample (where they might have been expected), nor could any peaks attributable to their glycosides. A trace component with spectral characteristics similar to those of an anthraquinone, eluting at about 154 min, is present, however. No constituents resembling anthocyanidins could be detected, which might have been expected in the case of the *R. cathartica*-type dyestuff on wool. The constituents eluting around 150 min were not the same as those in the *Lo-kao* samples. The major peaks centring around 150 min in the raw *Lo-kao* dyestuff seem to have little, if anything, to do with any component seen in the dyed silk. The peaks at 180 and *c.*200 min are impurities in the system and should be ignored.

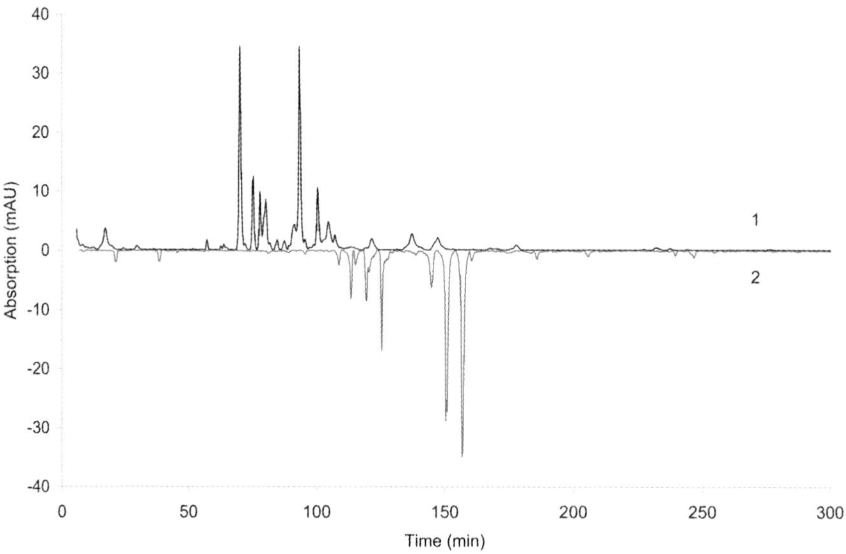

Figure 11 HPLCs of sap green (sample 8; reacted with BF_3/methanol) compared with *Lo-kao* (sample 1) after a short reaction time with BF_3/methanol. 1: sap green; 2: *Lo-kao*, plotted inverted.

The results obtained from the *Lo-kao* samples are rather surprising, given that the dyestuff is quite insoluble in normal solvents. The strength of boron trifluoride/methanol used usually reacts slowly with both lake pigments and dyed textiles (anthraquinone or flavonoid dyes), breaking up any Al or Ca bonds or complexes present and methylating vulnerable groups; any methylation that occurs is relatively gradual. The immediate changes observed on the addition of the reagent to *Lo-kao* were unexpected. It appears that the dyestuff is very vulnerable to this reagent, possibly breaking up into smaller fragments; simple methylation alone cannot be the explanation as this would lead to an increased retention time. Few late-eluting components were observed in these chromatograms.

Second set of analyses

It is clear that several components of *Lo-kao* with an absorption at 670–80 nm and various other features are present in the initial raw dyestuff, or on the silk, or are formed as the result of the reaction with the boron trifluoride/methanol reagent, or combinations of these. These are transitory and are already changing further minutes after the reagent is added to the sample. The constituents eluting at about 111–112 and 124–125 min occur both in the raw dyestuff and on the dyed silk; the earlier constituents on the silk (eluting at about 81, 86, 91 and 95 min, of which that at 91 min gives the strongest absorption), have similar UV-VIS features, but are absent in the raw dyestuff. The component eluting at about 119–20 min in the raw dyestuff is absent on the silk; that at about 120 min on the silk has different UV-VIS characteristics (λ_{max} 232, 264–6, *c*.366, *c*.466 (very weak) nm), somewhat reminiscent of an anthraquinone. The three components in the raw dyestuff, together with that at 106 min, also occur in the chromatograms obtained from the samples dissolved in 1-methyl-2-pyrrolidinone. The strongest absorption is given by the constituent at 124–125 min in all cases. It is thus clear that these four constituents are due to the original *Lo-kao* dyestuff and do not derive from the treatment with BF_3/MeOH. It seems likely that the constituent eluting at 124–125 min is the most significant. The features at 146–156 min in all the chromatograms obtained from the raw dyestuff after treatment with boron trifluoride-methanol, which have very different spectral characteristics and are not green, are absent from the chromatograms obtained from the 1-methyl-2-pyrrolidinone solutions. It is therefore conceivable that these constituents derive from the methylation process.[12]

In the case of sample 6 on wool, no compounds of this sort appear to have been produced; the chromatogram is comparable with that run earlier, although much feebler. There is no similarity between any of the chromatograms from the wool and those given by solutions of sap green, obtained from ripe *R. cathartica* berries, in water/methanol or BF_3/MeOH: the constituents are different.

Conclusion

All samples are multicomponent. HPLC chromatograms of the blue raw *Lo-kao* dye on cotton and the green colour on alum-mordanted silk in 4% boron trifluoride-methanol and appear substantially different, although they do have at least two constituents in common.

Use of 4% BF_3/MeOH reveals short-lived species with very similar HPLC retention times, but having a visible absorption maxima at *c*.670–680 nm. Examination of the solution of the raw dyestuff in 1-methyl-2-pyrrolidinone shows the presence of constituents with similar retention times and spectral characteristics; these, too, are transitory, but slightly less so. The general insolubility of the dyestuff and the transitory nature of the species present in both the diluted methylating reagent and 1-methyl-2-pyrrolidinone render its analysis difficult. It is hoped that LC–MS will give further information on the structure of these short-lived species and thus the *Lo-kao* dyestuff itself.

The purple and green dyes extracted from wool thought to have been dyed with bark from *Rhamnus cathartica* or another *Rhamnus* species are identical to each other, but different to the

Lo-kao dyes. The anthraquinones emodin and chrysophanol and related glycosides reportedly present in this bark were not detectable, perhaps because the amount present was below the detection limit of the detector, or perhaps because they are indeed absent. The constituents that were present could not be identified. A further examination of the constituents of *Rhamnus* spp. bark is necessary before any conclusions can be drawn.

All the components of the samples of sap green are different to those observed in any of the dyestuff samples. They have shorter HPLC retention times and are principally anthocyanidin and flavonoid glycosides, with small amounts of the free flavonoids.

Acknowledgements

The *Lo-kao* samples were kindly supplied by Su Grierson. Thanks are due to Colin Harvey, The National Gallery Photographic Department for photography of the samples shown in Plates 49 and 50.

Editor's note

This paper was presented as a poster at the Meeting. *Author for correspondence: 59 Swiss Avenue, Watford, Herts, WD18 7LL, UK (dha@chriscooksey.demon.co.uk). Paper received 14 August 2003; revised version received 31 May 2005.

Notes and references

1. Cooksey, C.J., Dronsfield, A.T. and Kirby, J. (2005) 'Chinese green, an enduring mystery', *Dyes in History and Archaeology* 20, pp. 155–64; Cloez, S. and Guignet, E. (1872) 'Recherches sur la composition chimique du vert de Chine (Lo-Kao)', *Comptes Rendus Hebdomadaires des Séances de l'Académie des Sciences* 74, pp. 994–7; Kayser, R. (1885) 'Ueber das Lokao Chinesische grün', *Berichte der Deutschen Chemischen Gesellschaft* 18, pp. 3417–29; Rüdiger, A. (1914) 'Beiträge zur Kenntnis des Lokao-Farbstoffes', *Archiv der Pharmazie* 252, pp. 165–86; Michel, A.-F. (1858) *Recherches sur la matière colorante des nerpruns indigènes*, Paris; Rommier, M. (1860) 'Recherches sur les matières colorantes vertes contenues dans certains nerpruns de France comparées à celles des nerpruns de Chine', *Comptes rendus de l'Académie des Sciences* 50, 2, pp. 113–15; Bridel, M. and Charaux, C. (1925) 'Le rhamnicoside, glucoside nouveau, générateur du vert de Chine, retiré de l'écorce de tige du Nerprun purgatif (*Rhamnus cathartica* L.)', *Bulletin de la Société de Chimie Biologique* 7, pp. 811–21 and briefly summarised in Bridel, M. and Charaux, C. (1925) 'Le rhamnicoside, glucoside nouveau, générateur du vert de Chine, retiré de l'écorce de tige du nerprun purgatif (note)', *Comptes Rendus Hebdomadaires des Séances de l'Académie des Sciences* 180, pp. 1047–9; Rauwald, H.-W. (1983) 'A new investigation on the composition of *Rhamni cathartici* cortex, IV[1]. The main anthraglycosides of the stem bark of *Rhamnus catharticus* L.: Emodin–8–O–gentiobioside, –glucoside and –primveroside', *Zeitschrift für Naturforschung* 38c, pp. 170–78; Rauwald, H.-W. and Just, H.-D. (1983) 'Isolierung und Charakterisierung der blau fluoreszierenden Leitstoffe aus der Rinde von *Rhamnus catharticus* L.' *Archiv für Pharmazie (Weinheim)* 316, pp. 399–408; Rauwald, H.-W. and Just, H.-D., (1983) 'Ein neues lactonisches Naphthalinglycosid aus der Rinde von *Rhamnus catharticus* L.', *Archiv fur Pharmazie (Weinheim)* 316, pp. 409–12.
2. Perkin, A.G. and Everest, A.E. (1918) *The Natural Organic Colouring Matters*, London, pp. 617–20. Specimens produced by Su Grierson.
3. Supplied by the late Dr Robin Hill (formerly Department of Biochemistry, University of Cambridge) in 1974.
4. Prepared in the National Gallery by J. Pilc, September 1989 and 1990.
5. Cooksey *et al.* (see note 1 above), citing J. Rutty, 1772, who describes the pigment consisting simply of expressed juice. Recipes for the other form he mentions, in which a little alum is added to the juice, are very common from the early 15th century onwards. The formation of a blue or purple complex between anthocyanidins present in the ripe berry and the aluminium in alum gives the final pigment a more pronounced green colour. The addition of alkali allows the

pigment to be precipitated and dried as a lake, but does not alter the hue much. For the recipe used for this pigment see Merrifield, M.P. (1849) *Original Treatises dating from the XIIth to XVIIIth Centuries on the Arts of Painting*, 2 vols, London (reprinted New York, Dover 1967), vol. II, *Ricette per far ogni sorte di colore* (the Paduan Manuscript), 17th century, no.127, pp. 706–7. For examples of recipes where alum alone is added see no. 31, pp. 662–5; 137, pp. 708–11, and, in the same volume, *Segreti per colori* (the Bolognese Manuscript), 15th century, no. 89, pp. 420–21.

6. Kirby, J. and White, R. (1996) 'The identification of red lake pigment dyestuffs and a discussion of their use', *National Gallery Technical Bulletin* 17, pp. 56–80, esp. p. 60.
7. Dr Hill's sample.
8. Ferreira, E.S.B., Hulme, A.N., McNab, H. and Quye, A. (2003) 'LC–Ion trap MS and PDA–HPLC – complementary techniques in the analysis of different components of flavonoid dyes: the example of Persian Berries (*Rhamnus* sp.)', *Dyes in History and Archaeology* 19, pp. 19–23.
9. Kirby, J. (2003) 'Sir Nathaniel Bacon's "Pinke"', *Dyes in History and Archaeology* 19, pp. 37–50, esp. p. 42.
10. Ferreira *et al.* 2003 (see note 8 above). The pattern of flavonoid constituents present in sap green prepared from ripe berries shows quantitative differences, however.
11. Ferreira *et al.* 2003 (see note 8 above); Schweppe, H. (1993) *Handbuch der Naturfarbstoffe: Vorkommen; Verwendung; Nachweis*, Landsberg/Lech, p. 379. The anthraquinone isoemodin and the emodin glycoside rhamnicoside – defined by Schweppe as emodin primeveroside – and frangulin (perhaps) are also present.
12. It is worth noting that, for all these constituents and in both solvents used, the chromatographs recorded at 491 and 540 nm give negative peaks, while that recorded at 330 nm is variable: usually negative, sometimes not. It is not clear why this should be the case: possibly a solvent absorption effect or fluorescence, or it may be connected with the choice of detection and reference wavelengths used.

Logwood under the Microscope

Brian H. Davies

Abstract Logwood extracts, from the imported heartwood of *Haematoxylon campechianum* L., were used as dark blue, violet and black dyes in western Europe from the 16th century and for over 350 years, eventually being replaced by synthetic organic dyestuffs. They were also used as components of violet and black writing and copying inks during the 19th century. With the development of the optical microscope and the necessity for staining to visualise detail in thin, translucent sections of tissue, and with the knowledge of its polychromic properties and the effects of different mordants, logwood became the basis of many staining procedures used in microscopy. The haematoxylin-eosin (H-E) stain is still regarded as an important general tissue stain in histology and histopathology.

Introduction

Logwood is the heartwood of a leguminous tree, *Haematoxylun campechianum* L., native to Central America. Discovered in Campeche on the west coast of the Yucatan peninsula during the conquest of Mexico in the early 16th century, and furnishing a water extract which would dye fabrics dark blue, purple or black, the wood was exported to Spain.[1] Transported as logs or blocks, this Campeachy wood later acquired the names of logwood or blockwood. Mark Catesby (1771) described how the British had first acquired their supplies of logwood, by 'getting it from the Spaniards ... either by Force or Stealth' and in the course of 'The bloody Disputes which this useful Tree has occasioned between the Spaniards and English'.[2] After an Anglo-Spanish treaty in 1667, such activities were discouraged and many of the British privateers settled as logwood cutters in what is now Belize on the east coast of the Yucatan peninsula; the country's coat-of-arms shows the logwood and mahogany cutters. They also established logwood plantations in Jamaica from 1715.[3] The quality of the cultivated logwood from Jamaica, though, was never as good as that from the natural resources of the Yucatan peninsula.[4]

Introduced into Britain as a dye by the middle of the 16th century, logwood was first used for dyeing woollens dark blue or black. Without the use of proper mordanting procedures to fix the dye to the fibres permanently, the colours it produced were fugitive and unreliable.[5] Although its introduction must have affected the interests of the woad growers of Elizabethan England, it was its unreliability that was given as the reason for its import and use being prohibited by an Act of Parliament in 1581: 'Whereas of late Years there has been brought into this Realm of England from beyond the Seas a certain Kind of Ware or Stuff called Logwood ... Forasmuch as the Colours made with the said Stuff called Logwood, *alias* Blockwood, is false and deceitful ... Logwood shall be forfeited, openly burned, and no Cloth or Wool shall be dyed therewith'.[6] Nearly 100 years later, in the reign of Charles II, this Act was repealed by another, which accepted that 'ingenious Industry of these Times hath taught the Dyers of England the Art of fixing the Colours made of Logwood *alias* Blockwood so that by Experience they are found as lasting and serviceable as the Colours made with any other sort of Dying-Wood whatsoever'.[7] The principle of mordanting, the 'Art of fixing the Colours', was now understood and, from 1661 and for some 200 years thereafter, logwood was one of the most important dyestuffs used in Britain.

The logwood dyestuff and its use in dyeing

The colouring principle of logwood was investigated in 1810 by M.E. Chevreul, who crystallised it and named it 'haematin'.[8] Once it was realised that this name was already in use for a derivative of haemoglobin, it was renamed 'haematoxylin'.[9] Its structure was determined, along with that of its coloured oxidation product, haematein, in 1908.[10] Also investigated at the same time were the structures of brazilin, from the heartwood of species of brazilwood, *Caesalpinia* spp., and its oxidation product, brazilein. Of these brazilwoods, Pernambuco wood from *C. echinata* Lam., was the best source of brasilin.[11] Scheme 1 shows the structures of these compounds and the close relationship between haematoxylin and brazilin. In each case, it is the oxidation product (i.e. haematein, brazilein) which is the strongly coloured component, the oxidation having produced a series of conjugated olefinic bonds running through the molecule.

In order for the logwood to be used effectively as a dye, the haematoxylin had to be oxidised to haematein. The imported logs, stripped of their bark and outer layers, were converted mechanically into small chips. The rasped or chipped wood, yellow in colour, was spread, a few inches deep, on the floor and made uniformly damp with water before being collected into heaps.[12] Fermentation commenced and caused a rise in temperature, which had to be controlled by turning the heaps; too high a temperature would destroy some of the haematein, leading to 'burnt' or 'over-aged' wood.[13] The fermented wood became dark red in colour and acquired a greenish encrustation of haematein crystals and a metallic lustre.[14] At this stage it was spread out again and dried to stop further reactions.

As an alternative to oxidising the haematoxylin to haematein during this fermentation process, oxidation would take place during the actual dyeing if an oxidising mordant were used. The use of such a mordant with fermented logwood extract, however, could lead to a deleterious over-oxidation of the haematein to brown products. As another alternative to fermentation, the fresh logwood could be extracted and the extract boiled to oxidise the haematoxylin.

(a) Logwood (*Haematoxylum campechianum* L.)

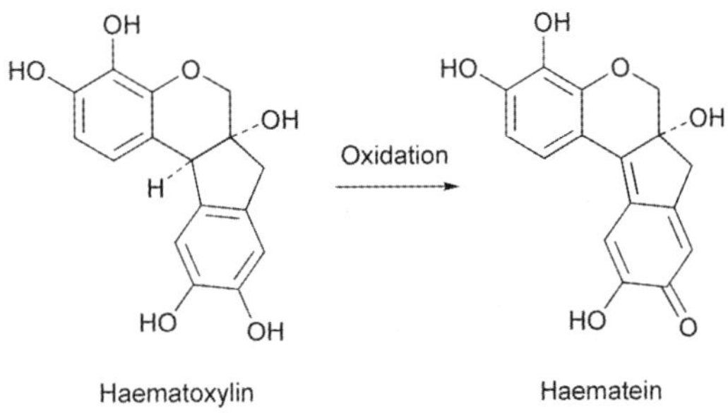

(b) Brazilwood (*Caesalpinia* spp.)

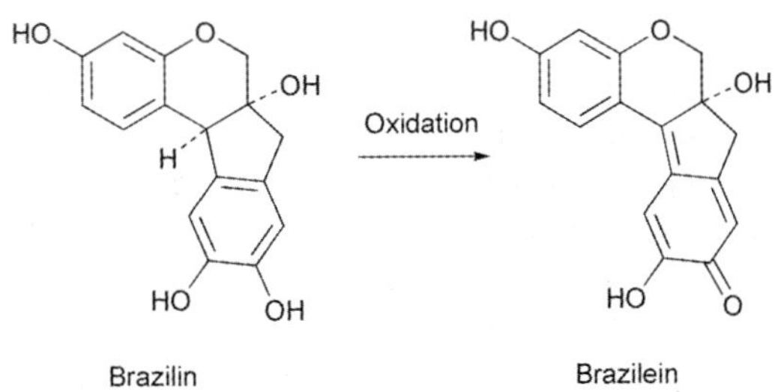

Scheme 1 Dyestuffs from the oxidation of wood extracts.

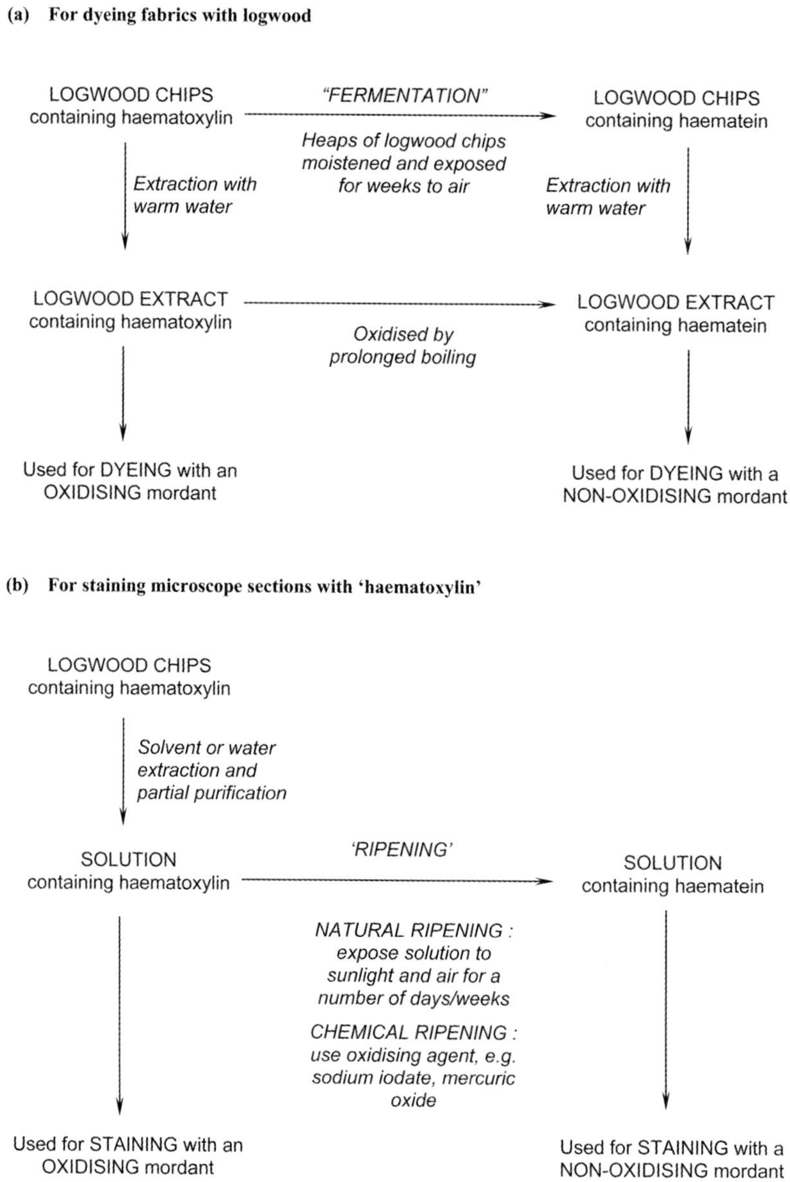

Scheme 2 Logwood preparations for dyeing and staining

The extraction of the logwood chips was generally done with warm (80 °C) water over a period of a few hours. A more rapid extraction was possible, by boiling with water for half an hour; this method gave a higher yield of dyestuff, but it was of a quality inferior to that produced with warm water.[15] These oxidation and extraction procedures are summarised in Scheme 2(a).

Logwood was largely used for dyeing dark blues, greys and black on silk, wool and cotton. With chromium and aluminium mordants, logwood dyes a dark blue, and even black; with tin, a dark purple and with iron, black.[16] Its particular importance was for the black-dyeing of wool, but it had considerable use in the dyeing of navy blues and for the darkening (saddening) of brown, olive and other colours on chrome-mordanted wool. For union fabrics, those composed of two or more different materials woven together, logwood dyed the cotton claret or purple shades with a tin mordant, and was useful for producing black cotton warps.[17]

Two of the principal logwood preparations for dyeing wool black were Iron Black and Chrome Black. The Iron Black process, the oldest known method for dyeing wool black, used iron(II) and copper sulfates as mordants. Sometimes known as Salzburg Black because of its use of 'Salzburg vitriol', a mixture of iron(II) and copper sulfates, it gave a bluish black which could be modified (to a purer black) by dyeing with fustic, rich in tannins.[18]

Chromium was discovered by L-N. Vauquelin in 1797 and its compounds soon became available and were used in dyeing. The Chrome Black process used potassium dichromate as an oxidising mordant, with sulfuric acid and copper sulfate. Acid dichromate alone gave a bluish black, but the presence of copper yielded a darker and more lightfast colour. Blues were produced with alum, copper and dichromate together, with the copper and dichromate at lower concentrations than for black dyeing.[19]

The dyeings with Iron Black and Chrome Black were both two-step procedures, with mordanting preceding the dyeing stage. There was also a third method, often referred to as Single-Bath Black. This originally involved dyeing with logwood and sumach (a source of tannins) followed by a darkening with iron(II) sulfate, but it evolved into a procedure using a solution containing logwood extract and the sulfates of iron and copper.[20]

Inks

The ready availability of chromium compounds by the mid-19th century, and knowledge of their use in dyeing, had considerable influence on the development of writing and other inks. Although the use of logwood in making a violet ink was known in the early 17th century,[21] it was William Lewis who, in carrying out the first systematic study of the composition of writing inks in 1763,[22] described an iron-gall ink which contained logwood in addition to the usual components of iron(II) sulfate, an extract of galls and gum arabic.[23] Later modifications to this formula resulted in logwood inks which contained only some of the usual constituents of iron-gall ink. Eisler's ink (1770) used logwood and gall extract,[24] while Reinige's ink (1857)[25] contained logwood and the iron salt, but no tannin.[26]

Reinige's ink was typical of the logwood inks of the 19th century in evolving away from the use of the natural tannins of iron-gall ink, and advantage was taken of the developing knowledge of the reaction of logwood with the salts of chromium and other metals. It was Runge (1847) who made a chromium ink based on logwood and potassium chromate,[27] while Böttger modified this (1848) by the addition of soda to prevent precipitation.[28] Subsequently in 1857 he developed an entirely different ink containing alum and copper sulfate in addition to logwood.[29] Another copper-logwood ink, that of Viedt (1875), contained no alum.[30]

The compounds of the various metals included in these preparations had much the same influence on the colour of the ink as they had when used in dyeing. Fresh logwood-chromate ink was reddish violet as it came off the pen, but this darkened on the paper to black; in combination with iron-gall ink it was violet or blue-violet.[31] The logwood-alum ink was also violet, while the addition of chromate to this gave a blue-black. Logwood-alum-copper acetate inks had a reddish appearance.[32]

A particular use of the logwood-chrome preparation was as copying ink, an ink which would yield, under pressure, an impression of the original on a sheet of dampened tissue paper. The inclusion of components such as sugar and glycerine ensured that the ink would remain moist for long enough to allow a good copy to be made. Although the earliest copying inks (e.g. that of James Watt, iron-gall ink plus gum arabic and alum) were modified iron-tannin inks, later copying inks, such as that of Adolph Ott, contained no tannin, but were based on logwood with added sugar and glycerine, and with potassium chromate and the sulfates of aluminium, iron and copper to ensure a good colour.[33] Other, less complicated, copying inks contained fewer of the metal components.[34] Some logwood copying inks were capable of making many copies. A procedure developed by Underwood and Burt, in which a particularly concentrated logwood ink was applied to paper pre-prepared with potassium chromate, would furnish up to 20 copies.[35] Logwood copying inks were important commodities of international commerce; the European companies of Antoine and of Faber supplied the US Patent Office with violet logwood copying ink for some 40 years from the 1850s.[36] Copying pencils, too, could be based on the logwood-chromate preparation,[37] but the introduction of the typewriter and carbon paper into business practice in the 1870s inevitably lessened the use of manual copying techniques.

Logwood was little used in inks for printing, but there are two notable examples. William Savage suggested in 1832 that the colour change with acid to which logwood is prone might be used in the prevention of fraud, and he devised a logwood-iron-gall ink which could be used for printing bank drafts, bills of exchange and receipts.[38] Some years later, Thomas de la Rue (1854) patented a logwood printing ink containing glycerine.[39]

Although logwood was used successfully in both dyeing and in the preparation of ink, there is very little evidence from manuscript sources of its use in graphic works of art, although a few used it in paint.[40] Its relative instability as a colour was recognised by Edward Norgate who, in his 17th-century treatise on painting miniatures, dismissed it, along with a number of other natural dyestuffs, as being unsuitable materials for his art. 'In this Catalogue of Colours I doe purposely omit Brazill Verdigreece Orpiment Rosett Turnsole Litmus Logwood and others not worthy the naming, nor by reason of their course and grosse Body, their dangerous qualitie (some beinge mineralls) nor their Cheapness and Community to be used at all in this cleanly, neat and exquesite Art, being indeed fitter for those that wash prints or colour Mapps then to be admitted into our Company'.[41]

This brief history of the uses of logwood is more or less complete now. Although it never found much popularity with graphic artists it was, for centuries, a highly prized fabric dye which, once its mordanting was understood, gave dark violets and blues and a particularly strong black. With the advent of the synthetic dyestuffs, it was soon replaced as a black by aniline black, naphthol black or naphthylamine black.[42] It had a history, too, as a component in inks, with a particular significance in copying ink, a preparation which lost its importance after the advent of the typewriter in 1734 and the increased use of carbon paper. But there is one area, not mentioned so far, where the usefulness of logwood persists to this day, and that is in the staining of biological specimens for examination by microscopy. The haematoxylin stains continue to be of the utmost importance in histology and histopathology.

Logwood as a histological stain

The history of logwood runs parallel with the history of the science and technique of optical microscopy, the first microscopes being developed in the second half of the 17th century and refined ever since. Insofar as Robert Hooke used his microscope to observe the appearance of hair and wool before and after dyeing,[43] he might be described as the first (in 1665) to use a dyestuff to stain specimens for microscopy, but he was not as deliberate in this as was Antoníe van Leeuwenhoek. In a letter of 1714, the latter described how, on comparing the muscle fibres of a lean and a fat cow, he found the sections too transparent for effective observation, so soaked them in an extract prepared by macerating *Crocus* in wine.[44] The transparency of thin sections continued to be a problem for later microscopists. As Quekett (1848) explained, 'Very thin objects ... became so indistinct ... some mode of making them dark becomes necessary; this may be effected ... either by charring or dyeing; ... dyeing ... may be effected by soaking them for a time in a decoction of fustic or logwood.'[45]

The opening sentence of Conn's classic *Biological Stains*, in response to the question 'What are biological stains and how do they differ from dyes?', states that, 'The answer is simply that they are dyes, but are adapted for very special purposes'.[46] Of course, before synthetic organic dyestuffs became readily available, microscopists were dependent on the natural dyes, so indigo, madder, saffron, fustic, phytolacca and cochineal all figured in early studies.[47]

In the context of the history of staining, Clark and Kasten (1983) suggest that von Waldeyer had been the first to use haematoxylin in 1863, albeit not very successfully because his use of a plain logwood extract involved no mordanting.[48] Böhmer (1865) knew about the effect of alum as a mordant in dyeing, so used alum in combination with a logwood stain. He also found that microscope preparations treated with oxidising materials such as chromic acid, potassium dichromate or copper sulfate would also stain effectively with logwood.[49]

Although the logwood microscope stains are known as the 'haematoxylin stains', as with its use in dyeing, the natural haematoxylin has to be oxidised to haematein to be effective. This is not surprising, nor is the requirement for mordanting, for wool and silk are natural proteins, taking up the logwood colour just as well in biological tissues as they do as components of fabrics. The essential chemistry of the two processes is the same; the only differences are in details of procedure.

The steps involved in the use of logwood as a microscope stain are shown in Scheme 2(b). An extract of logwood, not necessarily made with water, but perhaps with an organic solvent, undergoes some purification, yielding a higher quality of haematoxylin than was generally used in dyeing. This purified haematoxylin can be used directly as a stain as long as the mordant is a metal salt which has oxidising properties. Otherwise, the haematoxylin has to be oxidised to haematein in a process known as 'ripening'; this may be accomplished by 'natural' or chemical means. 'Natural ripening' is a gentle process in which the haematoxylin solution is exposed to air and light for a period of days, weeks or even longer. The haematein solution produced is very stable. 'Chemical ripening' involves a chemical oxidising agent; oxidation, carried out by reagents such as sodium or potassium iodate, mercuric oxide or iodine is immediate, but the haematein solution, prone to over-oxidation, is less stable than that produced by 'natural ripening'. The haematein produced by either form of 'ripening' is used as a stain with a non-oxidising mordant, typically alum.

A number of the haematoxylin stains used today were first developed in the late 19th century.[50] The alum-haematoxylin stains attributed to Delafield (1885) and Ehrlich (1886) are based on natural ripening, while others involve chemical ripening, for example, those of Mayer (sodium iodate, 1891), Harris (mercuric oxide, 1900), Carazzi (potassium iodate, 1911) and Cole (iodine, 1943).[51] These alum haematoxylins all initially stain the cell nuclei red, but they become blue-black on washing ('blueing') with tap water, this normally being sufficiently alkaline. The common practice is to use alum-haematoxylin as a regressive stain, that is, to overstain initially and then to use alcohol selectively to extract excess stain after the 'blueing' stage.

The way in which alum haematoxylin stains are used is reminiscent of the process of metachrome dyeing, the dye and the mordant metal being in the same solution and applied simultaneously. Some of the ways in which salts of other metals have been used have more in common with the procedure of mordant dyeing, in which the fabric is treated first with the mordant and then exposed to the dye.[52]

Other mordants that have been used in haematoxylin stains have been based on iron, chromium, copper, lead, tungsten, molybdenum and vanadium. The iron haematoxylins and the tungsten haematoxylins are still in general use.[53] For the iron haematoxylins, the mordant is iron(III) chloride or iron(III) ammonium sulfate. In the case of molybdenum mordanting, the reagent is phosphotungstic acid, while potassium permanganate is used as oxidant. The staining is said to be more precise if the tissue section has been pre-treated with potassium dichromate. This Mallory (1897) phosphotungstic acid-haematoxylin method exploits the polychromic properties of haematein, some cell components appearing in different shades of blue while others are red or brown.

Many stains of a more complicated nature, such as the trichrome stains, are particularly good at showing different cell components in different colours. These more modern stains generally contain some of the synthetic organic dyes, such as acid fuchsin or celestine blue, in addition to an alum-haematoxylin. They are often used in histology for examining particular types of tissue, such as connective tissue. By far the most important synthetic dye used in combination with haematoxylin, however, is the xanthene dye, eosin.

The xanthene dyes represent some of the earliest achievements of synthetic chemistry in the production of dyestuffs.[54] While the condensation of phthalic anhydride with phenol yielded phenolphthalein, its condensation with resorcinol gave fluorescein. The substitution of the fluorescein molecule with halogens yielded a series of red xanthene dyes (eosin, phloxine, erythrosine, rose Bengal); of these, eosin is the tetrabromo-derivative of fluorescein, and it has the property of staining tissues red or pink. It was introduced as a microscope stain by Fischer in 1876.[55]

Eosin is used in histology and histopathology as a counterstain with alum-haematoxylin. The section is stained first with an alum haematoxylin, washed, and then stained with eosin Y. The nuclei of the cells appear blue-black, while the cytoplasm is in various shades of pink. The particular value of the haematoxylin-eosin stain is to distinguish between the cytoplasm of different types of cell, through the various shades of blue, red and pink which are produced.

In microscopy, the general tissue stains are those which differentiate nuclei from the cytoplasm of the cells, distinguish between the various types of tissue, and identify pathological abnormalities in the various types of tissue. The best known of these are the haematoxylin stains, sometimes used alone, but often counterstained with eosin. This haematoxylin-eosin combination is in such general use that it is commonly referred to as the 'H-E stain'.

While those uses which originally made logwood such a prize among privateers on the Caribbean Sea are long since gone, this, one of the oldest of the natural dyes, in combination with one of the oldest synthetic dyestuffs, provides the haematoxylin-eosin (H-E) microscope stain which, in the 21st century, still has important uses in science and in medicine.

Editor's note

Address for correspondence: Department of Information Studies, Aberystwyth University, Aberystwyth, SY23 3AS, UK (bhd@aber.ac.uk).
Paper received 30 January 2003; revised version received 17 September 2003.

Notes and references

1. Holmyard, E.J. (1958) 'Dyestuffs in the nineteenth century', in C. Singer, E.J. Holmyard, A.R. Hall and T.I. Williams (eds), *A History of Technology: Volume V – The Late Nineteenth Century, c.1850–c.1900*, Oxford, pp. 257–83.
2. Catesby, M. (1771) *The Natural History of Carolina, Florida and the Bahama Islands*, 2 vols, 3rd edn, London, vol. 2, p. 66 (first published in 1731).
3. Jenkin, R.N., Innes, R.R., Dunsmore, J.R., Walker, S.H., Birchall, C.J. and Briggs, J.S. (1976) *The Agricultural Development of the Belize Valley*, London (Ministry of Overseas Development), pp. 111–12; Ure, A. (1875) *Ure's Dictionary of Arts, Manufactures and Mines*, 7th edn, eds R. Hunt and F.W. Rudler, 3 vols, London, vol. III, p. 143.
4. Muspratt, J.S. (1860) *Chemistry, Theoretical, Practical and Analytical, and Relating to the Arts and Manufactures*, 2 vols, Glasgow, vol. 1, p. 601.
5. Adrosko, R.J. (1971) *Natural Dyes and Home*

Dyeing, New York (Dover Publications), p. 45. (Originally published as 'Natural dyes in the United States', United States Museum Bulletin 281, Washington, Smithsonian Institution Press, 1968.)
6. Anno 23 Elizabethæ, c.9, A.D. 1581, in J. Raithby (ed.) (1811) *The Statutes at Large of England and of Great Britain from Magna Carta to the Union of the Kingdoms of Great Britain and Ireland*, 10 vols, London, vol. 2, p. 526.
7. Anno 13 & 14 Car.II, c.11 (XXVI), A.D. 1662, in Raithby 1811 (see note 6 above), vol. 3, pp. 65–6.
8. Chevreul, M.E. (1810) 'De la seconde partie du memoire sur le bois de Campèche et sur la nature de son principe colorant', *Annales de Chimie* 82, pp. 126–47.
9. Berzelius, J.J. (1840) *Lehrbuch der Chemie ... nach des Verfassers schwedischer Bearbeitung* ... tr. F. Wöhler, 10 vols (1833–41), vol. IX, p. 62.
10. Perkin, W.H., Jr. and Robinson, R. (1908) 'Brazilin and haematoxylin. Part VIII. Synthesis of brazilinic acid, the lactones of dihydrobrazilinic and dihydrohaematoxylinic acids, anhydrobrazilic acid, &c. The constitution of brazilin, haematoxylin, and their derivatives', *Journal of the Chemical Society* 93, pp. 489–517; Engels, P., Perkin, W.H., Jr. and Robinson, R. (1908) 'Brazilin, haematoxylin and their derivatives. Part IX. On brazilein, haematein, and their derivatives', *Journal of the Chemical Society* 93, pp. 1115–73.
11. The taxonomy of the South American *Caesalpinia* group is under revision, which is helping to clarify the identities of the individual species: see Cardon, D. (2003) *Le monde des teintures naturelles*, Paris, pp. 198 and 217–25; Nowik, W. (2001) 'The possibility of differentiation and identification of red and blue "soluble" dyewoods: determination of species used in dyeing and chemistry of their dyestuffs', *Dyes in History and Archaeology* 16/17, pp. 129–44; Lewis, G.P. (2005) 'Caesalpinieae', in G. Lewis, B. Schrire, B. MacKinder and M. Lock (eds), *Legumes of the World*, Kew (Royal Botanical Gardens), pp. 127–61.
12. Stansfield, W.S. (1934) 'Logwood', *Journal of the Society of Dyers and Colourists*, Jubilee Edition, pp. 139–68.
13. Perkin, A.G. and Everest, A.E. (1918) *The Natural Organic Colouring Matters*, London, p. 380.
14. Georgievics, G. von (1920a) *The Chemical Technology of Textile Fibres*, tr. C. Salter, 2nd English edn London, p. 221.
15. Georgievics 1920a (see note 14 above), p. 221.
16. Hummel, J.J. and Knecht, E. (1910) 'Dyeing', in *Encylopædia Britannica*, 11th edn, Cambridge, vol. 8, p. 749.
17. Stansfield 1934 (see note 12 above).
18. Georgievics 1920a (see note 14 above), p. 223.
19. Georgievics 1920a (see note 14 above), p. 224.
20. Georgievics 1920a (see note 14 above), p. 224.
21. Harley, R.D. (1982) *Artists' Pigments, c.1600–1835: A Study in English Documentary Sources*, 2nd edn, London, p. 64.
22. Muspratt 1860 (see note 4 above), vol. 2, p. 372.
23. Lewis, W. (1763) *Commercium Philosophico-technicum; or The Philosophical Commerce of Arts: designed as an Attempt to improve Arts, Trades, and Manufactures*, London, p. 377.
24. Eisler, — (1770) *Das Eislerische Dintefass* (rev. edn of *Das wohl bereitete Tinten-Fass*), Helmstedt, p. 8, quoted in Mitchell, C.A. and Hepworth, T.C. (1937) *Inks: Their Composition and Manufacture*, 4th edn by C.A. Mitchell, London (cited below as Mitchell 1937).
25. Reinige, — (1857) *Dinglers polytechnisches Journal* 143, p. 240, quoted in Mitchell 1937 (see note 24 above).
26. Mitchell 1937 (see note 24 above), pp. 139–44.
27. Runge, F.F. (1847) *Grundriss der Chemie*, 2 vols, Munich, vol. 2, p. 205.
28. Böttger, — (1848) *Dinglers polytechnisches Journal* 109, p. 225, quoted in Mitchell 1937 (see note 24 above).
29. Böttger, — (1857) quoted in Mitchell 1937 (see note 24 above).
30. Viedt, — (1875) quoted in Mitchell 1937 (see note 24 above).
31. Cleveland, R.-R. and Rhodes, B. (1999) 'Copying inks', in B. Rhodes and W.W. Streeter (eds), *Before Photocopying: The Art and History of Mechanical Copying, 1780–1938*, Newcastle, Delaware (Oak Knoll Press), pp. 29–45.
32. Lehner, S. (1926) *Ink Manufacture – Including Writing, Copying, Lithographic, Marking, Stamping, Typewriter and Laundry inks*, 3rd edn, tr. C.A. Mitchell, London, pp. 69–73.
33. Cleveland and Rhodes 1999 (see note 31 above).
34. Lehner 1926 (see note 32 above), pp. 78–9.
35. Underwood, J. (1857) 'On the history and chem-

istry of writing, printing and copying inks, and a new plan of taking manifold copies of written and printed documents, maps, charts, plans and drawings', *Journal of the Society of Arts* 5, p. 70.
36. Carvalho, D.N. (1904) *Forty Centuries of Ink*, New York.
37. Schwanhausser, G. (1875), *Improvement in Colored Copying: Pencils*, US Patent 169, 195, Oct 26 1875.
38. Savage, W. (1832) *On the Preparation of Printing Ink both Black and Coloured*, London, described in *Printing Ink Manual* (1969), Society of British Printing Ink Manufacturers Technical Training Board, ed. F.A. Askew *et al.*, 2nd edn, Cambridge (Heffer).
39. Carvalho 1904 (see note 36 above).
40. Harley 1982 (see note 21 above), p. 65.
41. Norgate, E. (1997) *Miniatura or The Art of Limning*, eds J.M. Muller and J. Murrell, New Haven and London, p. 59. This edition is of Norgate's revised version of the original manuscript of 1627–8, dated about 1648.
42. Georgievics, G. von (1920b) *A Text-Book of Dye Chemistry*, 2nd edn. (from 4th German edn, ed. E. Grandmougin, tr. F.A. Mason), London, p. 527.
43. Hooke, R. (1665) *Micrographia*, London.
44. Lewis, F.T. (1942) 'The introduction of biological stains: employment of saffron by Vieussens and Leeuwenhoek', *Anatomical Record* 83, pp. 229–53.
45. Quekett, J. (1848) *A Practical Treatise on the Use of the Microscope, including the Different Methods of Preparing and Examining Animal, Vegetable and Mineral Structures*, London.
46. Conn, H.J. (1961) *Biological Stains*, 7th edn, Baltimore, p. 1.
47. Baker, J.R. (1943) *The Discovery of the Uses of Colouring Agents in Biological Micro-technique* (Monograph of the Quekett Microscopical Club), London (Williams and Norgate).
48. Clark, G.C. and Kasten, F.H. (1983) *History of Staining*, 3rd edn, Baltimore, p. 62.
49. Böhmer, F. (1865) 'Zur pathologischen Anatomie der Meningitis cerebromedullaris epidemica', *Ärztliches Intelligenzblatt* (Munich) 12, pp. 539–48.
50. Conn 1961 (see note 46 above), p. 4.
51. Stevens, A. (1990) 'The haematoxylins', in J.D. Bancroft and A. Stevens (eds), *Theory and Practice of Histological Techniques*, Edinburgh (Churchill Livingstone), pp. 107–18.
52. Clark and Kasten 1983 (see note 48 above), pp. 69–70.
53. Stevens 1990 (see note 51 above).
54. Georgievics 1920b (see note 42 above), pp. 218–26.
55. Fischer, E. (1876) 'Eosin als Tinctionsmittel für mikroskopische Präparate', *Archiv für mikroskopische Anatomie* 12, pp. 349–52.

Edward Schunck: Forgotten Dyestuffs Chemist?

Christopher J. Cooksey* and Alan T. Dronsfield

Introduction

Henry Edward Schunck – known as Edward – lived a long and productive life. Born in 1820, he had many personal recollections of the doyen of the atomic theory, John Dalton (1766–1844). He died in 1903. One of the authors' grandfathers (ATD) could have served as his laboratory assistant. His scientific research spanned one of the greatest periods of organic chemistry. At its start – before he was 20 – he would have known chemists whose earlier studies would have had phlogiston as their only unifying concept. At its conclusion, some 80 years later, there were equations, structural formulae, coal-tar dyes and a burgeoning industrial organic chemistry, developed principally from W.H. Perkin's discovery of his dye mauveine.

Schunck's grandfather, Johann Carl (1745–1800), was an officer serving in the army of the Elector of Hesse. This army was 'hired' by the British to fight in the War of American Independence.[1] His younger son, Martin (Edward's father) settled in Malta and later (from 1808) in Manchester where 'he amassed a considerable fortune'.[2] Certainly this arose from his activities in the calico printing industry. Speculatively we may assume he might have been involved in cotton importing; the choice of his Manchester location would, indeed, support this view. Martin was also a noted export shipping merchant.[3] Edward had no need to work for a living; however, for a time he 'engaged for some years in practical work, but finding this repugnant to his taste and inclination he gave it up, and devoted himself to pure science'.[4] The family fortune enabled him to live the life of a gentleman-scientist and philanthropist.

Schunck attended a fee-paying school in Manchester and he records, 'Dalton lived with the master of the school I attended and I had frequent opportunities of seeing and wondering at his ways. How well I remember his peering out of his bedroom window ... and looking at the thermometer outside. Very soon he would appear at the front door of the house, arranged in his snuff-coloured Quaker-dress, with knee breeches and broad-brimmed hat.'[5] Though noted as a teacher, Dalton does not appear to have actually taught the young Schunck. Initially Edward's chemistry was self-taught, learning the subject from books. However, his family was on good terms with the Henry family, also of Manchester; Thomas Henry was a manufacturing chemist (mainly of patent medicines) and his son, William, was a noted textbook author who continued the family business. Schunck was particularly close to the third generation Henry (William) Charles. He was university trained as a physician, but did not practice and devoted himself to the family business. Schunck records, 'I have good reason to remember Dr Charles

Henry for he, learning that I was learning chemistry entirely from books, very kindly offered to give me some practical instruction in science and I was introduced into the laboratory attached to his work – a somewhat dark and dingy apartment, but a paradise to me who had never seen its like before.'[6] The young Schunck was sent to Berlin to further his chemical studies, where he studied under Rose. Heinrich Rose (1795–1864) discovered niobium, diligently analysed minerals and other inorganic substances and studied the chemistry of titanium, phosphorus, arsenic, antimony, sulfur, selenium and tellurium. While in Berlin, Schunck also studied under Magnus. Heinrich Gustav Magnus (1802–1870) published over 80 papers on many diverse topics in chemistry and physics. Schunck then worked with Justus Liebig at Giessen, where he obtained a PhD degree in 1841. It was from here that, in the same year, he published his first research paper, in Liebig's famous journal *Annalen*, on the topic of the effect of nitric acid on aloes.

Chrysammic acid

Edward Schunck extended Liebig's research on the action of nitric acid on aloes,[7] a subject which was first investigated by Henri Braconnot (1781–1855) in 1809.[8] Schunck published his results in two papers in 1841 and 1848.[9] The aloes–nitric acid reaction gives, among other products, aloetic acid, which on further reaction is converted into chrysammic acid (Χρυσος, gold, ἄμμος, sand). The preparation of chrysammic acid was described by Glover in 1855:[10]

> Chrysammic acid ... is obtained by steeping 1 part of aloes in 8 of nitric acid, sp. gr. 1.37, and heating the mass in a porcelain capsule until the chief part of the action is over, then distilling off in a retort two-thirds of the nitric acid; three or four parts of nitric acid are afresh introduced into the retort, and the whole kept for two or three days at a temperature near to boiling point. After disengagement of gas has ceased, water is added to the residue, which forms a precipitate – the chrysammic acid. The mother liquid contains oxalic and chrysolipic acids, which the latter appears to be picric.

This is exciting chemistry: nitric acid is corrosive to clothes and skin – the initial reaction would give clouds of brown gas, nitrogen dioxide, and the yellow products are explosive and easily detonated. Schunck analysed samples of chrysammic acid and several of its metal salts, concluding that the formula of the acid was $C_{15}H_3N_4O_{12}$. This is very nearly the currently accepted formula of $C_{14}H_4N_4O_{12}$ which was obtained by Mulder (1848, 1849) a few years later.[11]

Chrysammic acid is 1,8-dihydroxy-2,4,5,7-tetranitroanthraquinone, which is structurally quite similar to picric acid, the other product of the reaction. In the mid-19th century, the sodium salt of chrysammic acid was employed as a dye for wool, giving a blue or blue-grey shade.[12] The origin of the chrysammic acid is aloe-emodin or its glucoside, present in aloes, and the modern chemical structures are shown in Figure 1.

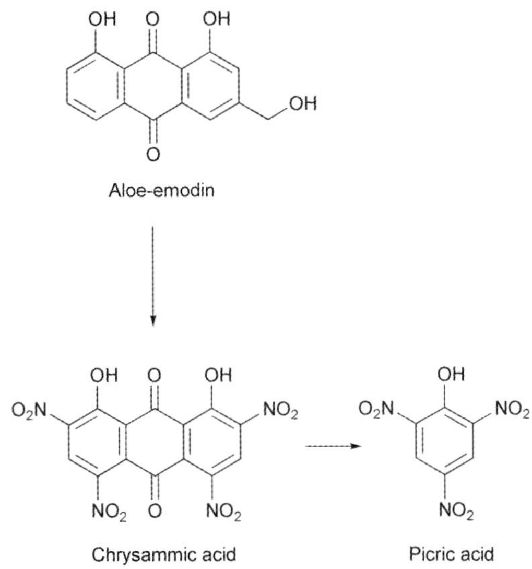

Figure 1 Chrysammic acid and related compounds.

Lichen purple

The purple from lichens was an important commercial product and came in a variety of forms, for example, orchil and cudbear. By the 1830s, the researches of Pierre Jean Robiquet (1780–1840), Friedrich Heeren (1803–1885), Jean-Baptiste Dumas and Robert John Kane (1809–1890) into the constituents of lichens had revealed three colour precursors – orcinol, erythrin and pseudoerythrin – but their constitution was not precisely known. Liebig encouraged Schunck to reinvestigate the subject using dye-producing lichens, 'of the genera *Lecanora*, *Urceolaria*, *Variolaria*, &c', that grow on the basalt rocks of the Vogelsberg in Upper Hessia (Oberhessen). In 1842, he discovered a new compound which he called lecanorin:[13]

> The lichens were reduced to a coarse powder and then treated with æther, in an apparatus of displacement, until the æther dissolved nothing more. The æthereal extract, which had acquired a green tinge from chlorophyll in solution, was distilled off; leaving as a residue a greenish yellow mass consisting for greater part of lecanorin. This mass was brought into a glass funnel and washed with small quantities of æther, until it had lost its green colour in part. It was then treated with boiling water in order to remove every trace of pseuderythrin, and, lastly, purified by dissolving it in a small quantity of boiling alcohol, which deposited on cooling a snow-white crystalline mass, consisting of lecanorin in a state of purity. The dark green æthereal fluid obtained by washing the impure lecanorin, contained besides lecanorin the greatest part of the pseuderythrin which had been extracted by the æther. The fluid was evaporated to dryness and the residual mass treated with boiling water, which deposited on cooling a mass of shining plates and needles of pseuderythrin, which was purified by recrystallization.

Lecanorin is now recognised to be a carboxylic acid and named lecanoric acid. Elemental analysis resulted in a formula $C_{18}H_8O_8$ and the products of decomposition were orcin (orcinol), water and carbonic acid (carbon dioxide), leading to the following 'equation' for this reaction:

1 atom* of anhydrous orcin	C16	H6	O2
2 atoms of water		H2	O2
2 atoms of carbonic acid	C2		O4
1 atom of lecanorin	C18	H8	O8

* In current usage, 'atom' is equivalent to 'molecule', 'orcin' to 'orcinol' and the above formulae are calculated with the old atomic weights in use before the 1860s: C = 6; H = 1; O = 8.

It all adds up, except that, as Schunck admits, 'In regard to the composition of orcin, I have been induced to replace the generally received formula [$C_{18}H_7O_3$] for its composition by a new one [$C_{16}H_6O_2$].'[14] Unfortunately, neither formula for orcin (orcinol) is correct and the first correct determination of the formula of orcinol, $C_7H_8O_2$, was by Charles Gerhardt three years later in 1845.[15] In addition, during Schunck's manipulations, his lecanoric acid had partially hydrolysed to give orsellinic acid, leading to an incorrect result. The true story was unravelled by Stenhouse (using old atomic weights) who presented his results like this:

$$\underset{\text{lecanoric acid}}{C_{32}H_{14}O_{14}} + \underset{\text{water}}{2HO} = \underset{\text{orsellinic acid}}{2C_{16}H_8O_8}$$

$$\underset{\text{orsellinic acid}}{C_{16}H_8O_8} = \underset{\text{carbon dioxide}}{C_2O_4} + \underset{\text{orcinol}}{C_{14}H_8O_4}$$

Later he discovered, in addition to lecanoric acid, another new compound, parellic acid, in *Lecanora parella* (L.) Ach.[16] The modern structural formulae are shown in Figure 2.

Madder

In 1846 Schunck started his extensive investigations into the colouring materials in madder.[17] Madder was an important dye and imports into the UK were valued at £1.25 million per year in

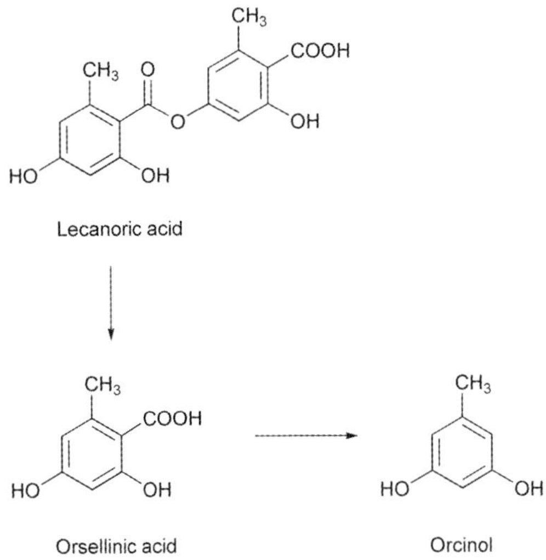

Figure 2 Lecanoric acid and related compounds.

the 1860s. The main colorant from madder was discovered by Robiquet and Colin in 1827 and called alizarin.[18] Their analysis gave a formula $C_{37}H_{48}O_{10}$. When purified by Schunck using sublimation and crystallisation,[19] he obtained a result which suggested $C_{14}H_8O_4$, but taking into account the analyses of metal derivatives as well, he chose $C_{14}H_{10}O_4$ as the best result. The modern formula is $C_{14}H_8O_4$. He found that oxidation of alizarin with nitric acid gave alizaric acid (phthalic acid), which, on heating, gave pyroalizaric acid (phthalic anhydride). This led to the suggestion that alizarin was a derivative of naphthalene, a C_{10} hydrocarbon, although Schunck pointed out that this did not explain the reactions of alizarin. He was vindicated when Graebe and Liebermann (1868) distilled alizarin with zinc dust to give anthracene, a C_{14} hydrocarbon, and subsequently (1869) synthesised alizarin from anthraquinone.

Schunck showed that alizarin was not the major colour precursor component of fresh madder root: this was a yellow, bitter water-soluble component, which he called rubian.[20] Rubian was obtained from the aqueous extract of madder root by adding bone-charcoal and extracting the bone-charcoal with ethanol. Rubian was an uncrystallisable gum, hydrolysable by acids or an enzyme contained in the madder root to give alizarin and a sugar. Exact details of the amounts of material used were often not disclosed in publications at this time, but Schunck does mention that very little rubian could be isolated: only 1000 grams from 1 hundredweight (cwt) [1 cwt = 50.80 kg]. In another context, showing that madder root contained sugar, he mentions extracting half a hundredweight of madder root with water. The solution obtained was fermented with yeast and the resulting liquor distilled to give the equivalent of 9 oz of absolute alcohol. Rubian itself could not be obtained in a pure form, but analysis of the lead salt of rubian led to a formula $C_{28}H_{34}O_{15}$ and the formation of alizarin was represented as

$$C_{28}H_{34}O_{15} = 2C_{14}H_{10}O_4 + 7H_2O$$

Following further work by Friedrich Rochleder (1819–1874),[21] who called rubian *Ruberythrinsäure*, Charles Gerhardt suggested a different formulation in 1854:[22]

$$C_{16}H_{16}O_8 + H_2O = C_{10}H_6O_3 + C_6H_{12}O_6$$

whereas the correct interpretation is:

$$C_{25}H_{26}O_{13} + 2H_2O = C_{14}H_8O_4 + C_6H_{12}O_6 + C_5H_{10}O_5$$

Rubian was actually a mixture of glycosides of di- and tri-hydroxy anthraquinones, of which a major component was ruberythric acid, which is an alizarin 2–β–primeveroside. Many other 'compounds' derived from the hydrolysis of rubian were described and enthusiastically named by Schunck: rubiretin, verantin, rubiacine, rubiadine, rubiapin, rubiafin and others, but some of these are most likely to be impure alizarin, and best forgotten. Schunck summarised his results in 1860,[23] returning to the subject at intervals thereafter.[24] Some of Schunck's original samples were examined in 1975 by Wilfrid Farrar.[25] Rubiadine (rubiadin) was shown to be 1,3-dihydroxy-2-methylanthraquinone, which Schunck assigned as the 4-methyl isomer.[26]

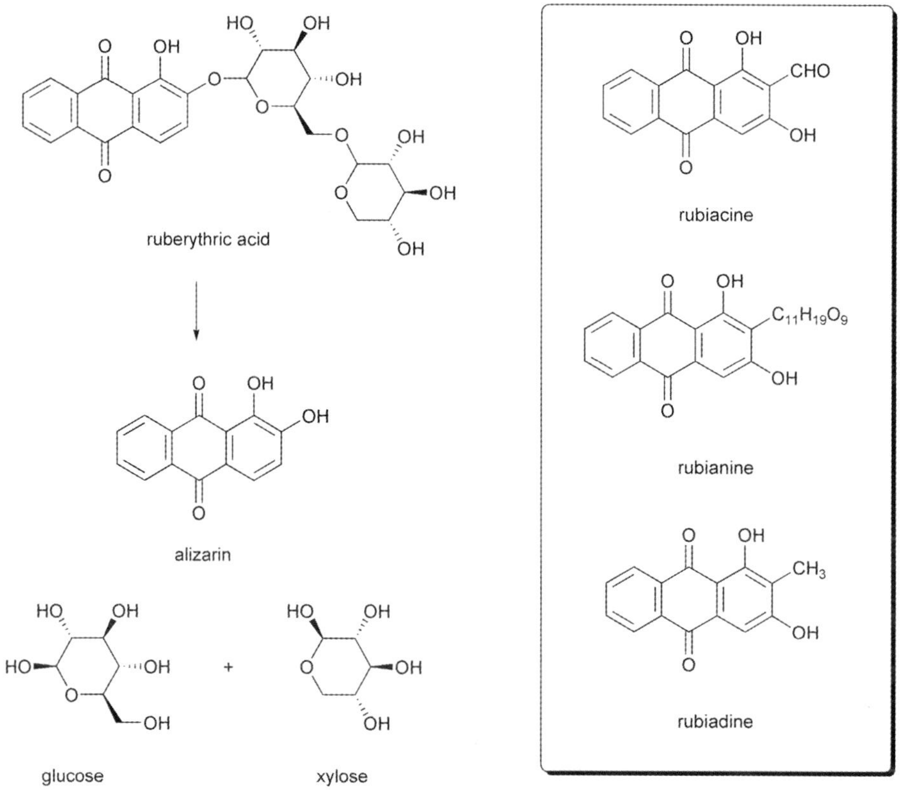

Figure 3 Alizarin and other compounds derived from rubian.

Rubiacine was identical to nordamnacanthal (1,3-dihydroxyanthraquinone-2-aldehyde) and rubianine was an unusual C-glucoside of unknown constitution. The modern chemistry of ruberythric acid and alizarin is shown in Figure 3.

Indigo

In 1855, Schunck tackled the subject of indigo, preferring the name indigo-blue to the alternative name, indigotine.[27]

> Having procured 10 lbs. of good French woad seed, I sowed it at the commencement of the spring last year on about half an acre of land. It was sown in drills about two feet apart, each drill being previously well supplied with farmyard manure. In a short time the young plants appeared, and grew vigorously during the summer months.

As was the case with madder, he found that the dye itself was not present in the woad; instead it was contained in a bitter, water-soluble precursor. His preferred extraction method was:

> The pounded woad leaves are extracted in a displacement apparatus, as before, with cold alcohol. To the green alcoholic extract there is added an alcoholic solution of acetate of lead, which produces a pale green precipitate, the precipitation being then completed by the addition of a little ammonia. The precipitate, which is bulky, is placed on a filter and washed with cold alcohol, until the percolating liquid instead of being dark green is only light green, and the excess of acetate

of lead and ammonia have been removed. It is then suspended in water, and a current of carbonic acid gas is passed through the liquid. The precipitate gradually becomes paler in colour and at last almost white, and loses considerably in bulk, while the liquid acquires a yellow colour. The latter being filtered, sulphuretted hydrogen is passed through it to precipitate a little oxide of lead contained in it, and being again filtered is evaporated as before, either in the air or *in vacuo* over sulphuric acid.

He proposed the name 'Indican' for the clear brown syrup obtained and suggested that the termination '–an' should be reserved for 'substances of a complex constitution, such as rubian, as are direct products of the vital energy of plants or animals'. After finding that the product was unstable he devised a means of evaporation of the solutions at low temperature. When treated with strong acid and heated to near boiling, pure indigo-blue was precipitated at first and later, as the reaction proceeds, the product also contained a red component: the indigo-red of Berzelius, which he named 'Indirubine'. In his usual way, further compounds are isolated, indiretine, indihumine, and so on and, like those rubian products, are best forgotten. The production of indigo from indican was represented as

$$C_{52}H_{33}NO_{36} + 2HO = C_{16}H_5NO_2 + 3C_{12}H_{10}O_{12}$$

The correct formula for indican using modern atomic weights is actually $C_{14}H_{17}NO_7$, and the need for oxygen in the indigo production process was not recognised. Later he investigated *Polygonum tinctorium* Ait. (Japanese indigo) as a source of indigo, believing that it contained the same indican.[28] He was unconvinced by much later reports,[29] which suggested that different precursors were involved.[30] The current belief is that *Isatis tinctoria* L. (woad) contains mainly isatan B, with indican as a minor component,[31] and indican is the major precursor in *Indigofera* spp. and *Polygonum tinctorium*. More recently a third precursor, isatan C, has been proposed in woad.[32] The structures of these precursors are shown in Figure 4.

The presence of indigo in urine has long been the subject of much curiosity and in 1857 Schunck tried to show that indican was the origin. Believing that the occurrence of indigo was more common than generally supposed at the time, he examined the urine of 40 individuals, all apparently healthy, with ages between seven and 55 years, mostly of the working class. In all but one case the result was positive. The largest amount of indigo was obtained from a man above the age of 50, a publican by trade. In his own case, the amount 'varied most capriciously from a tolerable quantity to a mere trace'. He thought this variation might be due to different kinds of diet, but, after many experiments, found only one which worked: 'I took on the next night, before going to bed, a mixture of treacle and arrowroot boiled with water in as

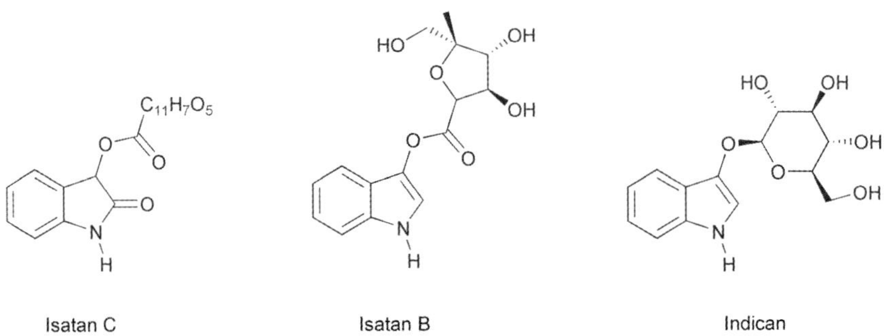

Figure 4 Precursors of indigo.

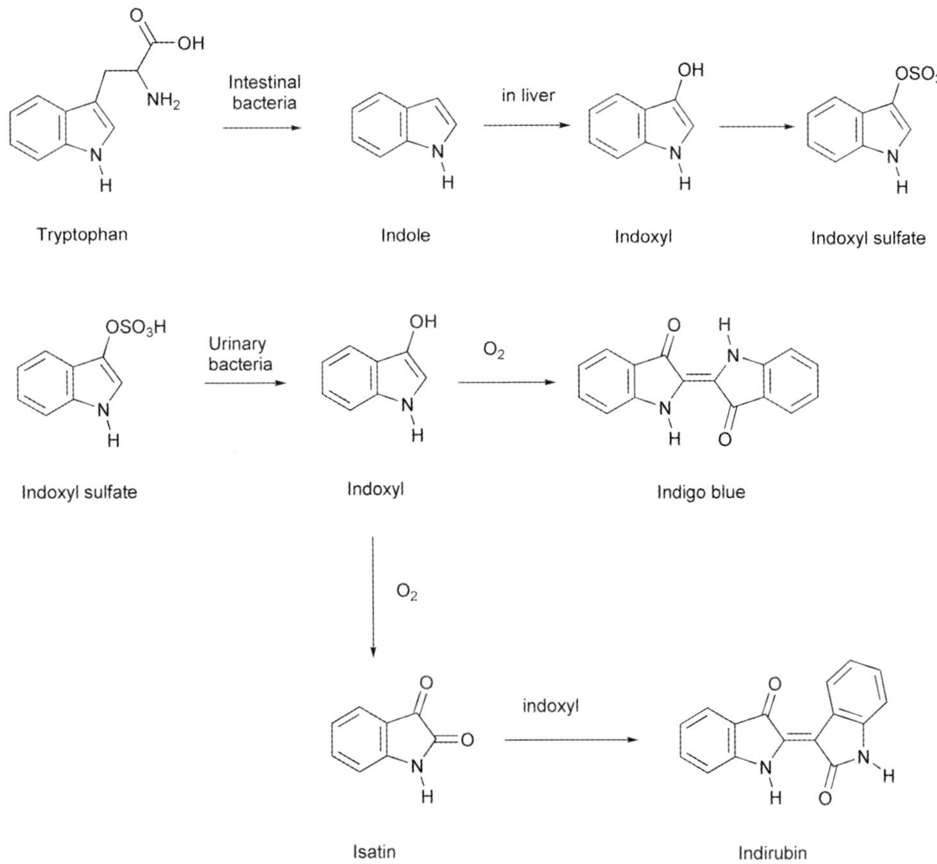

Figure 5 The chemistry of indoxyl sulfate.

large a quantity as the stomach could bear, and the effect was that the urine of the following night gave a large quantity of indigo-blue'.[33]

Over 130 years later, the subject was still being promoted as worthy of interest.[34] The indigo precursor in urine turns out not to be indican, although it was called 'medical indican', but indoxyl sulfate or gluconurate. The chemistry is shown in Figure 5.

Tyrian purple

Schunck rounded off his studies of ancient dyes with Tyrian purple obtained from 'Purpura Capillus' (*Nucella lapillus* (Linnaeus, 1758)) from Hastings and he was able to confirm the observations of many others since 1683: the formation of the purple from the mollusc required light and generated a garlic-like smell.[35] Since he only obtained 7 mg of the pigment from 400 molluscs, there was not sufficient material for elemental analysis, but he could confirm that it had some indigo-like properties. It dissolved in alkaline tin(II) solution, like indigo, but, unlike indigo, a solution in sulfuric acid became green and had no dyeing properties. He named the material he isolated 'Punicin', but that name, like many others he coined, never caught on. Two years later, he was able to examine the purple colour from *Purpura patula* (Linnaeus, 1758) from the west coast of Nicaragua.[36] The dyed yarn was extracted with boiling aniline, which on cooling precipitated the pigment; this

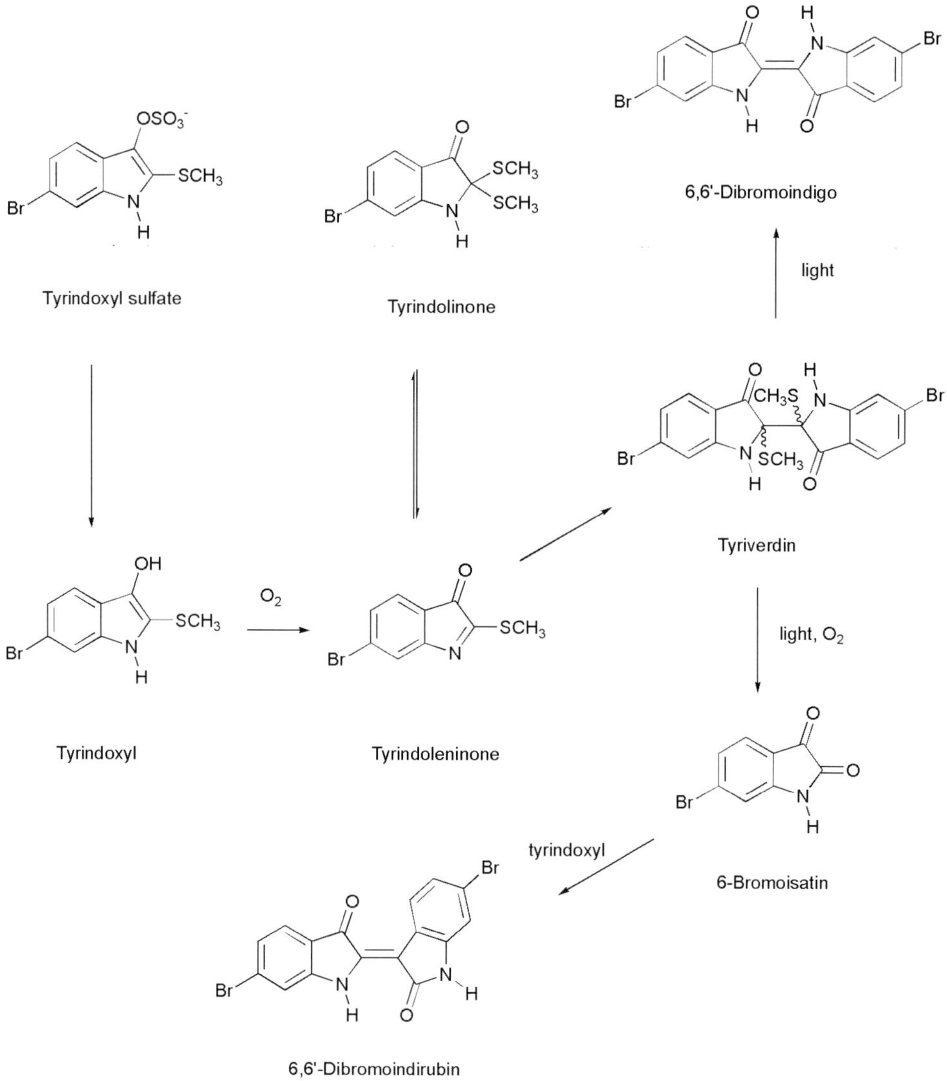

Figure 6 The formation of 6,6′-dibromoindigo.

he pronounced to be identical to that obtained from *Nucella lapillus*. As an additional observation, he notes the sublimation temperatures of indirubin (140 °C), indigo-blue (170 °C) and punicin (190 °C). The true nature of the purple colour from shellfish was not established until 30 years later. The heroic investigations of Paul Friedländer involved about 12,000 *Murex brandaris* (*Bolinus brandaris* (Linnaeus, 1758)) and revealed that the colouring matter was 6,6′-dibromoindigo. Details of the chemistry were only elucidated in the second half of the 20th century, as shown in Figure 6.

Chlorophyll

During the last 20 years of his life, Schunck investigated chlorophyll and published many papers.[37] But he had an almost impossible task

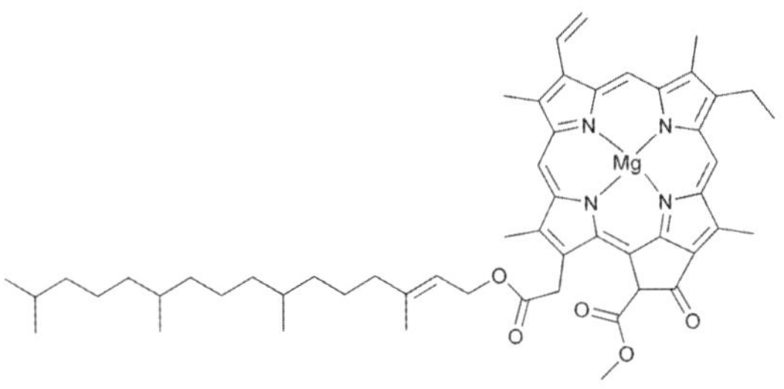

Figure 7 The structure of chlorophyll-*a*.

with such a large molecule, which is shown in Figure 7.

Schunck's laboratory

In 1851 Schunck married Judith Brooke and subsequently sired five sons and two daughters. For most of his life he lived in a large house at Kersal, some 10 km from Manchester city centre. It was in his private laboratory here that he conducted the majority of his researches. Some six years before his death, an anonymous writer describes the scene:[38]

> Imagine him in his laboratory, the finest private laboratory perhaps, in the world: stocked with bookcases filled with old volumes. The rooms are decorated with ornaments as beautiful as the skill of men accustomed to work in inlaid wood can make them. Around the room are busts of great philosophers who are gone; the whole building is one which invites study and reflection ... In his laboratories he has every kind of apparatus at his disposal; an ardent student of colour, he has invented many new colours, and explained the methods by which colours are built up in plants.

Honours and appointments

By choice, Edward Schunck held no university position, choosing to work independently. In the 1881 British Census he describes himself as 'Scientific Man Chemistry'. In his lifetime he published some 200 papers, although perhaps only 100 are unique since he often published essentially the same material in different journals. He joined the Chemical Society in 1841, the year of its foundation. In 1850, not yet 30, he was elected a Fellow of the Royal Society.[39] The citation reads, in part, 'Distinguished for his acquaintance with the science of Chemistry.' However, it was in the scientific community of Manchester that he made his greatest mark. He was elected into the Manchester Literary and Philosophical Society in 1842 and served as Secretary (1855–60) and President (1866–67, 1874–75, 1890–91 and 1897–98). He published 26 papers from 1854 to 1896 in its journal *Memoirs and Proceedings*. Schunck took an interest in applied chemistry; he was a founder member of the Society of Chemical Industry and was chairman of the Manchester section in 1888–9, and President in 1896–7. The Victoria University of Manchester (formerly Owen's College) conferred on him the honorary degree of Doctor of Science in 1899.

Interest and philanthropy

As to his interests outside colour and natural product chemistry, Harold Bailey Dixon recollects that he was 'fond of travelling, riding and billiards. He made many journeys on the Continent, and kept voluminous diaries. Botany was always a favourite study of his, and those who knew him in later life always found him hard at work in his laboratory, or busy in his greenhouses, or deep in a botanical book by way of relaxation ... His uniform courtesy and kindly but shrewd counsel, ever at the service of those who sought it, endeared him to a wide circle of younger men. He gave large sums in charity, but the help bestowed was given in secret. Few men have devoted their lives so unselfishly to the pursuit of knowledge for its own sake.'[40]

Edward Schunck died in 1903 at his home in Kersal, survived by his wife and four of his seven children. Earlier (1895) he had presented £20,000 (at 2002 values £1,250,000 or €1,875,000) to the University of Manchester to endow chemical research. In his will he bequeathed the university the building and contents of his laboratory, together with his library and some land in Kersal.

Reminders of Edward Schunck in the 21st century

Little remains today to remind Manchester citizens of one of their greatest 19th-century chemists. His house 'Oaklands', on the north-west edge of Manchester on the high land of Kersal Moor, has long been demolished. Its name survives in the name of a road and a 1960s nursery, which may have been built on the site of the house itself. Adjoining Oaklands Road is land which was until recently occupied by university halls of residence, but these, too, have been demolished and the area is rapidly becoming a modern housing estate.

Schunck was buried at St Paul's Church, Kersal, less than a mile from his house. This part of the churchyard is now overgrown and, through subsidence or vandalism, his headstone has toppled forward, rendering its inscription unreadable. His name survives as the Schunck Laboratory at the University of Manchester, where his plaque is on display, and as the name of the chemistry departmental library. The Schunck Library, for which there is a handwritten catalogue, still exists – not as an integral whole, but largely dispersed among the main chemistry book collection. His chemical stock of dyes and their intermediates, some 300 items, are preserved (in store) at the Manchester Museum of Science and Industry.

The published papers by Wilfred Farrar record some of his life and work.[41] There were obituaries published in the major scientific journals,[42] but only one portrait appears to have survived,[43] shown in Figure 8.

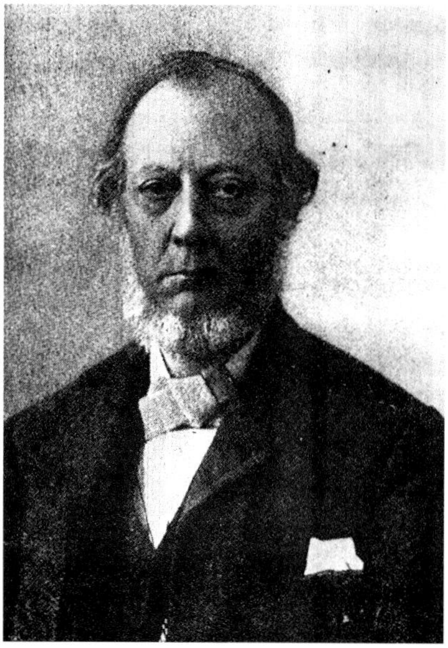

Figure 8 Edward Schunck, from an article published in 1897 (see text).

Editor's note

Author for correspondence: 59 Swiss Avenue, Watford, Herts, WD18 7LL, UK (dha@chriscooksey.demon.co.uk). Paper received 19 September 2003; revised version received 12 November 2004.

Notes and references

1. James, T.E. (1920) 'Schunck, Henry Edward', in S. Lee (ed), *Dictionary of National Biography, 1901–1911*, London, pp. 274–5.
2. 'Dr. Edward Schunck, F.R.S.' (1897) *Manchester Faces and Places* 9, 1, pp. 1–5.
3. James 1920 (see note 1 above).
4. 'Obituary. Henry Edward Schunck, Ph.D., D.Sc., F.R.S., &c' (1903) *Chemical News*, pp. 34–5.
5. 'Dr. Edward Schunck, F.R.S.' 1897 (see note 2 above).
6. 'Dr. Edward Schunck, F.R.S.' 1897 (see note 2 above).
7. Liebig, J. (1828) 'Ueber die bittere Substanz, welche durch Behandlung des Indigs, der Seide und der Aloë mit Salpetersäure erzeugt wird', *Annalen der Physik und Chemie* 13, pp. 191–207.
8. Braconnot, H. (1809) 'Examen des acides végétaux qui saturent la potasse et le chaux dans les plantes', *Annales de Chimie* 79, pp. 255–92.
9. Schunck, E. (1841) 'Ueber einige durch die

Einwirkung der Salpetersäure auf Aloë entstehende Producte', *Annalen der Chemie und Pharmazie* 39, pp. 1–25; Schunck, E. (1848) 'Ueber einige Zersetzungsprodukte der Chrysamminsäure', *Annalen der Chemie und Pharmazie* 65, pp. 234–44.

10. Glover, R.M. (1855) *Manual of Elementary Chemistry being a Practical Class Book*, London, p. 304.
11. Mulder, G.J. (1848) 'Chrysamminsäure', *Annalen der Chemie und Pharmacie* 68, pp. 339–40; Mulder, G.J. (1849) 'Zersetzungsproducte der Aloë durch Salpetersäure', *Annalen der Chemie und Pharmacie* 72, pp. 285–91.
12. Schweppe, H. (1993) *Handbuch der Naturfarbstoffe: Vorkommen; Verwendung; Nachweis*, Hamburg, p. 228.
13. Schunck, E. (1842) 'On some of the substances contained in the lichens employed for the preparations of archil and cudbear', *Chemical Society Memoirs* 1, pp. 71–7.
14. Schunck, E. (1842) 'Vorläufige Notiz über einige farbstoffgebende Substanzen der Flechten', *Annalen der Chemie und Pharmazie* 41, pp. 157–62.
15. Gerhardt, C. (1845) 'Nouvelle classe de composés organiques; Schunck – Principes colorants des lichens', *Comptes rendus mensuels des travaux chimiques* (1845), pp. 283–8.
16. Schunck, E. (1845) 'Ueber die Bestandtheile der Lecanora Parella', *Annalen der Chemie und Pharmazie* 54, pp. 257–84.
17. Dronsfield, A.T., Brown, T.M. and Cooksey, C.J. (1999) 'Alizarin, the forgotten dyestuff?' *Education in Chemistry* 36, pp. 20–22.
18. Robiquet, P.J. and Colin, J.J. (1827) 'Nouvelles recherches sur la matière colorante de la garance', *Annales de chimie et de physique* 2nd series, 34, pp. 225–53.
19. Schunck, E. (1848) 'Untersuchung der Krapps', *Annalen der Chemie und Pharmazie* 66 pp. 174–213.
20. Schunck, E. (1851) 'On rubian and its products of decomposition' *Philosophical Transactions of the Royal Society* 141, pp. 433–60.
21. Rochleder, F. (1852) 'Ueber die Wurzel der Rubia tinctorum' *Annalen der Chemie und Pharmacie* 82, pp. 205–16.
22. Gerhardt, C. (1854) *Traité de chimie organique*, 4 vols (1853–56), Paris, vol. 3, p. 49323; Rochleder, F. (1851) 'Untersuchung der Krappwurzel' *Annalen der Chemie und Pharmacie* 80, pp. 321–30. (A report from *Wiener Acad. Berichte* VI, 433, April 1851.)
23. Schunck, E. (1860) 'On the colouring matters of madder' *Journal of the Chemical Society* 12, pp. 198–221.
24. Schunck, E. and Roemer, H. (1877) 'Notes on madder colouring matters' *Journal of the Chemical Society* 31, pp. 665–75.
25. Farrar, W.V. (1975) 'Examination of some compounds isolated from madder 1845–1855', *Ambix* 22, 2, pp. 151–3.
26. Schunck, E. and Marchlewski, L. (1893) 'Supplementary notes on madder colouring matters' *Journal of the Chemical Society* 63, pp. 969–74; Schunck, E. and Marchlewski, L. (1894) 'Notes on madder colouring matters', *Journal of the Chemical Society* 65, pp. 182–7.
27. Schunck, E. (1855) 'On the formation of indigo-blue. Part 1', *Memoirs of the Manchester Literary and Philosophical Society* 2nd series, 12, pp. 177–208; Schunck, E. (1856) 'On the formation of indigo-blue. Part 2', *Memoirs of the Manchester Literary and Philosophical Society* 2nd series, 14, pp. 181–237.
28. Schunck, E. (1879) 'On indigo-blue from *Polygonum tinctorium* and other plants', *Memoirs of the Manchester Literary and Philosophical Society* 3rd series, 6, pp. 218–34.
29. Schunck, E. (1900) 'Recent researches on indican', *Chemical News* 82, pp. 176–7.
30. Schunck refers to Professor Hoogewerff and Mr ter Meulen (April 1900), *Koninklijke Akademie van Wetenschappen*: Schunck 1900 (see note 29 above).
31. Epstein, E., Nabors, M.W., Stowe, B.B. (1967) 'Origin of indigo of woad', *Nature* 216, pp. 547–9; Maier, W., Schumann, B. and Gröger, D. (1990) 'Biosynthesis of indoxyl derivatives in *Isatis tinctoria* and *Polygonum tinctorium*', *Phytochemistry* 29, 3, pp. 817–19. The chemical structure of isatan B, originally suggested by Epstein in 1967 and shown in Figure 4 (p. 194)

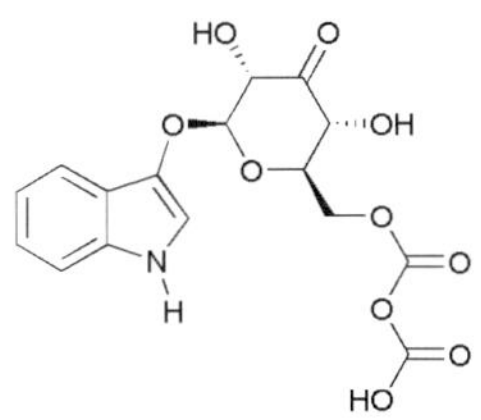

Isatan A

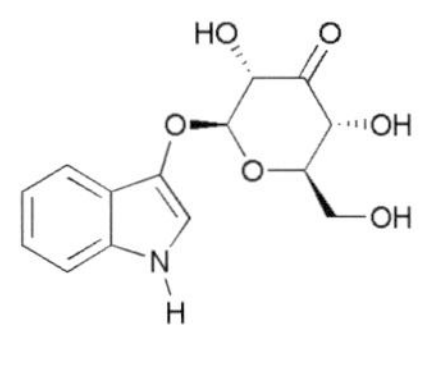

Isatan B

has recently been revised and a new precursor, isatan A, has been identified: Oberthür, C., Schneider, B., Graf, H. and Hamburger, M. (2004) 'The elusive indigo precursors in woad (*Isatis tinctoria* L.) – identification of the major indigo precursor, isatan A, and a structure revision of isatan B', *Chemistry and Biodiversity* 1, pp. 174–82.

32. Maugard, T., Enaud, E., Choisy, P. and Legoy, D. (2001) 'Identification of an indigo precursor from leaves of *Isatis tinctoria* (woad)', *Phytochemistry* 58, pp. 897–904.
33. Schunck, E. (1857) 'On the occurrence of Indigo-blue in urine', *Philosophical Magazine* 14, pp. 288–96.
34. Jackson, A.H., Jenkins, R.T., Grinstein, M., Ferramola de Sancovich A.M. and Sancovich, H.A. (1988) 'The isolation and identification of indigoid pigments from urine', *Clinica Chimica Acta* 172, 2–3, pp. 245–52; Blanz, J., Ehninger, G. and Zeller, K.P. (1989) 'The isolation and identification of indigo and indirubin from urine of a patient with leukemia', *Research Communications in Chemical Pathology and Pharmacology* 64, 1, pp. 145–56; Cooksey, C.J. and Dronsfield, A.T. (2008) 'Indigo in urine – the blue ring test', *Education in Chemistry*, 45, pp. 45–7.
35. Schunck, E. (1879) 'Notes on the purple of the Ancients. 1. The chromogen of Purpura Capillus. 2. Properties of the colouring matter formed by insolation from the chromogen of Purpura Capillus.', *Journal of the Chemical Society* 35, pp. 589–96.
36. Schunck, E. (1880) 'Notes on the purple of the Ancients. 3. Purple dyeing in modern times', *Journal of the Chemical Society* 37, pp. 613–17.
37. Schunck published a note on modified chlorophyll in 1880 and many papers on its constitution and chemistry from 1884, the last appearing in 1902, the year before his death. These are listed in the Appendix, a bibliography of Schunck's papers.
38. 'Dr. Edward Schunck, F.R.S.' 1897 (see note 2 above).
39. 'Henry Edward Schunck' (by C.L.B.) (1902–03) *Memoirs and Proceedings of of the Manchester Literary and Philosophical Society* 47, pp. xlix–liii.
40. 'Henry Edward Schunck, 1820–1903' (by H.B.D[ixon]) (1904) *Obituary Notices of the Fellows of the Royal Society*, pp. 261–6.
41. Farrar, W.V. (1975) 'Examination of some compounds isolated from madder 1845–1855', *Ambix*, 22, 2, pp. 151–3; Farrar, W.V. (1977) 'Edward Schunck FRS: a pioneer of natural product chemistry', *Notes and Records of the Royal Society of London* 31, pp. 273–96.
42. 'Obituary. Dr. Edward Schunck, F.R.S.' (1903) *Journal of the Society of Dyers and Colourists* 19, pp. 35–6; 'Obituary. Henry Edward Schunck, Ph.D., D.Sc., F.R.S., &c' (1903) *Chemical News*, pp. 34–5; 'Dr. H. E. Schunck, F.R.S.' (by A.G.P[erkin]) (1903) *Nature* 67, 1734, p. 275; 'Henry Edward Schunck, Ph.D., D.Sc., F.R.S.' (1903) *Journal of the Society of Chemical Industry* 22, p. 84; 'Edward Schunck (F.R.S.)' (1903) *Berichte der Deutschen chemischen Gesellschaft* 36, pp. 305–6; 'Henry Edward Schunck, 1820–1903' (by H.B.D[ixon]) (1904) *Obituary Notices of the Fellows of the Royal Society*, pp. 261–6.
43. The portrait appears in 'Dr. Edward Schunck, F.R.S.' 1897 (see note 2 above); Farrar, W.V. (1977) 'Edward Schunck FRS: a pioneer of natural product chemistry', *Notes and Records of the Royal Society of London* 31, pp. 273–96.

Appendix

Papers by Henry Edward Schunck (1820–1903)

1. Schunck, E. (1841) 'Ueber einige durch die Einwirkung der Salpetersäure auf Aloë entstehende Producte', *Annalen der Chemie und Pharmazie* 39, pp. 1–25.
2. Schunck, E. (1842) 'On some of the substances contained in the lichens employed for the preparations of archil and cudbear', *Chemical Society Memoirs* 1, pp. 71–7.
3. Schunck, E. (1842) 'On some of the substances contained in the lichens employed for the preparations of archil and cudbear', *Philosophical Magazine* 20, pp. 495–500.
4. Schunck, E. (1842) 'Vorläufige Notiz über einige farbstoffgebende Substanzen der Flechten', *Annalen der Chemie und Pharmazie* 41, pp. 157–62.
5. Schunck, E. (1842) 'Vorläufige Notiz über einige farbstoffgebende Substanzen der Flechten', *Journal de pharmacie et de chimie* 1, pp. 206–10.
6. Schunck, E. (1843) 'Pseudoerythrin ist lecanorsäure Aethyloxid', *Annalen der Chemie und Pharmazie* 45, p. 250.

7. Schunck, E. (1844) 'On some products of the decomposition of erythrin', *British Association Reports* (1844), pp. 31–2.
8. Schunck, E. (1845) 'Ueber die Bestandtheile der Lecanora Parella', *Annalen der Chemie und Pharmazie* 54, pp. 257–84.
9. Schunck, E. (1845) 'Ueber die Bestandtheile der Lecanora Parella', *Journal de pharmacie et de chimie* 8, pp. 395–400.
10. Schunck, E. (1846) 'On the substances contained in the *Roccella tinctoria*', *Journal für praktische Chemie* 38, pp. 449–61.
11. Schunck, E. (1846) 'On the substances contained in the *Roccella tinctoria*', *Philosophical Magazine* 34, pp. 261–71.
12. Schunck, E. (1846) 'On the colouring matters of madder', *British Association Reports* (1846), pp. 24–7.
13. Schunck, E. (1847) 'On the substances contained in the *Roccella tinctoria*', *Journal de pharmacie et de chimie* 11, pp. 403–7.
14. Schunck, E. (1847) 'On the colouring matters of madder', *Philosophical Magazine* 31, 205, pp. 46–9.
15. Schunck, E. (1847) 'On the colouring matters of madder', *Journal für praktische Chemie* 42, pp. 13–18.
16. Schunck, E. (1847) 'Ueber die in Rocella tinctoria enthalteren Stoffe', *Annalen der Chemie und Pharmazie* 61, pp. 64–80.
17. Schunck, E. (1847) 'On colouring matters', *British Association Reports* (1847), pp. 136–45.
18. Schunck, E. (1848) 'On colouring matters', *Journal für praktische Chemie* 45, pp. 286–304.
19. Schunck, E. (1848) 'Ueber einige Zersetzungsprodukte der Chrysamminsäure', *Annalen der Chemie und Pharmazie* 65, pp. 234–44.
20. Schunck, E. (1848) 'On the substances contained in the *Roccella tinctoria*', *Chemical Society Memoirs and Proceedings* 3, pp. 144–54.
21. Schunck, E. (1848) 'On colouring matters', *Philosophical Magazine* 33, 220, pp. 133–45.
22. Schunck, E. (1848) 'Untersuchung der Krapps', *Annalen der Chemie und Pharmazie* 66, pp. 174–213.
23. Schunck, E. (1848) 'Ueber einige Zersetzungsprodukte der Chrysamminsäure', *Journal de pharmacie et de chimie* 13, pp. 299–302.
24. Schunck, E. (1848) 'On colouring matters', *British Association Reports* (1848), pp. 57–74.
25. Schunck, E. (1848) 'Some remarks on the substances discovered by Mr. Stenhouse in *Roccella tinctoria* and *Evernia prunastri*', *Philosophical Magazine* 33, pp. 249–60.
26. Schunck, E. (1849) 'Some remarks on the substances discovered by Mr. Stenhouse in *Roccella tinctoria* and *Evernia prunastri*', *Journal für praktische Chemie* 47, pp. 18–30.
27. Schunck, E. (1849) 'On colouring matters', *Philosophical Magazine* 35, 235, pp. 204–25.
28. Schunck, E. (1849) 'On colouring matters', *Journal für praktische Chemie* 48, pp. 299–328.
29. Schunck, E. (1851) 'On rubian and its products of decomposition', *Philosophical Transactions of the Royal Society* 141, pp. 433–60.
30. Schunck, E. (1851) 'On rubian and its products of decomposition', *Proceedings of the Royal Society of London Abstracts*, pp. 27–30.
31. Schunck, E. (1852) 'On rubian and its products of decomposition', *Philosophical Magazine* 3, pp. 213–26.
32. Schunck, E. (1852) 'On rubian and its products of decomposition', *Philosophical Magazine* 3, pp. 354–71.
33. Schunck, E. (1852) 'On rubian and its products of decomposition', *Annales de Chimie* 35, pp. 366–76.
34. Schunck, E. (1852) 'On rubian and its products of decomposition', *Annalen der Chemie und Pharmazie* 81, pp. 336–54.
35. Schunck, E. (1852) 'On rubian and its products of decomposition', *Journal für praktische Chemie* 55, pp. 490–99.
36. Schunck, E. (1852) 'On rubian and its products of decomposition', *Journal de pharmacie et de chimie* 22, pp. 134–7.
37. Schunck, E. (1853) 'On rubian and its products of decomposition part II', *Philosophical Transactions of the Royal Society* 143, pp. 67–107.
38. Schunck, E. (1853) 'On rubian and its products of decomposition', *Journal für praktische Chemie* 59, pp. 453–88.
39. Schunck, E. (1853) 'On rubian and its products of decomposition', *Philosophical Magazine* 5, pp. 410–36.
40. Schunck, E. (1853) 'On rubian and its products of decomposition', *Philosophical Magazine* 5, pp. 495–514.
41. Schunck, E. (1853) 'On rubian and its products of decomposition', *Annalen der Chemie und Pharmazie* 87, pp. 344–64.
42. Schunck, E. (1853) 'On the constitution of the colouring matters of madder', *Philosophical*

Magazine 6, pp. 187–96.
43. Schunck, E. (1854) 'On the constitution of the colouring matters of madder', *Journal für praktische Chemie* 61, pp. 65–73.
44. Schunck, E. (1854) 'On the action of the ferment of madder on sugar', *Memoirs of the Manchester Literary and Philosophical Society* 2nd series, 12, pp. 109–28.
45. Schunck, E. (1854) 'On the action of the ferment of madder on sugar', *Philosophical Magazine* 8, pp. 161–75.
46. Schunck, E. (1854) 'On the action of the ferment of madder on sugar', *Journal für praktische Chemie* 63, pp. 222–40.
47. Schunck, E. (1855) 'On the formation of indigo-blue. Part 1', *Memoirs of the Manchester Literary and Philosophical Society* 2nd series, 12, pp. 177–208.
48. Schunck, E. (1855) 'On the formation of indigo-blue. Part 1', *Journal für praktische Chemie* 66, pp. 321–34.
49. Schunck, E. (1855) 'On rubian and its products of decomposition', *Philosophical Transactions of the Royal Society* 145, pp. 389–419.
50. Schunck, E. (1855) 'On the formation of indigo-blue. Part I', *Philosophical Magazine* 10, 64, pp. 73–95.
51. Schunck, E. (1856) 'On the formation of indigo-blue. Part 2', *Memoirs of the Manchester Literary and Philosophical Society* 2nd series, 14, pp. 181–237.
52. Schunck, E. (1856) 'On rubian and its products of decomposition', *Journal für praktische Chemie* 67, pp. 154–7.
53. Schunck, E. (1856) 'On rubian and its products of decomposition. – Part III', *Proceedings of the Royal Society of London* 7, pp. 477–82.
54. Schunck, E. (1856) 'On rubian and its products of decomposition', *Philosophical Magazine* 12, pp. 200–220.
55. Schunck, E. (1856) 'On rubian and its products of decomposition', *Philosophical Magazine* 12, pp. 270–83.
56. Schunck, E. (1857) 'On rubian and its products of decomposition', *Journal für praktische Chemie* 70, pp. 154–80.
57. Schunck, E. (1857) 'On the occurrence of indigo-blue in urine', *Memoirs of the Manchester Literary and Philosophical Society* 2nd series, 14, pp. 239–49.
58. Schunck, E. (1857) 'On a yellow colouring matter obtained from the leaves of the *Polygonum fagopyrum* or common buckwheat (Abstract)', *Proceedings of the Manchester Literary and Philosophical Society* 1, p. 1.
59. Schunck, E. (1857) 'On the occurrence of indigo-blue in urine', *Philosophical Magazine* 14, pp. 288–96.
60. Schunck, E. (1858) 'On a yellow colouring matter obtained from the leaves of the *Polygonum fagopyrum* or common buckwheat', *Memoirs of the Manchester Literary and Philosophical Society* 2nd series, 15, pp. 122–9.
61. Schunck, E. (1858) 'On the occurrence of indigo-blue in urine', *Journal für praktische Chemie* 75, pp. 376–90.
62. Schunck, E. (1858) 'On the formation of indigo-blue. Part 2', *Philosophical Magazine* 15, pp. 29–45.
63. Schunck, E. (1858) 'On the formation of indigo-blue. Part 2', *Journal für praktische Chemie* 73, pp. 268–75.
64. Schunck, E. (1858) 'On the formation of indigo-blue. Part 2', *Journal für praktische Chemie* 74, pp. 99–102.
65. Schunck, E. (1858) 'On the formation of indigo-blue. Part 2', *Journal für praktische Chemie* 74, pp. 174–81.
66. Schunck, E. (1858) 'On the formation of indigo-blue. Part 2', *Philosophical Magazine* 15, pp. 117–33.
67. Schunck, E. (1858) 'On the formation of indigo-blue. Part 2', *Philosophical Magazine* 15, pp. 183–92.
68. Schunck, E. (1859) 'On the phenomena of fermentation', *Philosophical Magazine* 18, p. 340.
69. Schunck, E. (1860) 'On the colouring matters of madder', *Journal of the Chemical Society* (1860), pp. 198–221.
70. Schunck, E., Roscoe, H.E. and Smith, R.A. (1861) 'On the recent progress and present condition of manufacturing chemistry in the South Lancashire District', *British Association Reports* (1861), pp. 108–28.
71. Schunck, E. (1862) 'On a new source of indigo', *Proceedings of the Manchester Literary and Philosophical Society* 2, p. 181.
72. Schunck, E. (1862) 'On sugar in urine', *Philosophical Magazine* 23, pp. 179–81.
73. Schunck, E. (1865) 'On some products derived from indigo blue', *Proceedings of the Manchester Literary and Philosophical Society* 4, pp. 70–76.
74. Schunck, E. (1865) 'On some products derived from indigo blue', *Memoirs of the Manchester*

Literary and Philosophical Society 3, pp. 66–91.

75. Schunck, E. (1865) 'On some products derived from indigo blue', *Philosophical Magazine* 30, pp. 293–311.
76. Schunck, E. (1866) 'On some products derived from indigo blue', *Journal of the Chemical Society* (1866), pp. 462–76.
77. Schunck, E. (1867) 'On the colouring and extractive matters of urine', *Proceedings of the Royal Society of London* 15, pp. 1–4.
78. Schunck, E. (1867) 'On a crystalline fatty acid from human urine', *Proceedings of the Royal Society of London* 15, pp. 258–9.
79. Schunck, E. (1867) 'On oxalurate of ammonia as a constituent of human urine', *Proceedings of the Royal Society of London* 15, pp. 259–60.
80. Schunck, E. (1868) 'On a crystalline fatty acid from human urine', *Proceedings of the Royal Society of London* 16, pp. 135–9.
81. Schunck, E. (1868) 'On a crystalline fatty acid from human urine', *Proceedings of the Royal Society of London* 16, pp. 140–43.
82. Schunck, E. (1868) 'On some constituents of cotton-fibre', *Memoirs of the Manchester Literary and Philosophical Society* 3rd series, 4, pp. 95–128.
83. Schunck, E. (1868) 'On some constituents of cotton-fibre (Abstract)', *Proceedings of the Manchester Literary and Philosophical Society* 7, pp. 91–7.
84. Schunck, E. (1868) 'On the colouring and extractive matters of urine', *Proceedings of the Royal Society of London* 16, pp. 73–125.
85. Schunck, E. (1868) 'On the colouring and extractive matters of urine', *Proceedings of the Royal Society of London* 16, pp. 126–35.
86. Schunck, E. (1868) 'On some constituents of cotton fibre', *Chemical News* 17, pp. 118–19.
87. Schunck, E. (1869) 'On the chemical formula of alizarine', *Proceedings of the Manchester Literary and Philosophical Society* 8, pp. 173–5.
88. Schunck, E. (1869) 'On the chemical formula of alizarine', *Chemical News* 19, p. 193.
89. Schunck, E. (1869) 'On the chemical formula of alizarine', *Moniteur Scientifique* 11, pp. 463–4.
90. Schunck, E. (1869) 'On Vogelsang and Geissler's experiments on the nature of the liquids enclosed in certain minerals', *Proceedings of the Manchester Literary and Philosophical Society* 4, p. 4.
91. Schunck, E. (1870) 'On artificial alizarine', *Proceedings of the Manchester Literary and Philosophical Society*, 4, p. 106.
92. Schunck, E. (1870) 'On the chemical composition of cotton', *British Association Reports* (1870), p. 63.
93. Schunck, E. (1871) 'On some constituents of cotton-fibre', *Memoirs of the Manchester Literary and Philosophical Society* 4, pp. 95–128.
94. Schunck, E. (1871) 'On anthraflavic acid, a yellow colouring-matter accompanying artificial alizarin', *Proceedings of the Manchester Literary and Philosophical Society* 10, pp. 133–41.
95. Schunck, E. (1871) 'On anthraflavic acid', *Chemical News* 23, pp. 157–9.
96. Schunck, E. (1873) 'On methyl-alizarin and ethyl-alizarin', *Proceedings of the Manchester Literary and Philosophical Society* 12, pp. 86–91.
97. Schunck, E. (1873) 'On the colour of Nankin cotton', *Proceedings of the Manchester Literary and Philosophical Society* 13, pp. 22–9.
98. Schunck, E. (1874) 'On the colour of Nankin cotton', *Chemical News* 29, pp. 5–6.
99. Schunck, E. and Roemer, H. (1875) 'Ueber eine neue Säure isomer mit Alizarin. (Vorläufige Mittheilung)', *Berichte der Deutschen chemischen Gesellschaft* 8, pp. 1628–9.
100. Schunck, E. (1876) 'On anthraflavic acid, a yellow colouring-matter accompanying artificial alizarin', *Memoirs of the Manchester Literary and Philosophical Society* 3rd series, 5, pp. 227–35.
101. Schunck, E. and Roemer, H. (1876) 'Ueber Anthraflavinsäure und Isoanthraflavinsäure', *Berichte der Deutschen chemischen Gesellschaft* 9, pp. 379–83.
102. Schunck, E. (1876) 'On methyl–alizarin and ethyl–alizarin', *Memoirs of the Manchester Literary and Philosophical Society* 3rd series, 5, pp. 236–41.
103. Schunck, E. (1876) 'On the colour of Nankin cotton', *Memoirs of the Manchester Literary and Philosophical Society* 3rd series, 5, pp. 362–9.
104. Schunck, E. and Roemer, H. (1876) 'On some isomerides of alizarine', *Proceedings of the Manchester Literary and Philosophical Society* 15, pp. 142–9.
105. Schunck, E. and Roemer, H. (1876) 'On some isomerides of alizarine', *Chemical News* 33, pp. 233–4.
106. Schunck, E. and Roemer, H. (1876) 'Ueber Anthrapurpurin und Flavopurpurin', *Berichte der Deutschen chemischen Gesellschaft* 9, pp. 678–82.

107. Schunck, E. and Roemer, H. (1876) 'Ueber Anthrapurpurin und Flavopurpurin', *Chemical News* 34, pp. 131–2.
108. Schunck, E. and Roemer, H. (1877) 'Notes on madder colouring matters', *Journal of the Chemical Society* (1877), pp. 665–75.
109. Schunck, E. and Roemer, H. (1877) 'Ueber Purpuroxanthincarbonsäure, eine neue, das natürliche Purpurin begleitende Substanz', *Berichte der Deutschen chemischen Gesellschaft* 10, pp. 172–5.
110. Schunck, E. and Roemer, H. (1877) 'Ueber den Nachweis geringer Mengen von Alizarin im Purpurin', *Berichte der Deutschen chemischen Gesellschaft* 10, pp. 175–6.
111. Schunck, E. and Roemer, H. (1877) 'Zur Kenntniss des Purpurins. Verwandlung des Purpurins in Chinizarin', *Berichte der Deutschen chemischen Gesellschaft* 10, pp. 550–56.
112. Schunck, E. and Roemer, H. (1877) 'Zur Kenntniss des Purpurins. Verwandlung des Purpurins in Chinizarin', *Journal of the Chemical Society* (1877), pp. 624–5.
113. Schunck, E. and Roemer, H. (1877) 'Ueber Munjistin, ε–Purpurin und Purpuroxanthincarbonsäure', *Berichte der Deutschen chemischen Gesellschaft* 10, pp. 790–92.
114. Schunck, E. and Roemer, H. (1877) 'Ueber Munjistin, ε–Purpurin und Purpuroxanthincarbonsäure', *Journal of the Chemical Society* (1877), pp. 788–9.
115. Schunck, E. and Roemer, H. (1877) 'Ueber Anthraflavon und über ein neues Bioxyanthrachinon', *Berichte der Deutschen chemischen Gesellschaft* 10, pp. 1225–7.
116. Schunck, E. and Roemer, H. (1877) 'Zur Kenntniss des Flavopurpurins', *Berichte der Deutschen chemischen Gesellschaft* 10, pp. 1821–3.
117. Schunck, E. and Roemer, H. (1878) 'Notes on madder colouring matters', *Journal of the Chemical Society* (1878), pp. 422–5.
118. Schunck, E. and Roemer, H. (1878) 'Ueber den Vergleich des ε–Purpurins mit der Purpuroxanthincarbonsäure und über Anthraflavon', *Berichte der Deutschen chemischen Gesellschaft* 11, pp. 431–4.
119. Schunck, E. and Roemer, H. (1878) 'Ueber Metabenzbioxyanthrachinon und Anthraflavinsäure aus Metaoxybenzoësäure', *Berichte der Deutschen chemischen Gesellschaft* 11, pp. 969–73.
120. Schunck, E. and Roemer, H. (1878) 'Ueber Anthrarufin, ein neues Bioxyanthrachinon aus der Metaoxybenzoësäure', *Berichte der Deutschen chemischen Gesellschaft* 11, pp. 1176–9.
121. Schunck, E. (1878) 'On indigo-blue from *Polygonum tinctorium* and other plants (Abstract)', *Proceedings of the Manchester Literary and Philosophical Society* 17, pp. 157–9.
122. Schunck, E. and Roemer, H. (1879) 'On some isomerides of alizarine', *Memoirs of the Manchester Literary and Philosophical Society* 3rd series, 6, pp. 37–48.
123. Schunck, E. (1879) 'On indigo-blue from *Polygonum tinctorium* and other plants', *Memoirs of the Manchester Literary and Philosophical Society* 3rd series, 6, pp. 218–34.
124. Schunck, E. (1879) 'Notes on the purple of the Ancients. 1. The chromogen of Purpura Capillus; 2. Properties of the colouring matter formed by insolation from the chromogen of Purpura Capillus.', *Journal of the Chemical Society* (1879), pp. 589–96.
125. Schunck, E. (1879) 'On indigo-purpurin and indirubin', *Journal of the Chemical Society* (1879), pp. 528–30.
126. Schunck, E. (1879) 'On indigo-blue from *Polygonum tinctorium* and other plants', *Chemical News* 39, pp. 119–20.
127. Schunck, E. (1879) 'On indigo–blue from *Polygonum tinctorium* and other plants', *Chemical News* 39, pp. 129–30.
128. Schunck, E. (1879) 'On indigo–blue from *Polygonum tinctorium* and other plants', *Chemical News* 39, pp. 143–4.
129. Schunck, E. (1879) 'On indigo–blue from *Polygonum tinctorium* and other plants', *Moniteur Scientifique* 21, pp. 442–9.
130. Schunck, E. and Roemer, H. (1879) 'Ueber die Zersetzung der Rubiansäure durch Salzsäure in Gegenwart von Quecksilber', *Berichte der Deutschen chemischen Gesellschaft* 12, pp. 2312–13.
131. Schunck, E. and Roemer, H. (1879) 'Ueber die Zersetzung des Pflanzenindicans bei Abschlus der Luft', *Berichte der Deutschen chemischen Gesellschaft* 12, pp. 2311–12.
132. Schunck, E. and Römer, H. (1879) 'Ueber Nitroalizarin', *Berichte der Deutschen chemischen Gesellschaft* 12, p. 1008.
133. Schunck, E. and Roemer, H. (1879) 'Ueber α- und β-Nitroalizarin und β-Amidoalizarin', *Berichte der Deutschen chemischen Gesellschaft* 12, pp. 583–9.

134. Schunck, E. (1880) 'Note on modified chlorophyll from the leaves of Eucalyptus globulus', *Proceedings of the Manchester Literary and Philosophical Society* 19, pp. 157–60.
135. Schunck, E. (1880) 'Note on modified chlorophyll from the leaves of *Eucalyptus globulus*', *Chemical News* 42, pp. 31–2.
136. Schunck, E. and Roemer, H. (1880) 'Ueber die Erkennung des Alizarins, des Isopurpurins sowie des Flavopurpurins, wenn sie neben einander Vorkommen, und über die quantitative Bestimmung des Alizarins', *Berichte der Deutschen chemischen Gesellschaft* 13, pp. 41–3.
137. Schunck, E. (1880) 'Notes on the purple of the Ancients. 3. Purple dyeing in modern times', *Journal of the Chemical Society* (1880), pp. 613–17.
138. Schunck, E. (1882) 'Remarks on the terms used to denote colour, and on the colours of faded leaves', *Chemical News* 21, pp. 43–56.
139. Schunck, E. (1882) 'Remarks on the terms used to denote colour, and on the colours of faded leaves', *Chemical News* 45, pp. 17–20.
140. Schunck, E. (1884) 'Remarks on the terms used to denote colour, and on the colours of faded leaves', *Memoirs of the Manchester Literary and Philosophical Society* 3rd series, 8, pp. 26–40.
141. Schunck, E. (1884) 'Note on the constitution of chlorophyll', *Proceedings of the Royal Society of London* 36, pp. 183–5.
142. Schunck, E. (1884) 'Supplementary note on the constitution of chlorophyll', *Proceedings of the Royal Society of London* 36, pp. 183, 285–6.
143. Schunck, E. (1885) 'Memoir of Robert Angus Smith, Ph.D., LL.D., F.R.S., F.C.S., &c.', *Memoirs of the Manchester Literary and Philosophical Society*, 3rd series, 10, p. 90.
144. Schunck, E. (1885) 'Memoir of Robert Angus Smith, Ph.D., LL.D., F.R.S., F.C.S., &c.', *Proceedings of the Manchester Literary and Philosophical Society* 24, p. 97.
145. Schunck, E. (1886) 'Contributions to the chemistry of chlorophyll. I', *Proceedings of the Royal Society of London* 39, pp. 348–61.
146. Schunck, E. (1887) 'Contributions to the chemistry of chlorophyll. II', *Proceedings of the Royal Society of London* 42, pp. 184–8.
147. Schunck, E. (1887) 'Chief characteristics of the chemistry of 1837 as compared with that of 1887', *British Association Reports* (1887), pp. 624–35.
148. Schunck, E. (1887) 'Remarks on dyeing with chlorophyll', *Journal of the Society of Chemical Industry* 6, p. 413.
149. Schunck, E. (1888) 'On the supposed identity of rutin and quercitrin', *Journal of the Chemical Society* (1888), pp. 262–7.
150. Schunck, E. (1888) 'Contributions to the chemistry of chlorophyll. III', *Proceedings of the Royal Society of London* 44, pp. 448–54.
151. Schunck, E. (1888) 'Some remarks on the theory of dyeing', *Journal of the Society of Chemical Industry* 7, pp. 815–19.
152. Schunck, E. (1889) 'On the green colouring matter from leaves found in one of the cuttings for the Manchester Ship Canal', *Memoirs of the Manchester Literary and Philosophical Society* 4th series, 2, pp. 231–3.
153. Schunck, E. (1890) 'On Dr. Gerhard Kruss's alleged discovery of a foreign substance in commercial nickel', *Memoirs of the Manchester Literary and Philosophical Society* 4th series, 3, p. 170.
154. Schunck, E. (1890) 'On a new compound of nitrogen', *Memoirs of the Manchester Literary and Philosophical Society* 4th series, 4, p. 16.
155. Schunck, E. (1892) 'Notes on some ancient dyes', *Memoirs of the Manchester Literary and Philosophical Society* 4th series, 5, pp. 158–61.
156. Schunck, E. (1892) 'On a copy of Colonel Drinkwater's "History of the Siege of Gibraltar"', *Memoirs of the Manchester Literary and Philosophical Society* 4th series, 5, pp. 139, 141.
157. Schunck, E. (1892) 'On a specimen of woollen cloth of Egyptian production of the fifth century', *Memoirs of the Manchester Literary and Philosophical Society* 4th series, 5, p. 172.
158. Schunck, E. (1892) 'Contributions to the chemistry of chlorophyll. IV', *Proceedings of the Royal Society of London* 50, pp. 302–17.
159. Schunck, E. and Brebner, G. (1892) 'On the action of aniline on green leaves and other parts of plants', *Annals of Botany* 6, pp. 167–84.
160. Schunck, E. and Marchlewski, L. (1892) 'Ueber die Einwirkung von Wasserstoffsuperoxide auf Aniline', *Berichte der Deutschen chemischen Gesellschaft* 25, pp. 3574–7.
161. Schunck, E. and Marchlewski, L. (1893) 'Zur Kenntniss des Phlorose', *Berichte der Deutschen chemischen Gesellschaft* 26, pp. 942–3.
162. Schunck, E. (1893) 'Extract from a letter by Prof. Flinders Petrie on some Egyptian fabrics',

Memoirs of the Manchester Literary and Philosophical Society 4th series, 7, p. 210.
163. Schunck, E. and Marchlewski, L. (1893) 'Supplementary notes on madder colouring matters', *Journal of the Chemical Society* (1893), pp. 969–74.
164. Schunck, E. and Marchlewski, L. (1893) 'Studien über Datiscin und seine Spaltungsprodukte', *Liebig's Annalen der Chemie* 277, pp. 261–76.
165. Schunck, E. and Marchlewski, L. (1894) 'Zur Kenntniss des Anthrachinonoxims', *Berichte der Deutschen chemischen Gesellschaft* 27, pp. 2125–6.
166. Schunck, E. and Marchlewski, L. (1894) 'Zur Kenntniss der Carminsäure', *Berichte der Deutschen chemischen Gesellschaft* 27, pp. 2979–85.
167. Schunck, E. and Marchlewski, L. (1894) 'Ueber die Einwirkung von Brom auf Datiscetin', *Liebig's Annalen der Chemie* 278, pp. 346–9.
168. Schunck, E. and Marchlewski, L. (1894) 'Studien über einige naturliche Zuckerarten', *Liebig's Annalen der Chemie* 278, pp. 349–59.
169. Schunck, E. and Marchlewski, L. (1894) 'Zur Kenntniss des Naphthazarins und über die Bildung der Naphtocyaninsäure', *Berichte der Deutschen chemischen Gesellschaft* 27, pp. 3462–5.
170. Schunck, E. and Marchlewski, L. (1894) 'Notes on madder colouring matters', *Journal of the Chemical Society* (1894), pp. 182–7.
171. Schunck, E. and Marchlewski, L. (1894) 'Dyeing with glucosides and rhamnosides', *Journal of the Society of Dyers and Colourists* 10, pp. 95–6.
172. Schunck, E., Knecht, E. and Marchlewski, L. (1894) 'Ueber einen in den Rebenblattern vorkommenden Farbstoffe. [Vorläufige Mittheilung]', *Berichte der Deutschen chemischen Gesellschaft* 27, pp. 487–8.
173. Schunck, E., Knecht, E. and Marchlewski, L. (1894) 'Ueber einen in den Rebenblattern vorkommenden Farbstoffe. [Vorläufige Mittheilung]', *Journal of the Society of Dyers and Colourists* 10, pp. 74–5.
174. Schunck, E. and Marchlewski, L. (1894) 'Zur Chemie des Chlorophylls', *Liebig's Annalen der Chemie* 278, pp. 329–45.
175. Schunck, E. (1894) 'Contributions to the chemistry of chlorophyll. V', *Proceedings of the Royal Society of London* 55, pp. 351–6.
176. Schunck, E. and Marchlewski, L. (1895) 'Zur Kenntniss der rothen Isomeren des Indigotins und uber einige Derivate des Isatins', *Berichte der Deutschen chemischen Gesellschaft* 28, pp. 539–47, 2525–31.
177. Schunck, E. (1895) 'The yellow colouring matter of *Sophora japonica*', *Journal of the Chemical Society* (1895), pp. 30–32.
178. Schunck, E. and Marchlewski, L. (1895) 'Zur Chemie des Chlorophylls', *Liebig's Annalen der Chemie* 284, pp. 81–107.
179. Schunck, E. and Marchlewski, L. (1895) 'Zur Chemie des Chlorophylls', *Liebig's Annalen der Chemie* 288, pp. 209–18.
180. Schunck, E. (1896) 'On carminic acid', *Memoirs of the Manchester Literary and Philosophical Society* 41, p. xiv.
181. Schunck, E. and Marchlewski, L. (1896) 'Zur Kenntniss des Isatins', *Berichte der Deutschen chemischen Gesellschaft* 29, pp. 194–203.
182. Schunck, E. and Marchlewski, L. (1896) 'Zur Chemie des Chlorophylls', *Liebig's Annalen der Chemie* 290, pp. 306–13.
183. Schunck, E. and Marchlewski, L. (1896) 'Zur Chemie des Chlorophylls', *Berichte der Deutschen chemischen Gesellschaft* 29, pp. 1347–52.
184. Schunck, E. (1896) 'Contributions to the chemistry of chlorophyll. VI', *Proceedings of the Royal Society of London* 57, pp. 314–22.
185. Schunck, E. and Marchlewski, L. (1896) 'Contributions to the chemistry of chlorophyll. VII. Phylloporphyrin and haematoporphyrin', *Proceedings of the Royal Society of London* 59, pp. 233–5.
186. Schunck, E. (1896) 'Some derivatives of anthraquinone', *Journal of the Chemical Society* 69, pp. 68–74.
187. Schunck, E. (1897) [on recollections of Dalton & others] *Journal of the Society of Chemical Industry* 17, p. 586.
188. Schunck, E. (1898) 'Alkoholische Gahrung ohne Hefezellung', *Berichte der Deutschen chemischen Gesellschaft* 31, p. 309.
189. Schunck, E. (1900) 'Recent researches on indican', *Chemical News* 82, pp. 176–7.
190. Schunck, E. (1901) 'The action of reagents on the leaves of *Polygonum tinctorium*', private publication.
191. Schunck, E. (1902) 'Contributions to the chemistry of chlorophyll. No. VIII. – Changes undergone by chlorophyll in passing through the bodies of animals', *Proceedings of the Royal Society of London* 69, pp. 307–12.

Papers about Schunck

192. Farrar, W.V. (1975) 'Examination of some compounds isolated from madder 1845–1855', *Ambix* 22, 2, pp. 151–3.
193. Farrar, W.V. (1977) 'Edward Schunck FRS: a pioneer of natural product chemistry', *Notes and Records of the Royal Society of London* 31, pp. 273–96.
194. 'Obituary. Dr. Edward Schunck, F.R.S.' (1903), *Journal of the Society of Dyers and Colourists* 19, pp. 35–6.
195. James, T.E. (1920) 'Schunck, Henry Edward', *Dictionary of National Biography, Second Supplement 1901–1911*, London, pp. 274–5.
196. Partington, J.R. (1970) *History of Chemistry*, 4 vols, London, vol. 4, pp. 397–400. (Reprint of 1964 edition.)
197. Farrar, W.V. (1970–76) 'Schunck, Henry Edward', *Dictionary of Scientific Biography*, New York, pp. 236–7.
198. 'Dr. Edward Schunck, F.R.S.' (1897), *Manchester Faces and Places* 9, 1, pp. 1–5 (with portrait).
199. 'Obituary. Henry Edward Schunck, Ph.D., D.Sc., F.R.S., &c' (1903), *Chemical News*, pp. 34–5.
200. 'Henry Edward Schunck' (by C.L.B.) (1902–1903) *Memoirs and Proceedings of of the Manchester Literary and Philosophical Society* 47, pp. xlix–liii.
201. Perkin, W.H., Jr. (1902–1903) 'The chemical researches of Edward Schunck', *Memoirs and Proceedings of of the Manchester Literary and Philosophical Society* 47, 6, pp. 1–8.
202. Dixon, H.B., 'Dr. Edward Schunck, 1820 – 1903', private publication.
203. 'Dr. H. E. Schunck, F.R.S.' (by A.G.P[erkin]) (1903) *Nature* 67, 1734, p. 275.
204. 'Henry Edward Schunck, Ph.D., D.Sc., F.R.S.' (1903) *Journal of the Society of Chemical Industry* 22, p. 84.
205. 'Edward Schunck (F.R.S.)' (1903) *Berichte der Deutschen chemischen Gesellschaft* 36, pp. 305–6.
206. 'Henry Edward Schunck, 1820 – 1903' (by H.B.D[ixon]) (1904) *Obituary Notices of the Fellows of the Royal Society*, pp. 261–6.
207. Kargon, R.H. (1977) *Science in Victorian Manchester*, Manchester, p. 95.
208. Farrar, W.V. (1997) *Chemistry and the Chemical Industry in the 19th Century: the Henrys of Manchester and other Studies ...*, eds R.L. Hills and W.H. Brock, Aldershot (Variorum).

Spectroscopic Studies (FTIR, SIMS, ES–MS) on the Structure of Anthraquinone–Aluminium Complexes

Jana Sanyova

Abstract The structures of the complexes of alizarin and other anthraquinones (purpurin and quinizarin) with aluminium, responsible for the properties of madder lakes, have been studied using different spectrometric methods: Fourier transform infrared spectroscopy (FTIR), secondary ion mass spectrometry (SIMS) and electrospray mass spectrometry (ES–MS). FTIR unfortunately gives very little information on the structure of the complexes. SIMS and ES–MS revealed new evidence in favour of the O-1/O-9 complexation site proposed in the Kiel and Heertjes model, but suggest a different alizarin:Al stoichiometry (2:1 instead of 4:2) with possible further aggregation of these 2:1 complexes through various mechanisms.

Introduction

Alizarin (II, Fig. 1) and aluminium form the complexes which are the foundation of the colour and other properties of madder lakes. Their exact structure is still not ascertained, and, even though the model of Kiel and Heertjes with its bidentate O-1/O-9 (Fig. 2a) complexation

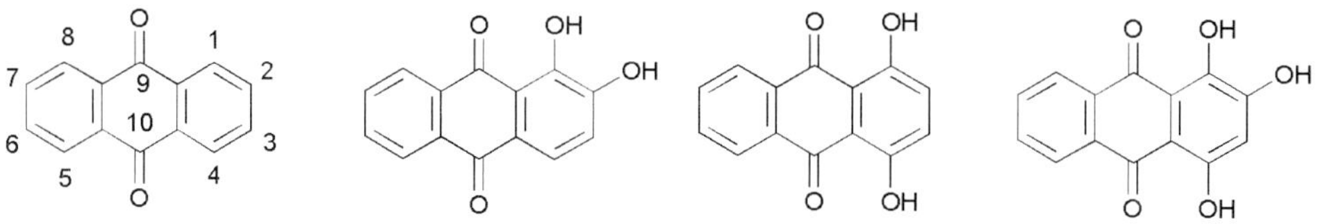

I II III IV

Figure 1 The structures of anthraquinone (I), alizarin (II), quinizarin (III) and purpurin (IV).

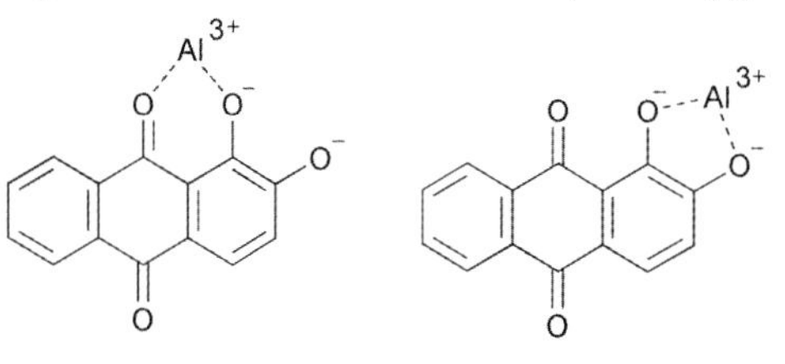

(a) (b)

Figure 2 Two possible coordination sites: (a) O-1/O-9, as in the Kiel and Heertjes model; (b) O-1/O-2, as in the Wunderlich model.

sites is often considered as the standard structure, experimental evidence from X-ray diffraction (XRD) patterns in favour of a different coordination site (O-1/O-2) (Fig. 2b) has more recently been published.[1]

One of the reasons for this continuing uncertainty is that, with many techniques, the observable complexes are necessarily different from those in real (solid) lakes: for example, crystalline for X-rays, ionised into the gas phase for mass spectrometry. In the work presented here, anthraquinone–aluminium complexes were studied using both 'solid state' (infrared spectroscopy) and 'extractive' (mass spectrometry) methods.

Materials and methods

Lakes were prepared from synthetic alizarin (Sigma, 98%), purpurin (Sigma, 98%) and quinizarin (Extrasynthèse, 97%) using methods adapted from 17th–18th-century madder lake recipes.[2]

FTIR spectra of anthraquinones and of their aluminium lakes in potassium bromide (KBr) discs were measured using a Nicolet 5DXC spectrometer. Mass spectra of complexes observed at the surface of solid lakes were obtained using a ToF-SIMS (time-of-flight secondary ion mass spectrometry) model instrument from Charles Evans & Associates. SIMS analyses the secondary ions ejected from a small spot at the surface of a solid by a high-energy primary ion beam. The experimental conditions are such as described by Poleunis *et al.* (1997).[3]

Mass spectra of complexes formed in alizarin–aluminium solutions which had gelled were obtained using an ES–MS 'LCQ-Deca' instrument from Finnigan (Thermo-Electron Corporation). ES–MS is a milder technique than SIMS and therefore gives a better chance of observing intact complexes rather than rearranged ions. The solution of 1:2:5 aluminium:alizarin:sodium hydroxide (NaOH), using 50 mM alizarin, which had turned into a loose gel some days after the preparation, was analysed twice: first in the gel-forming stage, secondly after dilution 1:1 with methanol and sonication. The sample was injected at 3 μL min^{-1} (nano-electrospray), and positive- and negative-ion mode spectra were recorded in the 200–4000 amu range.[4]

Results and discussion

Fourier transform infrared spectroscopy

FTIR has the advantage of allowing the study of complexes '*in situ*' (in the solid state), but unfortunately the differences observed between the spectra of pure anthraquinone (Fig. 3a), alizarin (Fig. 3b), quinizarin (Fig. 3c) and lakes of alizarin (Fig. 3d) are inconclusive. The IR spectrum of unsubstituted anthraquinone (I, Fig. 1) shows only one carbonyl absorption band near 1673 cm^{-1} (Fig. 3a): as expected, the two carbonyls are equivalent to each other and absorb at the same wavelength.

In the spectrum of alizarin, 1,2-dihydroxyanthraquinone (II, Fig. 1), two carbonyl bands are observed, near 1660 and 1634 cm^{-1} respectively (Fig. 3b): the two carbonyls now absorb at different wavelengths, due to the difference in their positions relative to the two phenols on C-1 and C-2. Some authors have attributed the larger of these two shifts (about –39 cm^{-1}) to

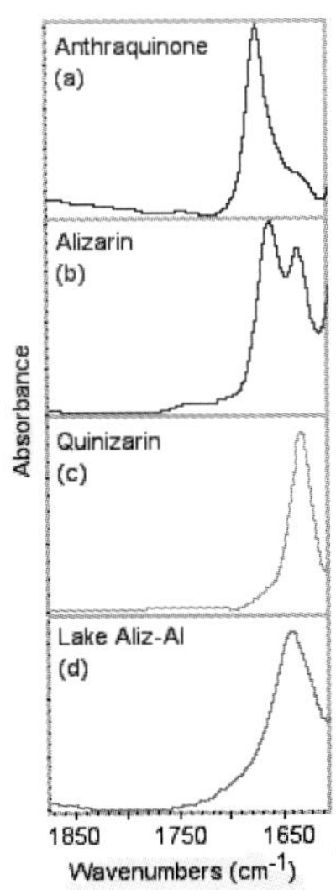

Figure 3 Infrared spectra of (a) anthraquinone; (b) alizarin; (c) quinizarin; (d) alizarin–aluminium lake, in KBr discs.

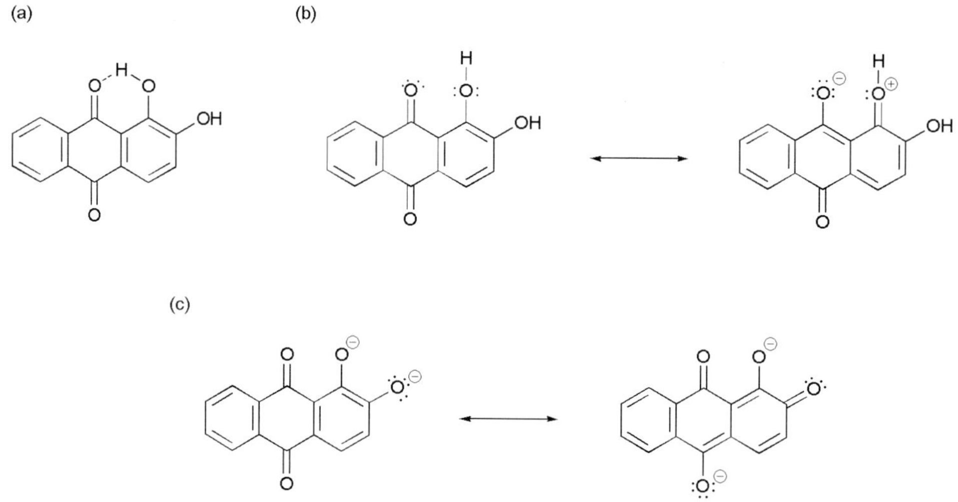

Figure 4 Possible causes for the IR band shifts: (a) 'sharing' of the proton on O-1 by the O-9 carbonyl; (b) mesomeric +M (electron donor) influence of O-1 hydroxyl on O-9 carbonyl; (c) in aluminium–anthraquinone lakes, mesomeric +M (electron donor) influence of deprotonated O-2 hydroxyl on O-10 carbonyl.

a hydrogen bond between the carbonyl on C-9 and the phenol on C-1, but Bloom *et al.* (1959) showed that the main contribution must be the mesomeric +M electron-donor effect of the C-1 phenol on the C-9 carbonyl (Fig. 4b).[5] Their argument is based on their observation that, in the spectrum of 2-amino-anthraquinone, a large shift (from 1975 to 1625 cm^{-1}) is also observed, even though the amine on C-2 is too far removed from any of the carbonyls for hydrogen-bonding. This shift must be due to the electron donor effect of the amino group, transmitted to the C-10 carbonyl through resonance. Amines are much stronger electron donors than hydroxyls, which explains why the shifts we observed with alizarin are smaller: about –39 for the effect of the C-1 phenol on the C-9 carbonyl, and –13 cm^{-1} for that of the C-2 phenol on the further removed C-10 carbonyl.

With quinizarin, 1,4-dihydroxyanthraquinone (III, Fig. 1) and purpurin, 1,2,4-trihydroxyanthraquinone (IV, Fig. 1), only one band is observed, at about 1621 cm^{-1} (Fig. 3c). Both carbonyls, on C-9 and C-10, have now shifted to lower wavelengths, because they are in resonance with the phenols on C-1 and C-4 respectively.

The wavelengths we observed for anthraquinone, alizarin and purpurin are within 2 cm^{-1} of those reported by Bloom.[6]

In the spectrum of alizarin lakes (Fig. 3d), only one band is observed between about 1634 cm^{-1} and about 1641 cm^{-1}. The discussion above shows that this second shift, relative to alizarin, must involve the second carbonyl on C-10 and must be related to the second phenol on C-2. This is best explained by the dissociation of the second phenol into phenolate. However, this does not permit a decision to be made between the Kiel and Heertjes or the Wunderlich models for complex formation, since the phenol on C-2 becomes a phenolate in both models.[7]

Secondary ion mass spectrometry

In SIMS, the sample surface is struck by high-energy ions, and the mass/charge ratio (m/z) of the secondary ions ejected is analysed. With alizarin lakes, the main ions detected in positive and negative modes have m/z of +505 and –503 respectively (Fig. 5a, b), indicating a 2:1 alizarin:Al ratio. The –503 ion is compatible with both models (Fig. 6b, c), whereas the +505

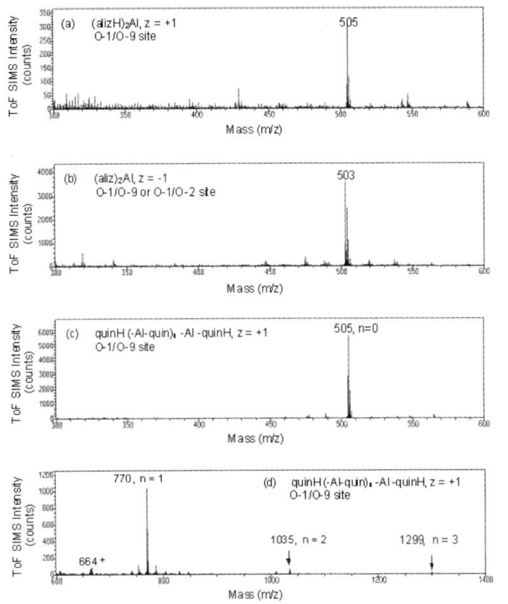

Figure 5 ToF-SIMS spectra of alizarin lakes in (a) positive and (b) negative ion mode, and of quinizarin lake in positive ion mode at (c) low and (d) medium mass ranges.

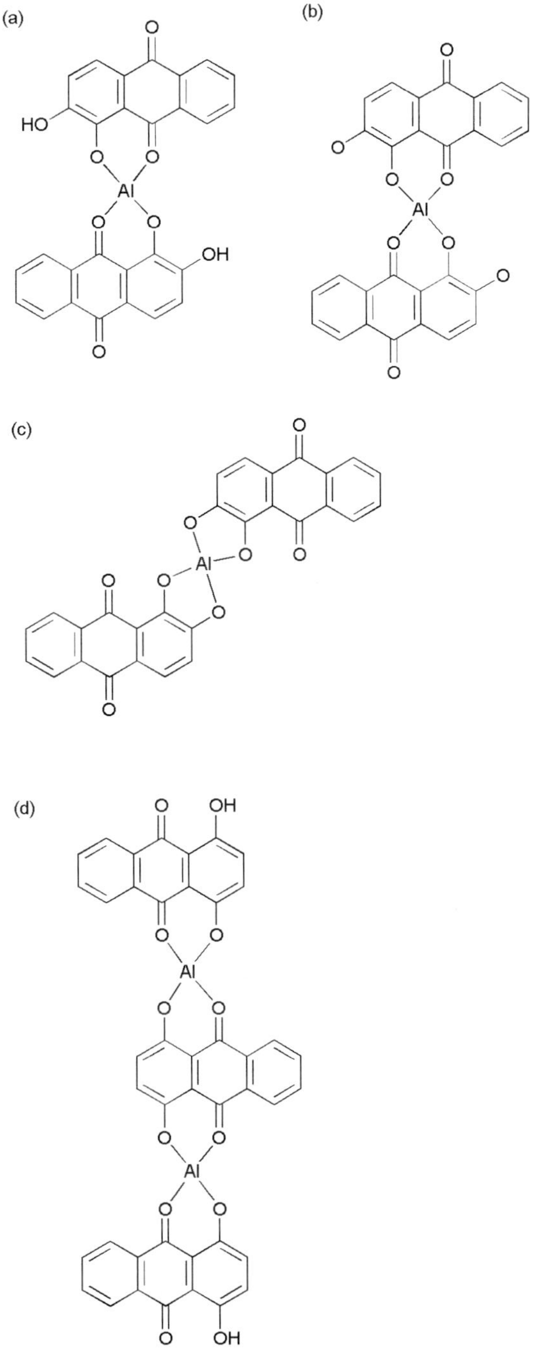

Figure 6 Possible structures for the ions detected by SIMS. (a) $(alizH)_2Al$, m/z = 505, z = +1, O-1/O-9 site; (b) $(aliz)_2Al$, m/z = 503, z = -1, O-1/O-9 site; (c) $(aliz)_2Al$, m/z = 503, z = -1, O-1/O-2 site; (d) (quinH-Al-quin-Al-quinH), m/z = 770, z = +1, O-1/O-9 site.

ion seems to imply that O-2 is still protonated and therefore the site for complex formation is O-1/O-9 (Fig. 6a); however both ions could also result from rearrangements under the harsh ionisation conditions. Quinizarin lakes (Fig. 6d) give a regular series of higher m/z ions (505 + n.265, Fig. 5c, d), attributed to 'oligomers' qu–(Al–qu)$_n$, made possible by the presence of two equivalent sites per molecule (O-1/O-9 and O-4/O-10). The net positive charge should normally increase with the degree of polymerisation, but unexpected charges are common with SIMS.

No significant peak was obtained from purpurin lakes. It is possible that the third phenol group binds to the alumina substrate, or allows the formation of a number of non-linear complexes which are more difficult to desorb from the alumina surface.

Electrospray mass spectrometry

The milder technique ES–MS is not suitable for solids, but has the advantage that it is more

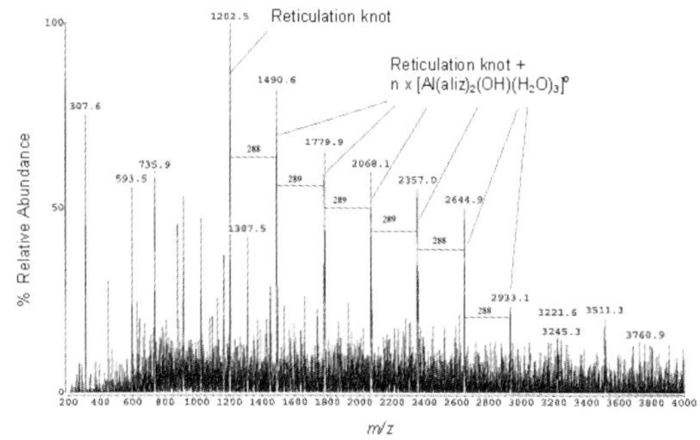

Figure 7 Positive ion mode ES–MS spectrum of a gelling alizarin–aluminium solution.

probable that the ions detected are such as they were in solution (Fig. 6). We applied it to (1:2) stoichiometric solutions of aluminium and alizarin, which were slowly turning into gels. After dilution and sonication, these solutions gave the same +505 and –503 ions as in SIMS. But unsonicated (gelling) solutions gave, in positive ion mode, a series of peaks, m/z = 1202.5 + n.288 (Fig. 7), interpreted as fragments of a three-dimensional gel network, with z = +2 and m = 2505 + n.576, where 576 is the mass of a 'monomer'. This monomer is composed of a $(alizH^-)_2Al^{3+}$ complex (m=505) and a hydroxyl – OH^- – group acting as a crosslink with the next complex, plus three water molecules. The first peak at m/z = 2505/2 would then be the basic 'reticulation knot' of the three-dimensional network. The structure that best fits the results involves two parallel oligomeric strands, each composed of two $(alizH^-)_2Al^{3+}$ complexes, which bond to each other by interpenetration of their aromatic side-rings (Plate 51).

Conclusion

Although the results presented here do not allow any definitive conclusion to be drawn about the coordination site, they are rather in favour of the O-1/O-9 site. They further suggest that, in contrast to the 4:2 stoichiometry proposed by Kiel and Heertjes, the basic stoichiometry of the complexes in the solid lakes and in concentrated solutions is 2:1, even though, in stoichiometric solutions, further aggregation of the complexes through Al–O–Al bonds and cross-linking through aromatic ring interactions are possible.

Acknowledgements

The author wishes to thank Professor Reiss (ULB, Service de Chimie Organique) and Dr Masschelein-Kleiner (Director of IRPA-KIK until 2003), who supervised the PhD thesis, part of which is presented in this poster. The help of Professor Bertrand and M. Poleunis (UCL-PCPM), for the acquisition and interpretation of the SIMS spectra, is also gratefully acknowledged.

Editor's note

This paper was presented as a poster at the Meeting. Address for correspondence: Institut Royal du Patrimoine Artistique, Parc du Cinquantenaire 1, B-1000 Brussels, Belgium (jana.sanyova@kikirpa.be). Paper received 17 March 2003; revised version received 23 February 2004.

Notes and references

1. Kiel, E.G. and Heertjes, P.M. (1963a) 'Metal complexes of alizarin. I. The structure of the calcium-aluminium lake of alizarin', *Journal of the Society of Dyers and Colourists* 79, pp. 21–7; Kiel, E.G. and Heertjes, P.M. (1963b) 'Metal complexes of alizarin. II. The structure of some metal complexes of alizarin other than Turkey red', *Journal of the Society of Dyers and Colourists* 79, pp. 61–4; Wunderlich, C.-H. and Bergerhoff, G. (1994) 'Konstitution und Farbe von Alizarin- und Purpurin-Farblacken', *Chemische Berichte* 127, pp. 1185–90.
2. Neri, A. (1612) *L'arte vetraria distinta in libri sette del R.P. Antonio Neri, fiorentino ...*, Florence, p. 102; Diderot, D. and D'Alembert, M. (1967) *Supplément à l'encyclopédie ou dictionnaire raisonné des sciences, des arts et des métiers, par une société de gens de lettres*, vol. 3, Amsterdam 1777 (facsimile reprint Stuttgart-Bad Cannstatt),

p. 689; Sanyova, J. (2001) *Contribution à l'étude de la structure et des propriétés des laques de garance*, PhD thesis, Université Libre de Bruxelles, pp. 199–208.

3. Poleunis, C., De Smet, F., Devillers, M. and Bertrand, P. (1997) 'ToF-SIMS study of Pr_6O_{11}-MoO_3 mixtures: effects of the ultrasonic treatment during the preparation', in G. Gillen, R. Lareau, J. Bennett and F. Stevie (eds), *Proceedings of the Eleventh International Conference on Secondary Ion Mass Spectrometry, SIMS XI, Orlando, USA, 7–11 September 1997*, Chichester, pp. 655–8.
4. De Hoffmann, E., Charrette, J. and Stroobant, V. (1994) *Spectrométrie de masse*, Paris-Milan-Barcelona, pp. 30–33.
5. Bloom H., Briggs L.H. and Cleverley, B. (1959) 'Physical properties of anthraquinone and its derivatives. Part I, infrared spectra', *Journal of the Chemical Society* (1959), pp. 178–85: see references 2, 3, 5 and 9.
6. Bloom *et al.* 1959 (see note 5 above).
7. Kiel and Heertjes 1963a and 1963b (see note 1 above); Wunderlich and Bergerhoff 1994 (see note 1 above).

Computation of a Visible Light Absorption Spectrum, a Semi-Empirical Quantum Mechanical Approach: Application to the Colour Determination of Alizarin Complexes in Solution or Chemisorbed onto Cellulose

François Delamare* and Bernard Monasse

Abstract Quantum mechanical modelling is used to compute the visible light absorption spectra of calcium-aluminium (and iron) complexes of alizarin. The methods we use, computing the energy difference between the fundamental and excited states of these complexes, enable us to determine the various conformations they can take as well as the corresponding visible light absorption spectra. The possibility of modifying each molecular parameter one at a time, which computation alone can offer, brings forth the respective roles of the mordant cation (Al^{3+}, Fe^{3+}), of the counter-ions (Ca^{2+}) and of the hydroxyls in the absorption spectrum. Our calculations show that these structures are stable and give not only spectra in the visible range, but also colours close to the expected values. It also lets us approach the chemisorption of these complexes onto a cellulose fibril. We observe the influence of the number of chemical bonds between complex and fibril on the ultimate absorption light spectrum of the latter.

Introduction

Although dyeing practices are rather well known, the understanding of the chemical bases of the process remains uncertain. We should all like to be able to answer these questions: Under what form does the dye exist in the vat bath? How does it react with the fibre? What exactly are the roles of the mordant and the other metallic cations? Which implies: what are the molecules present in the vicinity of the dye and what really is the composition of the fibre surface?

For this last issue, we have shown that a surface analysis technique, secondary ion mass spectrometry used in a static mode with time of flight mass selection (ToF–SIMS), could provide the relevant answers.[1] A different approach is presented here, based on computation.

Because the absorption of light by organic molecules is now explained classically through the molecular orbital theory, and because there exists software based on quantum mechanics to help chemists understand the stability and the properties of complex molecules, we decided to test its possibilities in the domain of dyeing processes[2] in order to observe the influence of the nature of the mordant, of other metallic counter-ions and the bonding to a fibre on colour. For this purpose, we selected the calcium-aluminium complex of alizarin (sometimes called Turkey red) bonding to a cellulose fibril.

Computational methods

Computations with quantum mechanical methods are carried out on a chemical structure. Each conformation is optimised by the minimisation of its energy. The electronic spectrum is then calculated.[3] To ensure the homogeneity of the results of our computations, the ZINDO/1 method is used for the two steps.[4] 'Semi-empirical' means that only the external electrons involved in the bonding are taken into account. All computations are done at absolute zero temperature.

Let us take an example with the alizarin molecule ($C_{14}H_8O_4$; Figure 1a/Plate 52a), hereafter named *Az*. Taking into account the electronic properties of each of its atoms, a first computation (ZINDO/1, PM3, HF) shows whether the structure can exist and the exact relative positions of the different atoms in the lowest energy conformation of the molecule. It therefore gives the optimal 3-dimensional geometry of the molecule. Thus the alizarin molecule is planar and the orientation of the hydroxyl groups is as indicated in Figure 1a, that is, in good accordance with the literature.[5]

The second step computes the different *energy levels* of the molecule (Figure 1b/Plate 52b). Below the reference line (0 eV), they are occupied by electrons; above, they are empty,

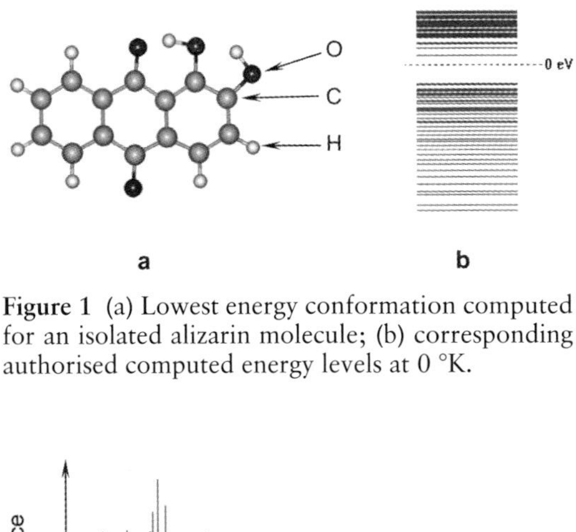

Figure 1 (a) Lowest energy conformation computed for an isolated alizarin molecule; (b) corresponding authorised computed energy levels at 0 °K.

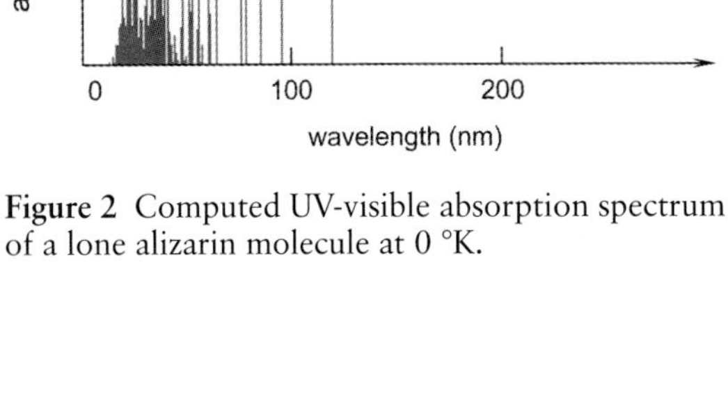

Figure 2 Computed UV-visible absorption spectrum of a lone alizarin molecule at 0 °K.

but they can be occupied if the molecule is excited. Figure 1b (Plate 52b) shows that, even for a simple molecule such as alizarin, the computed spectrum of the authorised energy levels is already complicated.

Absorption occurs when the energy of light matches an electron jump between these different authorised levels exactly. Absorption in visible light corresponds to jump energies ranging from 3.1 to 1.8 eV, (i.e. wavelengths of between 400 and 650 nm). With each energy level the value of the *oscillator strength* is computed, which gives absorbance. Absorbance versus wavelength can be plotted, in other words the absorption spectrum. Figure 2 shows that the absorption of energy by an isolated alizarin molecule occurs principally within the UV range. In a liquid, alizarin molecules are associated and the experimental absorption spectrum shifts toward higher wavelength values.

The computed spectrum (Fig. 2) contains only some peaks: it is a discontinuous spectrum. How can we obtain the continuous spectrum that we were hoping for? In reality, at room temperature, the absorbance spectrum is the result of the combination of numerous spectra corresponding to all the possible states (conformations and spin multiplicity) of the molecule. Therefore, in order to compute the absorbance, we take into account not only the lowest energy conformation already determined, but also several different stable conformations having a slightly higher energy. Then we determine these conformations and compute the corresponding spectra, which are added to the first. New peaks appear. The absorption spectrum is the envelope of all these peaks.

A problem remains: to find the stable conformations and determine their relative weight in order to take into account the most significant only. This problem can be solved by molecular dynamics, which could well be envisaged for a future project.

Only high resolution spectrometers are able to show up these discrete absorption peaks. They also agree with previous computations (see example below). The main advantage of the computational method is its predictive character. Another advantage is the possibility of modifying each molecular parameter one at a time, which only computation can offer. On the other hand, two problems can be noted. It is known that computations based on the Hartree-Fock model underestimate the absorption wavelengths. It is difficult to give an estimate of this error: let us say 10% as an order of magnitude. Secondly, there is an incompatibility between the fact that we compute at the atomic scale, but using parameters which are defined at a macroscopic scale, such as the acidity of the surrounding medium. We cannot therefore simulate a progressive variation of the pH of an aqueous surrounding medium: to add a proton to a given structure causes a drastic pH jump.

As far as the calcium-aluminium complexes of alizarin are concerned, the main difficulty does not lie in the number of atoms (120 up to 210) to be taken into account, but in the fact that the structure is a complex built around metallic cations.

The choice of a structure

From Kopp (1875) and Rosenstiehl (1875) to Soubayrol (1996), many scientists have tried to determine the structure of Turkey red, i.e. the calcium-aluminium complex of alizarin.[6] A literature review is given in Kiel (1963), and an extensive survey of the results in Kiel (1961)[7] but significant progress was made by Wunderlich (1993, 1994) and Soubayrol (1996) who propose not just planar, but 3-dimensional structures.[8] These two authors agree that this complex includes four alizarin (*Az*) molecules, two aluminium (Al^{3+}) and two calcium (Ca^{2+}) ions. They agree also that aluminium cations, which fill a very central position in the complex, are bonded to alizarin via the oxygen atoms of the hydroxyls. An oxygen bridge links the aluminium cations. But they differ drastically in their opinions about the role of calcium.

For Wunderlich, calcium plays the same role as aluminium: calcium cations are bonded to alizarin via the oxygen atoms of the hydroxyls. His proposition for the complex can be summarised in the formula: $Ca_2Al_2(O)_2Az_4$. According to nuclear magnetic resonance (NMR) and mass spectrometry (MS) experiments, Soubayrol proposes that the calcium cations should be directly linked to the aromatic carbon atoms of alizarin. Note that this differentiation of the roles of calcium and aluminium accords well with Kiel's experiments. In addition, Soubayrol's model places a hydrogen atom on each of the two oxygen atoms bonding the two aluminium cations. As a third difference, Soubayrol adds four external water molecules to clamp the structure of his complex, $Ca_2Al_2(OH)_2Az_4 \cdot 4H_2O$, strongly. Soubayrol proposes an ellipsoidal shape for this complex, the external alizarin molecules being stretched and bent, due to the presence of the metallic cations inside (Fig. 3).

This novel role proposed for the calcium cations seems to be a sufficient reason to choose Soubayrol's structure for testing by quantum mechanical modelling, in spite of the fact that the global electric charge of this complex is hard to determine. Following Soubayrol, we shall take it as 4+ in our computations.

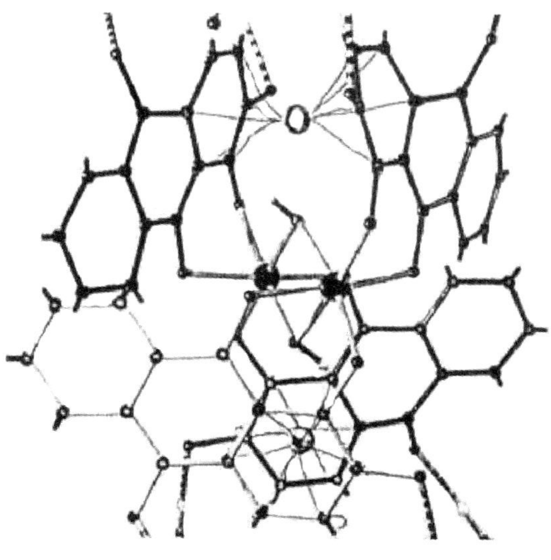

Figure 3 Structure proposed by Soubayrol for the calcium-aluminium complex of alizarin ($Ca_2Al_2(O)_2Az_4 \cdot 4H_2O$), deduced from NMR and MS experiments (half of a stereoscopic view).

Testing the structure proposed by Soubayrol

We begin our computations with the $Ca_2Al_2(OH)_2Az_4^{4+}$ structure. The first step of our computation shows that this structure is viable, but the spatial arrangement of the atoms is slightly different from that proposed by Soubayrol. Figures 4 and 5 (Plates 53 and 54) show the general shape of the complex. At first sight, alizarin molecules seem to be planar. It would be possible to define the angles between these planes. Figure 5 shows evidence for the role of the metallic cations in bonding the different parts of the complex. Note that the lines issuing from the calcium cations do not represent classical chemical bonds or, in other words, orbitals – they represent the equal influence between the carbon atoms of the aromatic structure and the metallic cations. Figure 6 (Plate 55), which uses a compact spheres model representation and respects the relative proportions of atomic diameters (d), emphasises the role of the bulkiness of the calcium cation (d = 0.198 nm), causing the inclination of the alizarin planes (for Al^{3+}, d = 0.100 nm). Figure 7 shows that alizarin molecules are, in fact, non-planar, i.e. bent, due to the presence of calcium. Note also that the small dimension of the complex (1.5 nm for the greatest dimension) offers great possibilities for it to diffuse inside non-crystalline areas of the fibres, or into the pores.

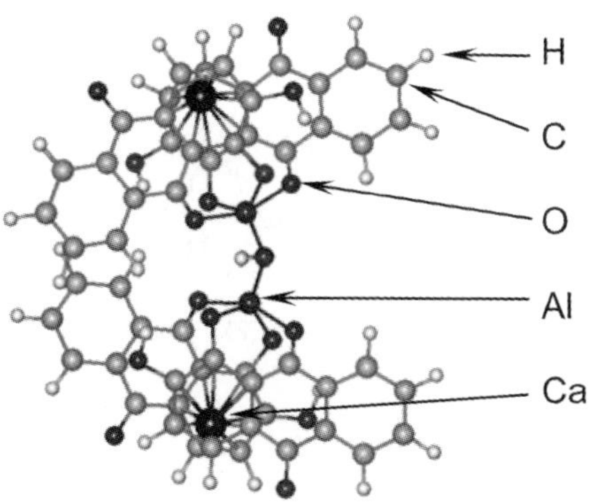

Figure 4 Computed top view of the structure of the $Ca_2Al_2(OH)_2Az_4^{4+}$ complex. Thick lines represent covalent bonds; thin lines, coordination bonds.

As a second step, we can compute the energy levels of the complex and look for the corresponding absorbance spectrum (Fig. 8). The absorbance is mainly in the visible range, with three maxima near 430, 500 and 540 nm.

In order to be more realistic, we now take into account the two other possible conformations of the complex related to the possible position of the central hydroxyl groups (Fig. 9/Plate 56). Each one has a specific absorption

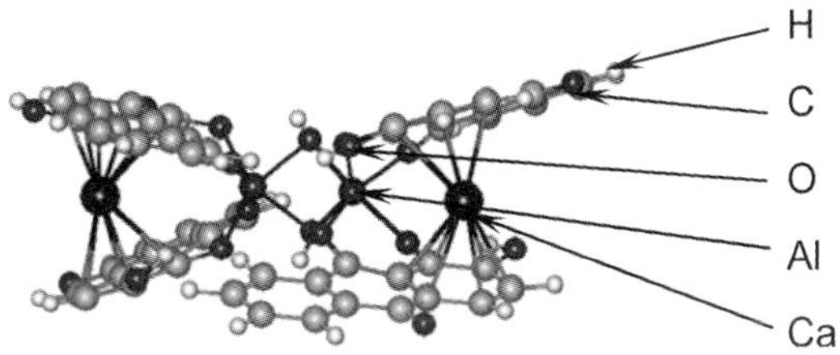

Figure 5 Computed side view of the structure of the $Ca_2Al_2(OH)_2Az_4^{4+}$ complex. Calcium cations are located in the median plane on the outside.

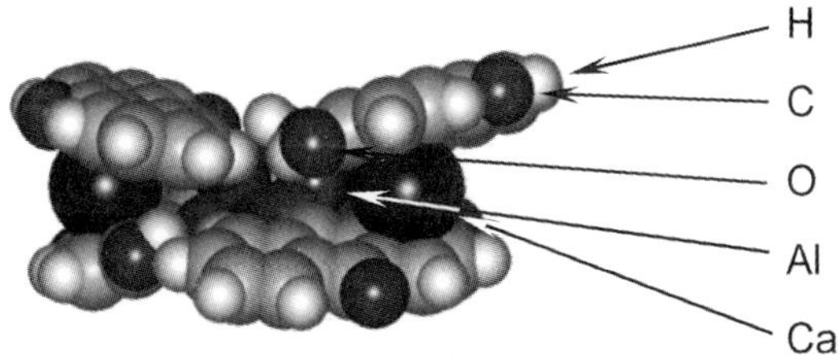

Figure 6 Same view as in Figure 5, but using a compact spheres model representation which respects the relative proportions of atomic diameters. It emphasises the role of the bulkiness of the calcium cation, causing the inclination of the alizarin planes. Note the aluminium ions hidden in the heart of the complex.

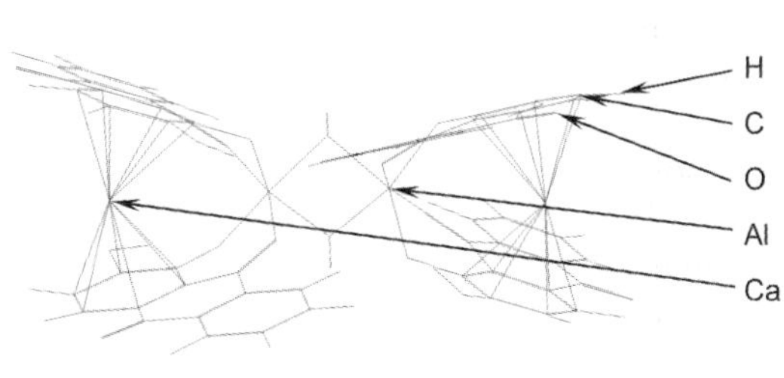

Figure 7 Same view as in Figure 5, evidencing the role of the calcium cation in the bending of the alizarin molecules.

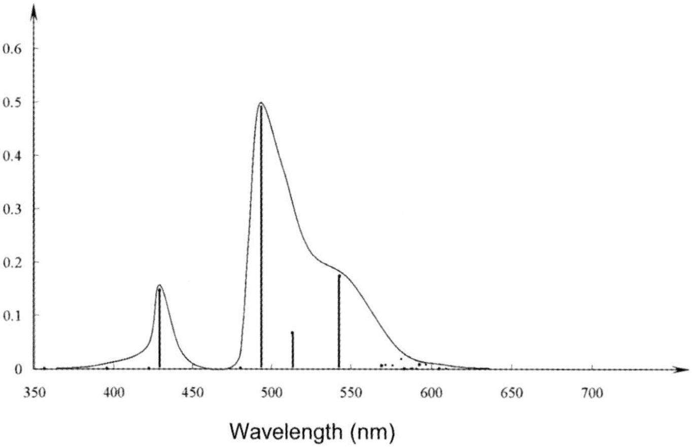

Figure 8 Visible light absorption spectrum of $Ca_2Al_2(OH)_2\textit{Az}_4^{4+}$ corresponding to the more stable conformation (symmetric) shown in Figures 4–7.

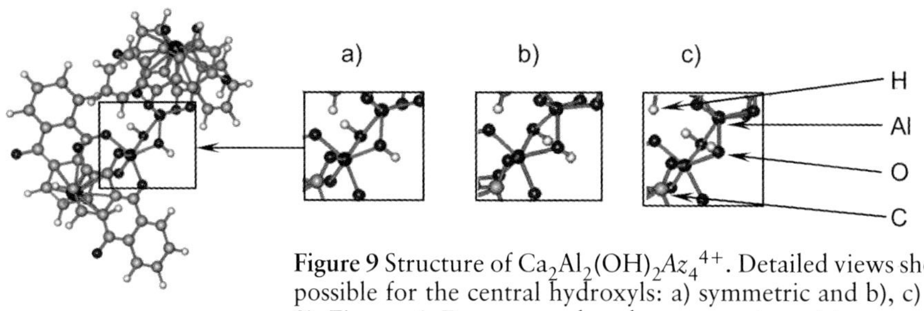

Figure 9 Structure of $Ca_2Al_2(OH)_2\textit{Az}_4^{4+}$. Detailed views show the three positions possible for the central hydroxyls: a) symmetric and b), c) anti symmetric 1 and 2). Figures 4–7 correspond to the symmetric position.

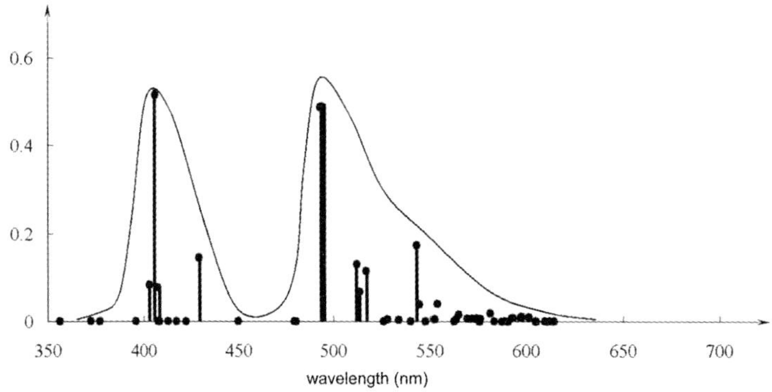

Figure 10 Visible light absorption spectrum of $Ca_2Al_2(OH)_2\textit{Az}_4^{4+}$ computed with the three conformations shown in Figure 9.

spectrum. They are rather similar, but they are sufficiently different to be characteristic of the conformation: slight variations in conformation produce significant variations in spectra.

How does the resultant spectrum appear? We don't know, because we ignore the question of which conformation is the most stable, present to the greatest extent in the aqueous solution. Figure 10 shows the spectrum computed if we

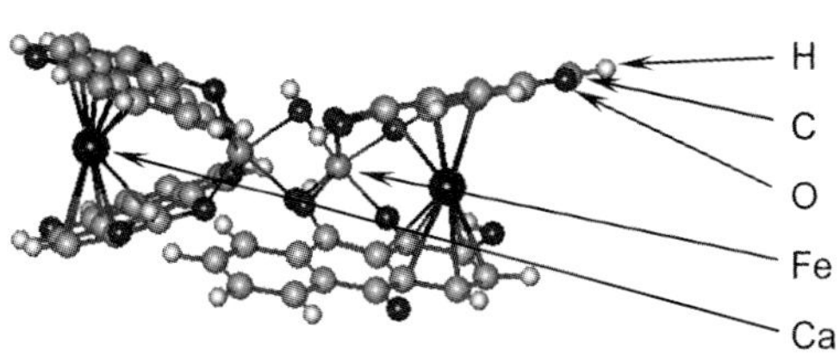

Figure 11 Computed structure of the $Ca_2Fe_2(OH)_2\textit{Az}_4^{4+}$ complex, where Fe replaces Al: compare with Figure 5.

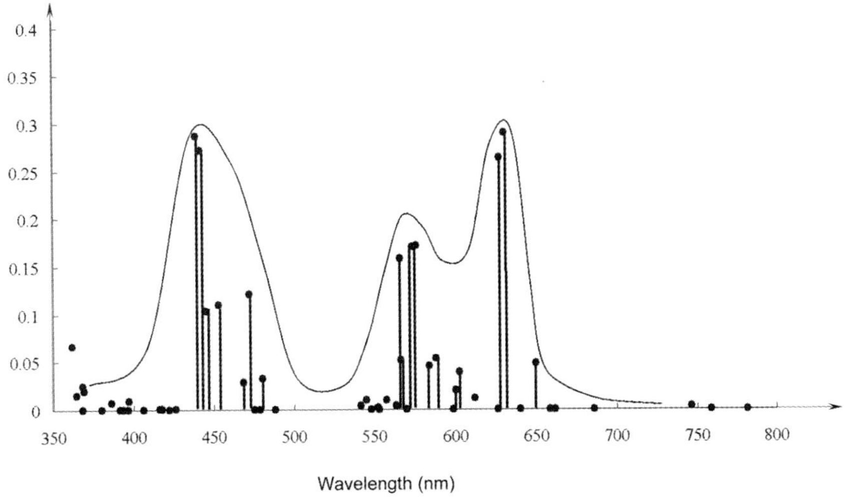

Figure 12 Computed absorption spectrum corresponding to the $Ca_2Fe_2(OH)_2Az_4{}^{4+}$ complex: compare with Figure 10.

suppose that the conformations are equally stable. It can be compared with the experimental spectrum published by Wunderlich (1994), which exhibits an absorbance in the ultraviolet B radiation range (280–320 nm), and another in the 460–600 nm range. Guineau (1994) has observed that the absorbance measured in the ultraviolet B range is an artefact of the measurement and that absorbance maxima at 500 and 540 nm are characteristic of alizarin complexes.[9] The computed colour is a purple.

Computing the effect of the nature of the mordant on the colour

Can we show evidence for the influence of the nature of the mordant on the colour? Let us replace aluminium by iron. Increasing the diameter of the mordant cation from 0.100 nm (Al^{3+}) to 0.128 nm (Fe^{3+}) does not modify the geometry of the complex, which remains imposed by the bulkiness of the calcium cation (0.198 nm) (Fig. 11/Plate 57). The structural differences with the reference complex are very small. On the other hand, the electronic structure of the mordant cation drastically changes, since iron (a transition metal) exhibits *d* orbitals. One might suppose that this simple fact ought to modify completely the energy levels, hence the absorption spectrum. As expected, in spite of the evident similarity of the structure, the absorption spectrum (computed taking into account the same three conformations) is completely different (Fig. 12).

The role of water molecules

Up to now we have computed the spectra of the $Ca_2Al_2(OH)_2Az_4{}^{4+}$ structure. Now, as Soubayrol does, we add four external water molecules to clamp the structure of the complex. They link two alizarin molecules previously bonded by a calcium cation, from a quinonic oxygen in the first alizarin to a hydroxyl group belonging to

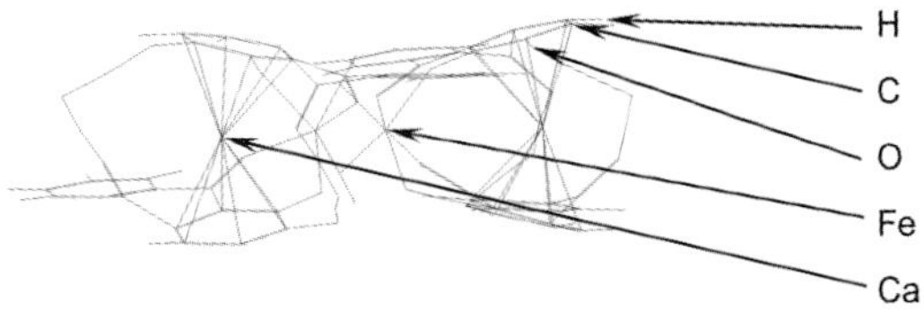

Figure 13 Structure of $Ca_2Fe_2(OH)_2Az_4{\cdot}4H_2O$. Four water molecules clamp the complex. Note the deformation of the alizarin molecule (left).

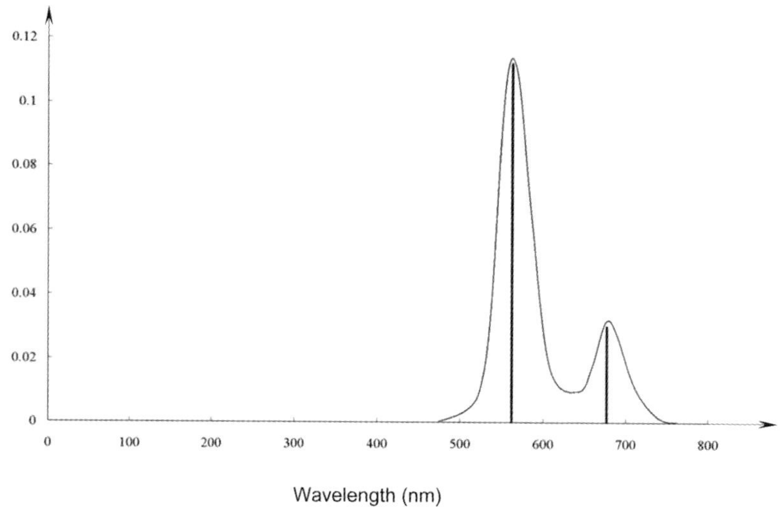

Figure 14 Visible light absorption spectrum of $Ca_2Al_2(OH)_2Az_4{\cdot}4H_2O$ (Fig. 13) computed with only one conformation.

the other (Fig. 13). The stresses on the alizarin molecules seem to be modified. The corresponding computed spectrum (Fig. 14) exhibits a weaker absorption with two maxima, near 560 and 680 nm.

The effect of the reaction of the complex with cellulose

What about the reaction of this complex with a cellulose fibre to obtain the Turkey red dyeing? In spite of the presence of many hydroxyl groups, cellulose is well known for its low level of chemical reactivity, probably due to the fact that most of the hydroxyl groups are involved in hydrogen bonds. But we are not directly concerned with this problem because we are not observing the bonds formed between the complex and the cellulose fibril in reality; instead we impose these bonds.

Generally speaking, authors believe that the role of the mordant is to link the dye to the fibre. In this calcium-aluminium complex of alizarin, aluminium cations are so completely hidden within the internal part of the complex that such a reaction seems unlikely. Soubayrol proposes a reaction mechanism with a cellulose fibril, a simple sequence of sugars (Fig. 15/Plate 58). He suggests the existence of a simultaneous bonding of the complex with two distinct fibres, by way of four peroxide bonds on each

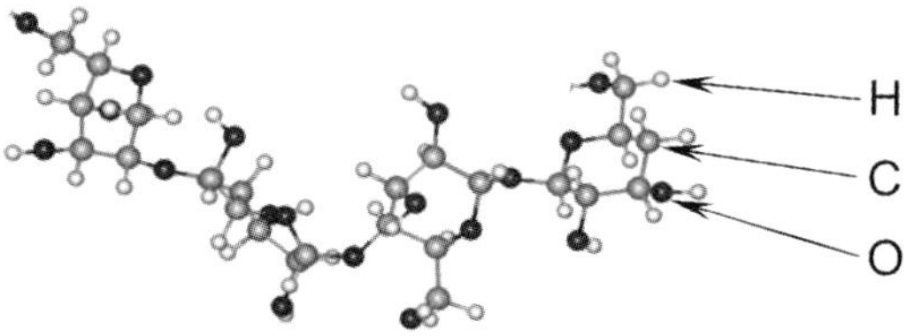

Figure 15 Non-planar structure of a cellulose fibril (here 81 atoms).

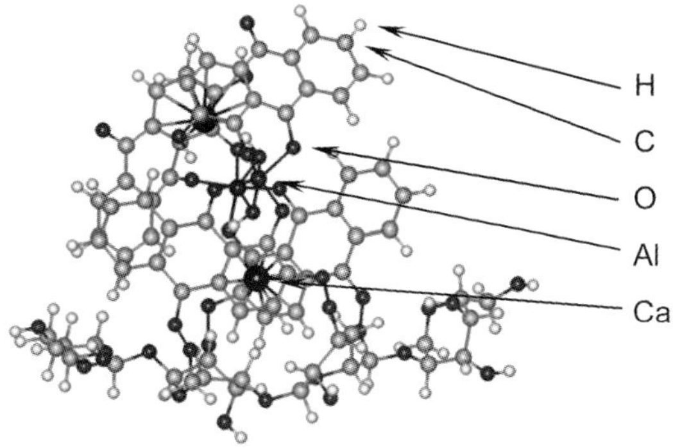

Figure 16 Chemisorption model of $Ca_2Al_2(OH)_2Az_4^{4+}$ on a cellulose fibril with creation of four peroxide bonds.

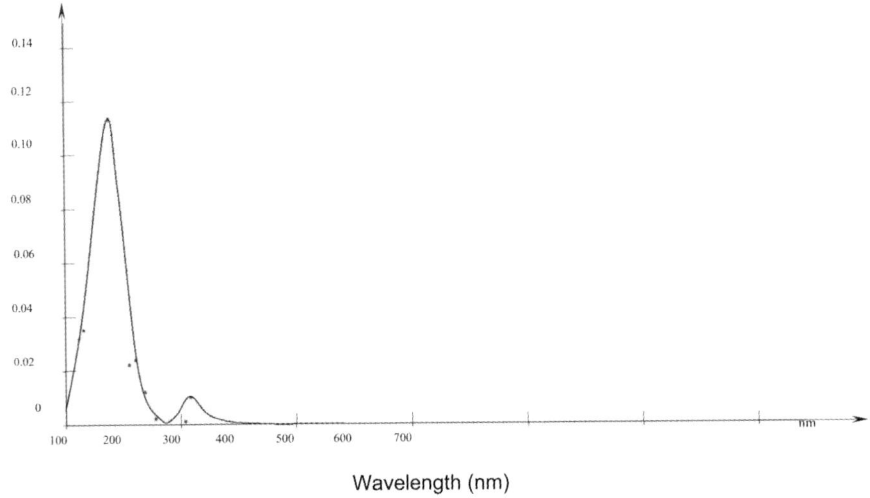

Figure 17 Computed absorption spectrum corresponding to the chemisorption mode presented in Figure 16. Light absorption occurs only in the UV region.

fibre. Because this seems to us very unlikely, let us try four peroxide bonds on one fibril only. Two alizarin molecules only are concerned with these four bonds (Fig. 16/Plate 59). For each, we bond one of the quinone oxygens (O-9) and the *m*-hydroxyl group (O-2) with the closest hydroxyl groups of the cellulose. The resulting absorption spectrum (Fig. 17) should be compared to Figure 10. The resulting absorption is not in the visible, but in the ultraviolet region. That is probably due to our choice of four bonds; too many bonds constrain the system so we must relax several.

If we relax two bonds in the O-9 position, as shown in Figure 18 (Plate 60), the effect of this relaxation is evident (Fig. 19). The colour shifts into the visible range and we obtain a purple. This spectrum is slightly different from our reference spectrum (Fig. 10). This is evidence for the fact that the linking process of the complex with the fibre plays an important role in the absorption of light in the visible range. We can also affirm that the chemisorption model proposed by Soubayrol is not the right one.

Does this mean that our model is close to reality? No; our proposition is possible, but not certain. The major reason is that, generally speaking, molecules – even simple ones – are dissociated during the chemisorption process, depending on the nature of the surface. And we have no idea of what really happens when this complex contacts the cellulose.

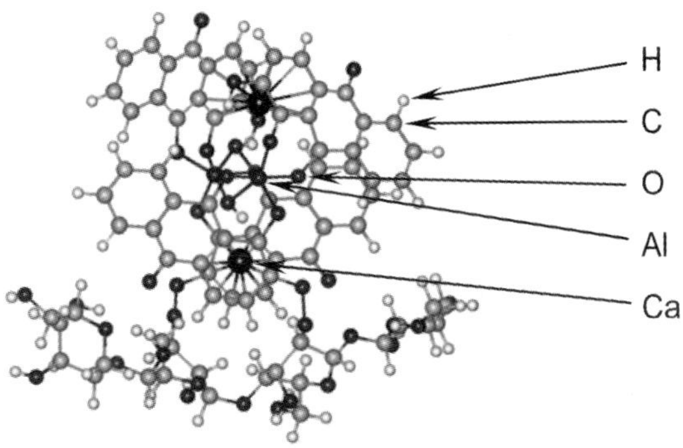

Figure 18 Chemisorption of $Ca_2Al_2(OH)_2Az_4^{4+}$ onto a cellulose fibril with creation of only two peroxide bonds.

Conclusion

To compute the visible light absorption spectrum associated with a structure, computational tools exist. They are sophisticated but due to the high complexity of the problem, we must make simplifications. The difficulty is to

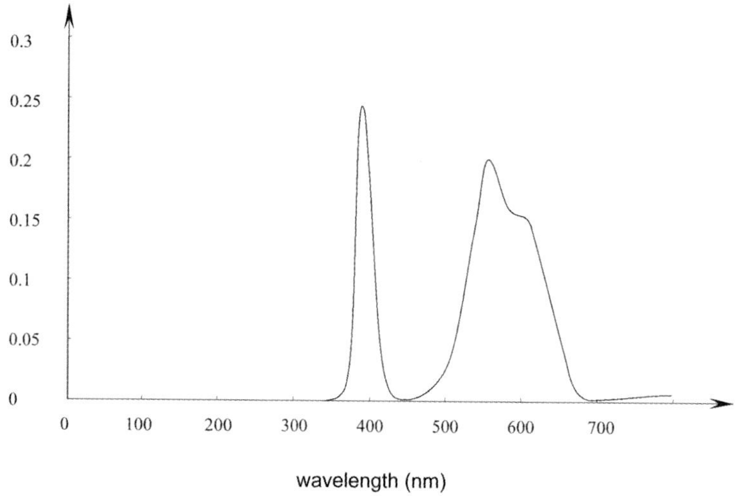

Figure 19 Computed absorption spectrum corresponding to the chemisorption mode presented in Figure 18. Light absorption occurs in the visible region.

do it at the right level to keep the precision necessary to compute a visible light absorption spectrum, which is related to a colour. To do so, we need a realistic situation suggested by powerful analytical tools. Then a comparison to the experimental colour can play the role of a touchstone to check our results.

We tested here the possibility of computing a complex build around metallic cations, namely Ca/Al/*Az* or Ca/Fe/*Az* complexes. Our calculations show that these structures are stable and give not only spectra in the visible range, but colours close to the expected values.

Editor's note

*Author for correspondence: École des Mines de Paris, CEMEF, BP 207, F-06904 – Sophia Antipolis Cédex, France (francois.delamare@gmail.com). Paper received 26 March 2003; revised version received 10 March 2004.

Notes and references

1. Delamare, F. and Repoux, M. (2005) 'Studying dyes by time-of-flight secondary ion mass spectrometry', *Dyes in History and Archaeology* 20, pp. 39–50.
2. Armstrong, D.R., Fortune, R., and Perkins P.G. (1972) 'Molecular orbital theory for the excited states of transition metal complexes', *Journal of the Chemical Society, Faraday Transactions 2* 68, pp. 1839–46.
3. Armstrong *et al.* 1972 (see note 2 above).
4. Armstrong *et al.* 1972 (see note 2 above); Stewart, J.J.P. (1985) 'MOSOL, MOPAC for solid-state physics', *Quantum Chemical Program Exchange* 5, pp. 62–3.
5. Hovaneissian, M. (2002) 'Étude des colorants et des pigments organiques rouges de garance par CLHP et IRTF. Application au domaine archéologique', *Mémoire de DEA*, Université d'Aix-Marseille III, pp. 1–26.
6. Kopp, E. (1875), *Berichte* 8, p. 980; Rosenstiehl, A. (1875), 'Recherches sur le rôle que joue dans la teinture les diverses matières colorantes de la garance', *Bulletin de la Société Industrielle de Mulhouse*, p. 56.
7. Kiel, E.G. (1961) *Metaalcomplexen van alizarinerood*, PhD thesis, University of Delft; Kiel, E.G. and Heertjes, P.M. (1963) 'Metal complexes of alizarin. I – The structure of the calcium-aluminium lake of alizarin', *Journal of the Society of Dyers and Colourists* 70, pp. 21–7.
8. Wunderlich, C.-H. and Bergerhoff, G. (1993) 'Crystal structure of di-μ-oxo-bis (bis (1,2-

dihydroxy anthrachinonato) aluminium-(aquabis (dimethylformamide) calcium)), $Al_2O_4(C_{14}H_6O_4)_4Ca_2(HCON(CH_3)_2)_4(H_2O)_2$, (alizarin complex)', *Zeitschrift für Kristallographie* 207, pp. 185–8; Wunderlich, C.-H. and Bergerhoff, G. (1994) 'Konstitution und Farbe von Alizarin- und Purpurin-Farblacken', *Chemische Berichte* 127, pp. 1185–90; Soubayrol, P., Dana, G. and Man, P.P. (1996) 'Aluminium-27 solid-state NMR study of aluminium coordination complexes of alizarin', *Magnetic Resonance in Chemistry* 34, 8, pp. 638–45; Soubayrol, P. (1996) *Préparation et étude structurale des complexes formés entre l'aluminium et l'alizarine*, PhD thesis, Université P. et M. Curie, Paris, pp. 1–176.

9. Guineau, B. (1994) 'Acquisition et traitement de données colorimétriques. Application à l'étude de la teinture des textiles', in M. Stefanaggi (ed.), *La conservation des textiles anciens : journées d'études de la SFIIC, Angers, 20-22 Octobre 1994*, Champs-sur-Marne, pp. 53–68.

Evaluation of Consolidants for Black Iron-Tannate-Dyed Maori Textiles

Rangi Te Kanawa, Gerald J. Smith,* Glenn A. Fenton, Ian J. Miller and Cara L. Dunford

Abstract The indigenous Maori people of New Zealand traditionally use a local iron-containing mud to obtain a black dye to decorate their textiles and garments made from *Phormium tenax* Forst. (*harakeke*) fibres. Ferric – iron(III) – ions in the mud form complexes with gallo-or condensed tannins extracted from the bark of *Elaeocarpus dentatus* Vahl (*hinau*) or *Leptospermum scoparium* Forst. (*manuka*) respectively to give the black dye. This dye causes a gradual weakening and embrittlement of the fibre. Eventually a consolidation treatment is required to maintain the integrity of the material. Also, as the dyed fibre ages, the colour fades and reflects less blue light; that is, it becomes browner.

Several polymers have been evaluated for their abilities to consolidate and inhibit further fading of the traditionally black-dyed fibres of Maori textiles. Among the polymers evaluated were the widely used consolidants, Klucel-G and Funori, a sulfated seaweed polysaccharide. Novel consolidants based on salts of an alginate from brown seaweed – specifically, sodium alginate, zinc alginate and zinc hydroxypropylalginate – were also evaluated. The treated fibre samples were subjected to accelerated thermal ageing for periods of 665 hours in sealed vessels and were compared for their strength, colour change and flexibility. The alginates were found to be effective in inhibiting the fading of the fibres. Also, the consumption of oxygen occurring as a result of the accelerated ageing was found to be substantial for the zinc alginates.

Introduction

Prior to contact with Europeans, the indigenous Maori people of New Zealand used a local iron-containing mud known as *paru* to obtain a black dye to decorate their textiles and garments, which are woven from *Phormium tenax* Forst. (*harakeke*) fibres. Indeed, it is still used in traditional Maori weaving today. The fibres are first coated with gallo-or condensed tannins extracted from the bark of *Elaeocarpus dentatus* Vahl (*hinau*) or *Leptospermum scoparium* Forst. (*manuka*) respectively. After drying, the fibres are immersed in *paru* and the iron(III) ions in the mud form black complexes with the tannin on the surfaces of the fibres.[1] As is the case with iron-tannate dyes on cellulosic substrates used by other cultures,[2] the *P. tenax* fibre is subject to degradation, including a gradual weakening, embrittlement and fading of the fibre over time.[3] For a number of years we have been involved in research designed to elucidate this process and to reduce damage to black-dyed *P. tenax* fibres that have been coloured black with the iron-tannate dye.[4]

When the degradation reaches an advanced stage, a consolidation treatment is required to maintain the physical integrity of the dyed fibres. The work described in this report has been concerned with evaluating the performance and understanding the chemistry of a number of consolidants in relation to fading that occurs with thermal ageing in moist air (relative humidity 80%). This fading involves a colour change from black to tan brown, resulting from a loss of yellow- and red-absorbing chromophores and hence an increased reflectance in the longer wavelength region of the visible spectrum.

Consolidation technique

The current conservation practice for fragile Maori textiles is to support them with a backing fabric or Japanese tissue. Textiles that cannot be handled, because of disintegration of the brittle black elements, are laid flat and cushioned within an Ethafoam (Dunlop Flexible Foams) surround to reduce movement of the elements relative to each other.

A different approach was employed in the present work: a supporting polymeric material was introduced, which impregnates, embeds and binds the damaged fibres in a flexible matrix. To the extent that the added material can be extracted, this process is reversible. However, only small amounts of a supporting polymer can be introduced without altering the appearance of the fibres and this limits the improvement in mechanical strength that can be achieved. Any flexible polymer can, in principle, be used in this way, provided it is soluble in a benign solvent, convenient to apply and, when it has been applied, the appearance of the material is close to that of the original. We have achieved this goal using alginates from brown seaweed. A number of other polymers have been used previously to consolidate black-dyed *P. tenax* fibres: Klucel G, a hydroxypropyl cellulose and Funori, a seaweed-derived polysaccharide.[5] The results of these treatments were compared with the alginate consolidants.

Methods

Fibre dyeing

Fibre bundles were extracted from leaves of *P. tenax* by stripping off the epidermal layer and soaking in water, followed by mechanical beating to release the individual fibres. Cleaned fibres were impregnated with tannin by immersing in an aqueous extract produced by boiling bark from the *hinau* tree (*Elaeocarpus dentatus*) in water. Because it is not possible to introduce reproducible and identical quantities of iron(III) ions using the iron-containing mud, the dried fibres were doped with iron(III) ions by soaking in aqueous 0.05 M iron(III) ammonium sulfate ($FeNH_4(SO_4)_2$) solution buffered with 0.05 M sodium acetate (CH_3COONa) adjusted to pH 5 by the dropwise addition of dilute hydrochloric acid (HCl) solution and dried. This solution has approximately the same acidity as the mud and it is therefore a satisfactory model system for introducing iron.

Consolidation

Solutions of sodium alginate in water with compositions of 2%, 1% and 0.5%, and 1% sodium hydroxypropylalginate were applied to bundles of fibres with an artists' paint brush and left to dry. The commercially available alginates that contain calcium to increase their viscosities should be avoided. This treatment was repeated several times to ensure the alginate had thoroughly penetrated into the fibre bundles and the excess alginate was carefully removed by sponging with tissues. A 3% solution of the acetate salt of a divalent metal ion such as zinc ($(CH_3COO)_2Zn$) dissolved in alcohol was then applied. The fibres were further sponged with alcohol to remove excess acetate solution and dried. Other bundles of fibres were similarly treated with 2% solutions of Klucel G and Funori.

Accelerated ageing

Air-dried samples (0.3 g) of dyed *P. tenax* fibres were subjected to accelerated ageing in sealed,

500 mL vessels for four weeks at 80 °C and a relative humidity of 80%, maintained by a small volume of a 20:80 mole ratio glycerol-water mixture.[6]

Gas/volatiles analysis

Periodically during the course of the accelerated ageing, the contents of carbon dioxide and oxygen in the head-space of the sealed vessels containing the fibres were measured using a quadrupole mass spectrometer (Dycor-Metek MA100/200). Gas samples were withdrawn from the head-space through a very fine glass capillary inserted through a silicone rubber septum.

Colour changes

The fading of the samples (bleaching of the dye chromophore/s) was quantified by reflectance spectroscopy of discs of compressed, 2–3 mm lengths of fibres, using a Perkin Elmer LS 55 spectrofluorimeter in coupled emission-excitation mode. The reflectance reference was a compressed mixture of Teflon powder and carbon black with a uniform reflectance of 10% from 350 to 700nm.

Results and discussion

Colour changes

The effects of the accelerated ageing regime on dyed fibres consolidated with Klucel G, zinc alginate and sodium alginate are shown in Plates 61–63. The colour changes for unconsolidated fibres and for Funori were similar to that shown for Klucel G. The colour change for fibres consolidated with zinc hydroxypropylalginate is similar to that for zinc alginate. The black colour was most stable with the zinc alginate and hydroxypropylalginate consolidants. While the black colour was well preserved with 2% sodium alginate, fading was apparent with 1% and 0.5% sodium alginate, indicating that the alginate polymer itself affords protection against the reactions responsible for colour

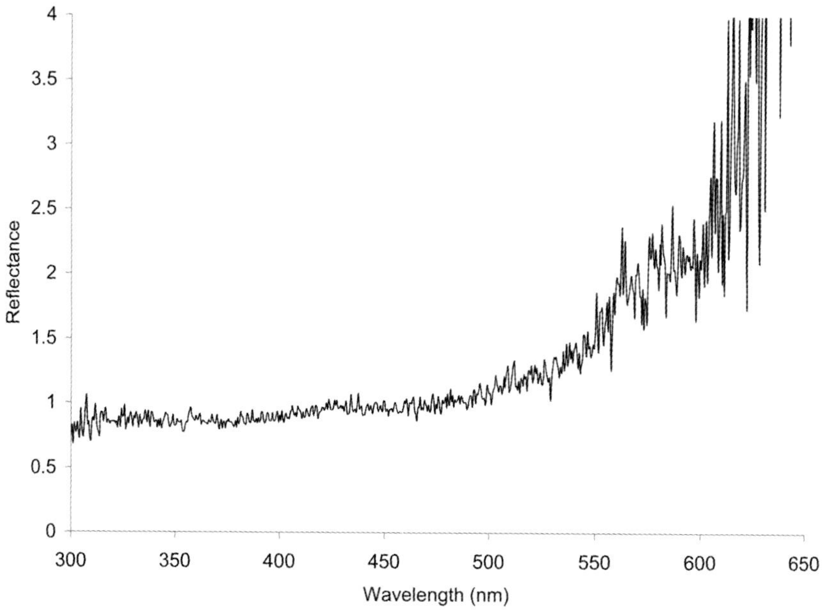

Figure 1 The reflectance spectrum, $^{age}I/^{unage}I$, of dyed *P.tenax* fibres consolidated with Klucel G as a result of accelerated ageing for 665 hours relative to unaged, dyed, consolidated *P. tenax*: that is, the change in reflectance as a result of accelerated ageing.

loss. There was substantial fading from black to tan or brown with the unconsolidated fibres and with those consolidated with Klucel G and Funori. This is associated with the loss of dye chromophores absorbing in the yellow–red part of the visible spectrum at wavelengths longer than about 550nm. This is evident in the reflectance spectrum for Klucel G-consolidated fibres shown in Figure 1.

Mechanical properties

Treatment with all the consolidants significantly reduced the flexibility of the unaged fibre bundles, but they were effective in binding the fibres together. After accelerated ageing, in all cases there was an embrittlement of the fibres that led to fragmentation of the fibre bundles when handled, in a way comparable to unconsolidated, aged fibres.

Oxygen consumption and carbon dioxide evolution

The oxygen consumption and carbon dioxide evolution by the fibres consolidated with zinc alginate during the accelerated ageing regime in sealed vessels are substantially greater than for unconsolidated fibres and those treated with other consolidants, as shown in Figure 2 and 3.

The increased oxygen consumption that attends the ageing of the samples consolidated with the zinc alginates suggests that the alginates are sacrificially oxidised, thereby protecting the black dye from bleaching. The evolution of carbon dioxide from unconsolidated fibre has been observed previously and has been attributed to the metal ion-induced decarboxylation of unesterified carboxylic (uronic) acid groups associated with the hemicellulose component of the fibres. Alginate polysaccharides also have a high abundance of free carboxylic acid groups. In the presence of metal ions such as iron(III) or zinc ions these will also be decarboxylated to produce carbon dioxide.

Alginates react irreversibly with acids, neutralising and removing them from the system. This involves an initial ion exchange between the metal ions associated with alginate and acid hydrogen ions. The ion exchange leads to the formation of an insoluble product, thereby rendering the hydrogen ion unavailable for participation in degradative reactions. This is followed by a decarboxylation reaction of the alginate to give initially the corresponding polysaccharide with a pentose residue and gaseous carbon dioxide.[7] The effect of this is to eliminate acids formed during the course of ageing. In this regard it is noted that Funori is a sulfated polysaccharide with the potential to form sulfuric acid.[8] In previous work, volatile carboxylic acids such as acetic acid have been shown to be produced by deacetylation of the hemicellulosic component of *P. tenax* fibres during ageing. It has also been

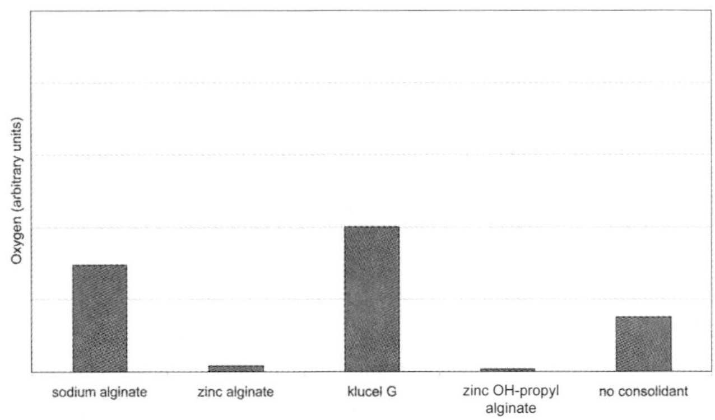

Figure 2 The consumption of oxygen as a result of thermal ageing of consolidated fibres for 665 hours.

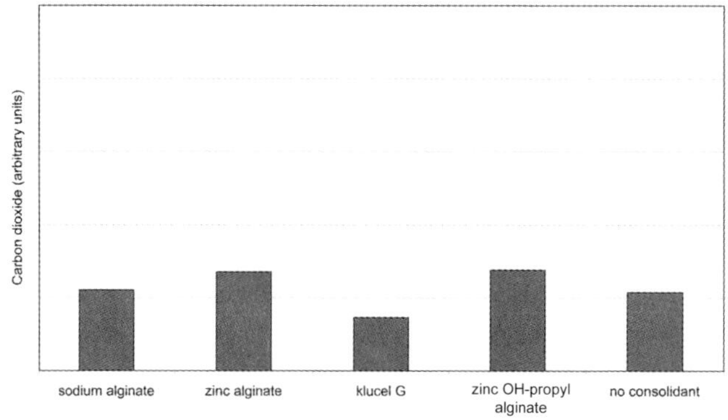

Figure 3 The evolution of carbon dioxide as a result of thermal ageing of consolidated fibres for 665 hours.

shown that such acids catalyse the degradation of the cellulosic substrate of iron-tannate dyes.[9]

The decarboxylated polysaccharide product is susceptible to reaction with free radicals and oxygen.[10] This latter reaction explains the high depletion of oxygen in the sealed vessels containing fibres consolidated with alginates.

The nature of the binding of divalent metal ions to alginate polymers dissolved in water is such as to cause the polymer to precipitate from solution as extended chains rather than as folded or coiled structures.[11] This confers a fibrillar form on the solid alginate matrix that is aligned with the *P. tenax* fibres and which provides added flexibility to the textile fibre–matrix composite.

Conclusion

A novel method for consolidating threads of badly degraded black-dyed *P. tenax* has been developed in this work. Divalent metal ion (for example, zinc) alginates provide the fibre with support and sufficient strength to be handled without further disintegration, while preserving the colour and appearance of the material even after one month of accelerated ageing.

Editor's note

This paper was presented as a poster at the Meeting. *Author for correspondence: New Zealand Institute for Industrial Research, PO Box 31310, Lower Hutt, New Zealand (G. Smith@irl.cri.nz). Paper received 31 July 2003; revised version received 19 November 2003.

Acknowledgements

This work was supported by the New Zealand Foundation for Research, Science and Technology, Contract C08X0215-4. We thank Dr Vincent Daniels, British Museum, for valuable discussions and advice and the New Zealand Maori Weavers group for their support, encouragement and interest.

Notes and references

1. Te Kanawa, D. (1994) *Weaving a Kakahu*, Wellington, NZ (Bridget Williams Books).
2. Wunderlich, C.H. (1994) 'Geschichte und Chemie der Eisengallustinte', *Restauro* 100, pp. 414–21; Bicchieri, M. and Pepa, S. (1996) 'The degradation of cellulose with ferric and cupric ions in a low-acid medium', *Restaurator* 17, pp. 165–83; Neevel, J.G. (1995) 'Phytate: a potential conservation agent for the treatment of ink

corrosion caused by iron gall ink', *Restaurator* 16, pp. 143–60.
3. Daniels, V. (1999) 'Factors affecting the deterioration of the cellulosic fibres in black-dyed New Zealand flax', *Studies in Conservation* 44, pp. 73–9; Daniels, V. (1996) 'Maori black mud dyeing of *Phormium tenax* fibre', *Dyes in History and Archaeology* 14, pp. 9–11.
4. Te Kanawa, R., Thomsen, S., Smith, G.J., Miller, I.J., Andary, C. and Cardon, D. (2002) 'Traditional Maori dyes', *Dyes in History and Archaeology* 18, pp. 47–50; More, N., Smith, G.J., Te Kanawa, R. and Miller, I.J. (2003) 'Iron-sensitized degradation of black dyed Maori textiles', *Dyes in History and Archaeology* 19, pp. 144–8; Smith, G.J., Te Kanawa, R., Miller, I.J. and Fenton, G.A. (2005) 'Stabilization of cellulosic textiles decorated with iron-containing dyes', *Dyes in History and Archaeology* 20, pp. 89–94.
5. Daniels, V. (1998) 'The deterioration and conservation of black-dyed New Zealand flax: Part 3 Mechanical properties', *The British Museum Report 1998/24*; Scott, G. (1988) 'A Maori cloak', *Conservation News* 35 (March), p. 10.
6. Miller, C.S. and Dalton, N.W. (1953) *Glycerol*, New York.
7. King, A. (1983) 'Brown seaweed extracts (alginates)', in M. Glicksman (ed.), *Food Hydrocolloids*, Boca Raton, pp. 115–88.
8. Penman, A. and Rees, D.A. (1973) 'Carrageenans Part IX. Methylation analysis of galactan sulphates: structure of carageenan', *Journal of the Chemical Society, Perkin Transactions* 1, pp. 2182–7.
9. Te Kanawa *et al.* 2002 (see note 4 above); More *et al.* 2003 (see note 4 above).
10. Penman and Rees 1973 (see note 8 above).
11. Penman and Rees 1973 (see note 8 above).

In the Circle of Colour: Natural Dyeing of Flax, Silk, and Wool

Katarzyna Schmidt-Przewoźna* and Witold Przewoźny

The exhibition

In the Circle of Colours was organised at the Ethnographical Museum, Division of the National Museum in Poznań, Poland, from 17 March until 28 April 2002 (Fig. 1).[1] The exhibition presented the results of experiments involving the reconstruction of old traditional dyeing techniques, developed at Studio Natural Art, the Institute of Natural Fibres, Poznań. A rich part of the exhibition consisted of exhibits from the collections of the Ethnographical Museum in Poznań. Fabrics, carpets, elements of clothes, Australian pictures and small utensils were produced by people of different ethnic groups living on different continents and showed people living in harmony with their own tradition, in isolation from the contemporary

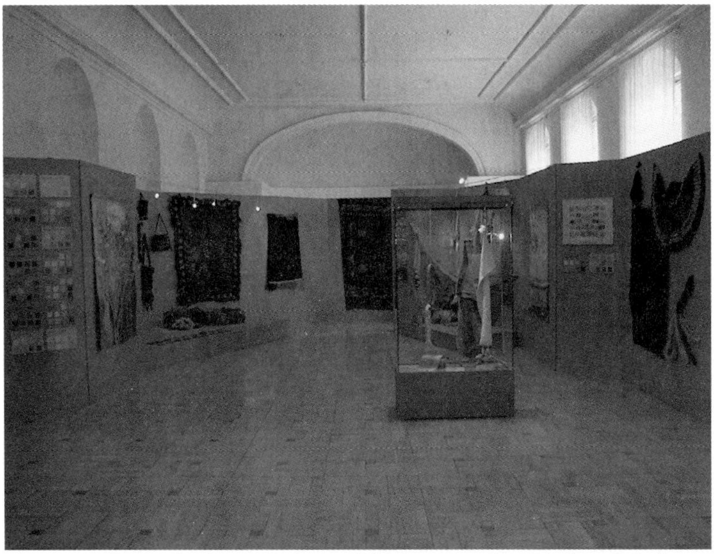

Figure 1 General view of the exhibition *In the Circle of Colours*, Ethnographical Museum, Division of the National Museum in Poznań, 2002. (Photo: Witold Przewoźny)

civilisation of the Western World. The natural ways of obtaining dyestuffs and methods of their application can be found among these cultures, which still persist in harmony with nature and admiration of its beauty.

Among the more interesting exhibits was the use of indigo in fabrics of the Dogon people from Mali and the lively, colourful skirts from the mountain regions of Laos (Plate 64). Red and shades of brown could be seen in woollen carpets from Afghanistan, Pakistan and India and the colourful bags of Pushtun (Pashtun) shepherds from Afghanistan were dyed with turmeric (*Curcuma longa* L.), among other dyes (Plate 65). The dyestuff from annatto (*Bixa orellana* L.) could be found in fabrics and clothes of the Panare Indians from Venezuela and the Tucuna from the Brazilian Amazon basin (Plate 66a and b).

The achievements of Polish dyeing mills of the 17th and 18th centuries were illustrated by a collection of waist sashes worn by Polish noblemen. These wide, silk sashes with characteristic oriental designs and lively, saturated colours, from several famous manufacturers, were real embellishments of the exhibition (Fig. 2).[2]

The exhibition also had an educational aspect, showing consecutive stages of dyeing with woad (*Isatis tinctoria* L), madder (*Rubia tinctoria* L.), alder buckthorn (*Rhamnus frangula* L.), annatto and logwood (*Haematoxylum campechianum* L.). The tapestry illustrated in Plate 67, for example, was made using yarns dyed with madder, weld (*Reseda luteola* L.), turmeric, tansy (*Tanacetum vulgare* L.), dandelion (*Taraxacum officinale* Weber), alder buckthorn, annatto and cutch (*Acacia catechu* Willd.).

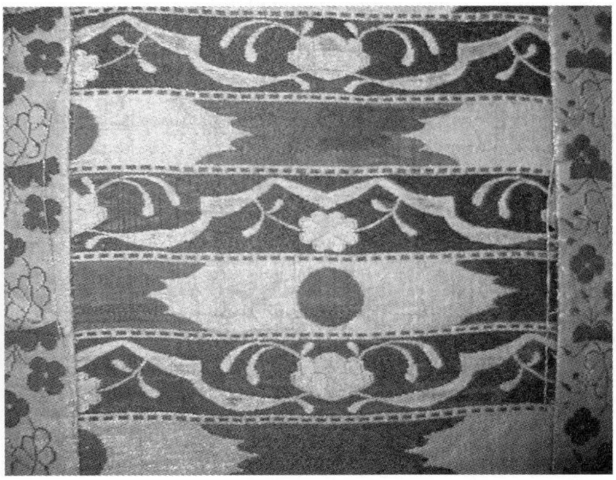

Figure 2 Detail of 18th-century waist sash of Słuck manufacture. (Photo: Katarzyna Schmidt-Przewoźna)

Colour trials illustrated the use of basic plant and animal dyestuffs on linen, silk and cotton fabrics. The scale of colours obtained was arranged thematically, creating blue–yellow–red axes for the exhibition (Plate 68). Blue, for example, is given by indigo, the king among dyestuffs, obtained from the indigo plant, *Indigofera tinctoria* L. and related species and other plants, including woad (*Isatis tinctoria* L.). Its use has strong correlations with aspects of human culture as blue emphasised social status in many societies. A very large number of plants, wild and cultivated, give yellow, often described as the symbol of sun, harvest and abundance. Examples in the exhibition included saffron (*Crocus sativus* L.), weld (*Reseda luteola* L.) and turmeric (*Curcuma longa* L.), which gives a particularly intense colour. Red in European culture was reserved for noblemen and in Poland an expensive red dyestuff was obtained from the cysts of the tiny insect Polish cochineal insect (*Porphyrophora polonica* L.). After the discovery of America and the growing use of imported New World cochineal (*Dactylopius coccus* Costa), the use of Polish cochineal, a less efficient source of dyestuff, gradually died out. Madder (*Rubia tinctorum* L.) is one of the oldest dyeing agents: it is thought to have been used in the Indus valley around 3000 BC. With madder one can obtain a wide palette of colours from warm rose, through red, to carmine and violet. The use of plants growing wild in Poland was also illustrated, including bilberry (*Vaccinium myrtillus* L.), which gives a non-permanent blue, safflower (*Carthamus tinctorius* L.) and nettle (*Urtica dioica* L.), which in autumn has red stems, appropriately used to dye red.[3]

Editor's note

This paper was presented as a poster at the Meeting. *Author for correspondence: Instytut Włókien Naturalnych (Institute of Natural Fibres), ul. Wojska Polskiego 71b, 60-630 Poznań, Poland (kasia@inf.poznan.pl). Paper received 7 May 2003; revised version received 5 January 2004.

Notes and references

1. Since this text was accepted for publication, a longer article by the authors on this exhibition has appeared: Schmidt-Przewoźna, K. and Przewoźny, W. (2004) 'In the Circle of Colour – exhibition in the Ethnographic Museum', *Journal of Natural Fibers*, 1, 2, pp. 93–8. See also Schmidt-Przewoźna, K. (2005) 'Historical dyes in Poland and their revival', *Dyes in History and Archaeology* 20, pp. 110–15.
2. Schmidt-Przewoźna, K. (2008) 'Natural dyes used in Polish workshops in the 17th–18th centuries', in this journal, pp. 148–53.
3. Schmidt-Przewoźna 2008 (see note 2 above).